Eastern Alpine Guide

Mount Clay at sunrise, New Hampshire. (JS)

Eastern Alpine Guide

Natural History and Conservation
of Mountain Tundra East of the Rockies

Written and Edited by Mike Jones and Liz Willey
Foreword by Laura Waterman

University Press of New England
Hanover and London

University Press of New England
www.upne.com
Manufactured in the United States of America
Designed and typeset by Matthew R. Burne
Cartography by Liz Willey

For permission to reproduce any of the material in this book, contact
Permissions, University Press of New England, One Court Street,
Suite 250, Lebanon NH 03766; or visit www.upne.com

All photographs are the copyright of the original photographer, as
noted. Photographs are labeled with photographer's initials throughout
the book (see page 343 for credit information).

Library of Congress Cataloging-in-Publication Data

Names: Jones, Mike, 1980– author, editor. | Willey, Liz, author, editor.
Title: Eastern alpine guide : natural history and conservation of mountain
    tundra east of the Rockies / written and edited by Mike Jones and Liz
    Willey.
Description: Hanover : University Press of New England, [2018] | Originally
    published: New Salem, Massachusetts : Beyond Ktaadn/Boghaunter Books,
    2012. | Includes bibliographical references.
Identifiers: LCCN 2017060704 | ISBN 9781512603026 (paperback) |
    ISBN 9781512603033 (ebook)
Subjects: LCSH: Mountain ecology—Northeastern States. | Mountain
    ecology—Canada, Eastern. | Mountain biodiversity conservation—
    Northeastern States. | Mountain biodiversity conservation—Canada,
    Eastern. | Mountains—Northeastern States. | Mountains—Canada, Eastern.
    | Natural history—Northeastern States. | Natural history—Canada, Eastern.
Classification: LCC QH541.5.M65 J66 2018 | DDC 577.5/3—dc23
LC record available at https://lccn.loc.gov/2017060704

5 4 3 2 1

The publishers gratefully acknowledge support of Beyond Ktaadn, the Waterman Fund, the
International Appalachian Trail, the Walden Woods Project, and Huntington Graphics.

The Waterman Fund

For Everett Oakes

and our parents, Betty & Tom and Lynn & Lloyd

and our sisters, Laura, Larisa, Laurie, McKenzie, Leslie, and Penelope

(*and* Addie)

In memory of Cary Girod, *who loved mountains.*

MTJ and LLW

"On the tops of mountains, as everywhere to hopeful souls, it is always morning."

H.D. Thoreau, *Walden* draft (1846–1847)

"For what canst thou pray here—but to be delivered from here."

H.D. Thoreau on Ktaadn, *Journal* (1846)

Additional contributions throughout—*by Matthew R. Burne, Iona Woolmington, Will Kemeza, & Charley Eiseman*

Contents

Foreword

Conservation is a state of harmony
between men and land.

— Aldo Leopold

This is, at heart, a book about wildness and the importance of preserving it.

Wildness can be found in the forests of the eastern mountains of North America, but, to my mind, it is most readily tracked in the alpine tundra, perhaps because tundra is old. Tundra plants date back to the ice age. The tundra is a landscape that has survived, possibly because much of it is relatively inaccessible, with little human impact. Looked at this way, tundra connects us to land that is wild. But as wildness is intangible—more a perception than anything else—it can be compromised, even erased by unbridled human contact.

In the northeastern United States, this ice age remnant of vegetation graces minimally scattered high points. But tundra, and therefore the possibility of wildness, becomes more prolific if we tramp due north, out through the top of Maine, and cross the international border into eastern Canada. In the provinces of Québec and Newfoundland & Labrador, authors Michael Jones and Lisabeth Willey and the other contributors to *Eastern Alpine Guide* have identified more than 40 alpine mountain ranges, many of which are little-known and support thousands of hectares of tundra. To those of us whose grandest encounters with alpine in eastern North America have occurred on the eight square miles of New Hampshire's Presidential Range (and to whom the 110 acres on Vermont's Mount Mansfield is precious beyond price), or on Québec's Mont Jacques-Cartier, or Newfoundland's Gros Morne—this expansion of our vision is nearly ungraspable. The very names, too, have a ring that can fill us with longing to shoulder our packs and set off: the Lewis Hills, home to the arctic hare and rock ptarmigan; Labrador's Mealy Mountains and the nearly unknown Otish Range of Québec that by the authors' estimate contains thousands of hectares of alpine terrain; the Monts Groulx in Québec where wolves roam, the very symbol of wildness, having never been extirpated as they have from Newfoundland, Gaspésie, or New England.

Many of the mountains in eastern Canada are barely half the height of Mount Washington (1916 m / 6288'), the top of the world in the northeastern United States. But loftiness counts for little here. These peaks dance on either side of the 50th parallel and gain the benefit of a lower treeline with every degree gained from their more northern location. Many rise swiftly and steeply in a curve of cirques, their massive shoulders rounding up to a windswept tableland. Here is found beauty.

Here are alpine plants familiar from the Presidentials, Katahdin, Mont Jacques-Cartier and Gros Morne: the cushions and mats of mountain cranberry, three-toothed cinquefoil, alpine bilberry, Labrador tea, and diapensia. Their roots are interlocked as the individual plants compete for moisture and nutrients. For shelter too, as the summers are short and damp and cold, and the winters long and icy. But there are dozens of other plants not seen on these mountains that are unique to the uplifts farther north with their own particular geology. Caribou can be common on the tablelands. From where the authors stand, they can see no roads, no buildings, no powerline corridors. Except for the caribou, they are alone. Remoteness is an attribute of wildness. They have walked few trails. Difficulty of access is an attribute of wildness. There is no safety net here in the event of accident. Uncertainty, too, is an attribute of wildness.

Remoteness is an attribute of wildness. Difficulty of access is an attribute of wildness. Uncertainty, too, is an attribute of wildness.

These mountains are unique in their combination of geology, climate, weather, animals, and plants. Yet, these authors know that looming threats (climate change, highways, logging to the extent of deforestation, iron mining, wind turbines, and dams—one 213 m [700'] high and nearly 1.6 km [1 mi] across) lie just beyond their vision, threatening the wildness, this seemingly endless expanse of eastern alpine—the total extent of which in their words, "has never really been articulated."

This is a landscape that those familiar with the Northeast's alpine zones would feel at home in, that is, in terms of recognizing some of the plants. The *Guide* authors have called the Mealy Mountains and Monts Otish "frightening, untamed ranges." Mystery is an attribute of wildness, and in size, the eastern Canadian alpine could overwhelm those of us whose main experience with above treeline tramping is the Presidential Range in New Hampshire or the more heavily developed peaks of Québec and Newfoundland. Because of its size, the alpine tundra of eastern Canada possesses wildness to a degree not present in the Northeast's mountains. That is why it matters. This eastern alpine, with its remnant of ice age plants, is relatively undeveloped and intact. Wolves still live here and the predator-prey cycles are intact also. It is a landscape worth preserving.

Every incursion into wildness diminishes us as humans. We can ensure its preservation. But first we must come to value and love wildness for its own sake, and as we do, ourselves.

New Hampshire's White Mountains saw early visitors, the botanists and the artists, at the turn of the 19th century. By the 1850s, Mount Washington boasted two hotels on its windy summit that were taking tourists who arrived by foot up the trails, or more significantly, by the 1860s, were transported by the Cog Railway up a western flank, or by the Carriage Road to the east. (Both rail and road are in active use today.) Mountain hotels sprang up on other summits. Hikers can see remnants of their foundations on Mounts Moosilauke and Lafayette. Great hotels filled the White Mountain notches from Pinkham to Crawford to Franconia, and those who visited arrived by rail from Boston and Providence, Philadelphia and New York. Most did not come to rough it or to experience its wildness, though by the 1870s Moses Sweetser's guidebooks detailed the trails and described the summit views.

This pattern of recreational development can be traced across the northeastern United States and southeastern Canada. It spread to the Green Mountains and the Adirondacks, south to the Berkshires in Massachusetts and further south into Connecticut, and north to the mountains of Gaspésie. Of the more prominent peaks of the northeastern United States only Katahdin escaped some sort of summit building. Its trail system, too, lagged behind the ranges more easily reached by rail or road.

No sooner was the trail system firmly in place in the White Mountains than the logging began. By the 1880s, hundreds of miles of railway tracks penetrated the deep wild heart of the mountains. Logging roads, laid out like contour lines on a map, switchbacked up the steep-sided mountains so as to get at more trees, all hauled to the mills by teams of horses. Men and horses lived in the logging camps, rarely coming out of the woods. Sparks, spewing from the trains' smokestacks, set fire to the forests, burning thousands of acres. Those in the mountain villages, where summer colonies devoted to tramping had taken root, complained of "the smokes." The air, even on the highest peaks, was thick and acrid, views obscured. Hikers had no sooner cleared their trails from the logger's slash than they would be covered again. By 1911, the Weeks Act, leading to the establishment of the White Mountain National Forest, ended this excessive despoliation. The forests grew slowly back.

"We abuse land because we regard it as a commodity belonging to us," wrote Aldo Leopold in *The Sand County Almanac*. "When we see land as a community in which we belong, we may begin to see it with love and respect." Seeing land only as commodity and without love, is what these authors hope to forestall. Once machines are turned loose upon the tundra, its recovery is unlikely. While our eastern forests can begin recovery from extensive logging in a generation or two, it takes centuries to grow these tundra plants. But there is precedent. It has been set by those who fought to save New Hampshire's White Mountain forests with the passage of the Weeks Act. Before that, in the Adirondacks, in 1892, a conservation effort established the Adirondack Park that today contains a unique landscape of towns and forest industries, yet is as wild as any comparable mountain landscape in the Northeastern United States.

The authors of the *Eastern Alpine Guide* are asking us to expand our vision for conservation northward into the alpine of eastern Canada. Industry and development are pushing inward. At the recreational level, some trails have already been built. This *Guide* itself will bring more people. These authors view this as a necessity in their call for a wise application of conservation; the landscape must be allowed to retain its wild character, and without a concerned constituency, the landscape will soon be overrun by the industrial interests. Wildness, as I see it, is perhaps the most threatened "commodity" on our planet today.

"Here was no man's garden...It was the fresh and natural surface of the earth."
—*H.D. Thoreau*

Captain Alden Partridge, on the summit of Camel's Hump in 1818, a Vermont mountain whose alpine expanse, even then, was measured in square feet, wrote that he found the view, "grand and sublime...The whole appeared to me so strangely illustrative of the original state of chaos."

Being in a wild spot can lift us out of ourselves. It *can* be frightening. Thoreau found it so on Katahdin in 1848. "Nature was here something savage and awful," he wrote. "Here was no man's garden...It was the fresh and natural surface of the planet earth." He reached the tableland, and found himself "deep within the hostile ranks of clouds." He was trapped in the "cloud factory...Vast, Titanic, inhuman Nature has got him at disadvantage." Thoreau turned back. He did not reach the summit of Katahdin. He desired only to regain his companions whom he found on the lower slopes, picking cranberries. Thoreau was overwhelmed by this mountain wildness, an entirely different degree of wildness from that found along the shores of his beloved and familiar Walden Pond. "...wind on our cheeks!" Thoreau wrote with exuberance, about his experience on Katahdin. "[T]he *solid* earth! the *actual* world! the *common sense*! *Contact*! *Contact*! *Who* are we? *Where* are we?" We can feel, as well, his fear.

Some will visit this eastern alpine wilderness with this guide in hand. Others will choose to stay beside their own Waldens in New York, New England, southern Québec and Maritime Canada. For them, it will be enough to know that here, because of its great size and isolation, there are opportunities for remoteness, uncertainty, and mystery.

Aldo Leopold wrote, "Many of the diverse wildernesses out of which we have hammered America are already gone; hence in any practical program the unit areas to be preserved must vary greatly in size and in degree of wildness." Wilderness, as a resource, *can* shrink. It cannot grow. The same can be said of wildness. These authors are calling our attention to a unique alpine wilderness in an unsuspecting region of North America... "that represents the last intact (if fraying) biome on the planet." Our time here is short; our impact large. We must be mindful of what we leave.

Laura Waterman
East Corinth, Vermont
April 1, 2012

Preface (Editors' Note)

The *Eastern Alpine Guide* has been many years in the making. On one hand, this book is simply the most recent in a long tradition of celebrating the striking significance of eastern alpine tundra. On the other hand, it may be the first to describe this particular ecosystem holistically, setting aside the major geopolitical boundaries that divide our region, including state and provincial boundaries, the U.S.–Canada border, and the border between Québec and the Atlantic provinces. This is not a technical account, nor is it a trail guide; it is a natural history guide to these wind-blistered landscapes, the relict alpine tundra of temperate eastern North America. This book is written for naturalists, hikers, wilderness advocates, mountaineers, and scientists who have already established some connection to alpine landscapes, and for those who are ready to do so. If you have never visited the alpine tundra, you must. Alpine summits are worlds apart from the grounded, overwrought forests below. If you already have firsthand experience of an alpine mountain—Mount Washington, say, or Mont Jacques-Cartier, or Gros Morne—we hope that you will intuitively pick up the proverbial "thread" that leads northward to less-traveled treeless mountains. If you don't have that firsthand experience, we hope that this book provides you with tools and information to appreciate the significant threshold, from forest to tundra, that occurs near the tops of our highest mountains.

Our primary purpose in writing this guide is to bring some color and depth to the dozens of underprotected and little-known alpine peaks of the East, with the hope of contributing in small ways to the development of a constituency for eastern alpine landscapes. We have avoided writing a trail guide, and we have tried to emphasize the responsibility that each visitor shares in maintaining the wildness of alpine summits. This really can't be overstated. We excluded detailed trail information for several important reasons. Most practically, many of the ranges profiled have few trails across them, and hiking in wild and remote regions requires a measure of self-reliance and responsibility. This book should not be the only preparation in attempting to embark on a traverse across some of the most rugged and lonely terrain in eastern North America. Secondly, in the seldom-traveled mountains, dispersed use is apparently the best way to minimize the impact of a few additional visitors each year. Finally, we feel the reader should be able to experience each range through a fresh lens.

The most southerly of these ranges are well known, well studied, well climbed, and well loved. Mount Washington, Katahdin, the Gaspé peaks, and Gros Morne all have heavily developed hiking trails and a dedicated following. But the other ranges, particularly those in the north, are, for the most part, isolated and remote. So a clear question, at least to us, as we began to develop accounts of the eastern alpine areas, was this: Why not leave them be? And part of the answer lies in the fact that as the global human footprint increases (by the day), these mountains become less remote and less wild, whether climbers scale their flanks or not. The issue is more complicated than whether or not to promote the existence of these places.

Certainly, the last thing we want for these ranges is for them to become any less wild. But the global pressures of the 21st century dictate that much of their alpine wildness is on a trajectory to be lost—whether by industry or temperature increases or precipitation change or bureaucracy will vary from mountain to mountain. For this reason, we posit that an informed constituency is an essential step toward permanent protection and ideal management. Furthermore, we hope that improved dialog between the local residents, mountain enthusiasts, land managers, and scientists across state and provincial boundaries will, by extension, improve the conservation outlook for all of the various mountains, protected and otherwise.

We've come to see this book as a very rough conservation blueprint for eastern alpine mountains, based on the two-tiered premise that alpine mountains, by virtue of their complex topography and relatively intact surroundings, will provide our best harbors for boreal and subarctic biodiversity in the uncertain years to come. Where alpine mountains have not been sufficiently protected, we must design an appropriate path to their long-term protection. Where alpine mountains have already been protected, we must continue to improve their management. Ideally, this book will contribute in a positive way toward the protection of these mountains in their wild state.

Thank you for reading this! We hope you enjoy it. See you on the mountain.

Mike & Liz

Acknowledgments

We first seriously imagined this book on a crisp night in the winter of 2008 when we were gathered in one of the hand-hewn, pre-Revolution barrooms in the Colonial Inn in downtown Concord, Massachusetts. Heavy, cold air seeped off of Monument Square, but the old tavern was full of warm light. The prospect of writing a book that unified what was known about the eastern alpine mountains was an exciting and intuitive notion. Whether or not we were on to a good idea, it seemed so at the time. We had not written a book before.

For much of the next three years, we rambled over the high alpine peaks of New York, New England, Newfoundland, Labrador, and Québec, exploring many of the alpine mountain ranges profiled in this book. Along the way met many people who were variously kind, brilliant, visionary, funny, and similarly enamored with the eastern alpine mountains. Several of these wonderful people eventually agreed to contribute their ideas or photographs to this book. We'd like to acknowledge their specific contributions here.

First, we'd like to gratefully acknowledge several scientists for their contributions of original chapters on the natural history and physical science of eastern mountains. Jean-Philippe Martin and Dr. Daniel Germain, from the Université du Québec à Montréal, co-authored Chapters 2 and 3, on the ancient mountain-building events that formed the foundations of today's eastern mountains and the processes during and after the Pleistocene "ice age" that sharply altered the face of the old mountains. Dr. Kenneth Kimball, Director of Research for the Appalachian Mountain Club, authored Chapter 4, an accessible overview of mountain weather and climate, with Neil Lareau and Liz Willey.[1] Marilyn Anions (NatureServe Canada) co-authored Chapter 5. Our good friend Dr. Noah Charney, at the University of Massachusetts, wrote the bulk of Chapter 6, a breakneck discussion of alpine fauna.[2]

The mountain range profiles found in Chapters 7 to 21 were greatly improved through the contributions of Marilyn Anions (NatureServe Canada); Dr. Guillaume de Lafontaine (Université Laval); Claudia Hanel (Newfoundland & Labrador Department of Environment and Conservation); Luise Hermanutz, Andrew Trant, and Laura Siegwart Collier (Labrador Highlands Research Group, Memorial University); Michael Lederer (American Alpine Club); Paul Wylezol (International Appalachian Trail—Newfoundland & Labrador); and Julia Goren (Adirondack Mountain Club). Their contributions made this collaborative work possible and the chapters to which they contributed are noted in the Table of Contents.

Four additional authors made critical contributions either in the conceptual stage or throughout the process. Matt Burne was the lead designer of the *Eastern Alpine Guide* and contributed greatly to the style and content throughout the book. Iona Woolmington made critical contributions to most of the chapters and throughout the text and made countless recommendations to improve the narrative flow. Will Kemeza, our good friend and cofounder of Beyond Ktaadn, was instrumental in the conception of this book and in exploring

---

1   Chapter 4 is based largely on Ken's essay "Wind: A Jekyll-and-Hyde Factor for the Northeastern Mountains," published in the Winter/Spring 2012 issue of *Appalachia*.

2   Noah coauthored the masterful and eclectic *Tracks and Signs of Insects and Other Invertebrates* (2010), which we highly recommend.

the ranges, and he contributed to Chapter 18 (Monts Groulx)[3] and Chapter 22 (Conservation). Charley Eiseman contributed many excellent images to the text.

This is a highly illustrated book, and we relied on photographs to shoulder the burden of our story. Most of the authors identified earlier contributed photographs, as did several other excellent naturalists. In particular, we'd like to acknowledge the photographic contributions of Ludovic Jolicoeur, Nathaniel Rayl, Mollie Freilicher, Jared Woodcock, Bill Deluca, Daniel Dubie, Monica Kopp, Pat Johnson, Brian Tang, Mark Rainey, W.L. Dickson, Jon Feldgajer, Joe Hardenbrook, Jerry Kobalenko, Mathieu Lemieux, Jonathan Mays, Daren Worcester, and Benjamin Griffiths. We'd especially like to acknowledge Jim Salge for his excellent cover photograph from Mount Washington's Great Gulf.

There are already a handful of excellent books detailing the patterns and processes of alpine regions around the world. We definitely wrote this book on the "shoulders" of Dr. Nancy Slack and Allison Bell's *AMC Field Guide to New England Alpine Summits*, and the *AMC Field Guide to Mountain Flowers of New England* by Dr. Stuart Harris and others. We have done our best to honor their tradition. We recommend both of these books highly and carry them with us on all of our trips into alpine backcountry. Additionally, we highly recommend the excellent, thorough, and thoughtful books of Laura and Guy Waterman, including *Forest and Crag* and *Wilderness Ethics*, two very different books that capture the historical and ethical complexity of mountain conservation. We also relied heavily on the work of alpine ecologists from around the globe including Laszlo Nagy, Georg Grabherr, Christian Körner, Ann H. Zwinger, and Charlie Cogbill. Last, we humbly acknowledge the early explorations and writings of Merritt Lyndon Fernald, William Oakes, Noel Odell, Arthur Stanley Pease, Olof Rune, Henry David Thoreau, Jacques Rousseau, Vero Copner Wynne-Edwards, Edward Tuckerman, and many others who made contributions to the field of eastern alpine ecology during its infancy.

This project found its feet because of the support of many people.

We are especially grateful for the early support of the **Guy Waterman Alpine Stewardship Fund** (watermanfund.org), and we are thankful for the encouragement of Laura Waterman, Val Stori, Julia Goren, Annie Bellerose, Matt Larson, Jeff Lougee, and Rick Paradis. We are also indebted to private donors who contributed very generously and early to the *Eastern Alpine Guide* project, including Sally Waisbrot, Steve Sauter, McKesson, Inc., Janet Chayes, Ruah Donnelly, David Small, the Athol Bird and Nature Club, Norcross Wildlife Foundation, Georgia Stanley, Laura Fallon, An Tang, Bryan Windmiller, Henry Woolsey, and anonymous donors.

We are also thankful for the time, resources, and dedication of Kathi Anderson and Juliet Trofi at the Walden Woods Project, a Concord, Massachusetts–based conservation organization that served as Beyond Ktaadn's fiscal sponsor since 2009. Beyond Ktaadn's board of directors—Steve Young, Val Stori, and Matt Burne—provided extremely useful feedback

---

3   Will's essay, "Monts Groulx, An Entire Earth" from the Summer/Fall 2010 issue of *Appalachia* greatly influenced the tone of Chapter 18.

throughout the process. Steve Young, the creative force behind the Center for Northern Studies, originally inspired us to explore the alpine barrens of Newfoundland more than a decade ago and has been a mentor to us in the years since.

Nancy Ritger and Doug Weihrauch of the Appalachian Mountain Club have been particularly important inspirations for more than a dozen years. Several other AMC folks provided encouragement and feedback along the way, including Chris Thayer, Walter Graff, and Eric Pedersen. Chris Woodside, editor of the AMC's *Appalachia*, provided welcome encouragement and feedback throughout the process of researching and writing this book. Three essays first published in *Appalachia* appear in partial or edited form in this book: Will Kemeza's "Monts Groulx, An Entire Earth" first appeared in the Summer/Fall 2010 issue and now appears as excerpts in Chapter 18. Our essay "Ten Peaks in the Great Eastern Alpine Zone: A List of Great Mountains Expands Old Ideas of Boundaries" first appeared in the Winter/Spring 2012 issue and now forms part of the introduction. Ken Kimball's "Wind: A Jekyll-and-Hyde Factor for the Northeastern Mountains" appeared also in that issue and has become Chapter 4.

Many excellent biologists provided thoughtful feedback and criticism throughout the process. Brendan Dunphy contributed his interpretation of mosquito species common throughout the eastern alpine region, and we have directly incorporated his expertise into Chapter 6. Ludovic Jolicoeur of the University of Québec at Sherbrooke provided encouraging feedback, as well as important observations of birds and insects, throughout the process. Sean Robinson graciously helped with the bryophyte identification. We found the online resource http://bugguide.net extremely helpful when identifying insects. Plant taxonomy follows VASCAN, the Database of Vascular Plants of Canada (http://data.canadaensys.net/vascan).

We found the website http://peakbagger.com an intuitive, fascinating, and information-rich source of information about mountain geography.

Jean Hoekwater (Baxter State Park), Leighlan Prout (White Mountain National Forest), Isabelle Thibault and Dominic Boisjoly (Québec provincial government), Claudia Hanel (Newfoundland Department of Environment and Conservation), and many others provided encouragement and support during the course of research activities intended to complement this book.

Rob Elliott, of Arizona Raft Adventures in Flagstaff, during two river trips through the Grand Canyon in 2008 drove home the significance of steady constituency-building in the long course of protecting and defending our stunning natural landscapes.

For companionship in the field, we wish to thank our many coauthors as well as Bill Aughton, Tom Seidel, Michel Denis, and Guillaume Fortin. For logistic help as well as camaraderie abroad, we are extremely thankful to Michel Denis, Isabelle Thibault, Dominic Boisjoly, J.P. Messier, Ludovic Jolicoeur, Marie-Pierre Clavette, Kevin Noseworthy, and Stewart and Marta Butt. For encouragement and support of all kinds, we thank our sisters Laura, Mckenzie (& Jake), Penelope, Larisa (& Jay), Laurie (& Eric), and Leslie (& Alex). Heartfelt thanks to Laurie and Eric for frequently dogsitting our stubborn and wonderful Labrador, Addie, who once demonstrated her herding abilities in a terrible, thunderous lightstorm on the high Groulx.

Jared Gange, of Huntington Graphics, saw merit in the "Eastern Alpine Guide" concept and worked thoughtfully with us for more than a year to see it through to completion—and introduced us to several of Vermont's best bakeries.

We'd like to thank our parents, Tom & Betty and Lynn & Lloyd, for their support through the past 10 years and for introducing us to New Hampshire's White Mountains 25-odd years ago. We know that they worried when we left for the mountains in snow and wind. Their support has made it possible for us to pursue the projects and places we love, and we owe them everything.

Thanks again to everyone who didn't ask "why?"

Mike & Liz

Introduction 1

Alpine tableland, Bay of Islands region, Newfoundland. (CE)

Map 1-1. Alpine mountain ranges of eastern North America.

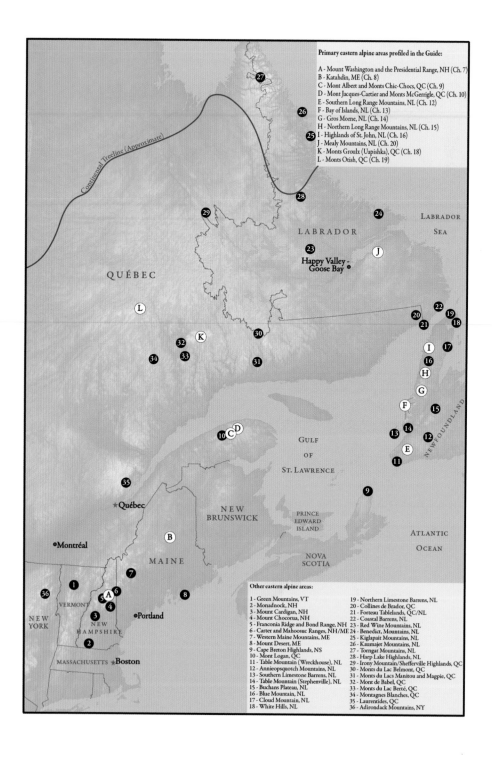

Primary eastern alpine areas profiled in the Guide:

A - Mount Washington and the Presidential Range, NH (Ch. 7)
B - Katahdin, ME (Ch. 8)
C - Mont Albert and Monts Chic-Chocs, QC (Ch. 9)
D - Mont Jacques-Cartier and Monts McGerrigle, QC (Ch. 10)
E - Southern Long Range Mountains, NL (Ch. 12)
F - Bay of Islands, NL (Ch. 13)
G - Gros Morne, NL (Ch. 14)
H - Northern Long Range Mountains, NL (Ch. 15)
I - Highlands of St. John, NL (Ch. 16)
J - Mealy Mountains, NL (Ch. 20)
K - Monts Groulx (Uapishka), QC (Ch. 18)
L - Monts Otish, QC (Ch. 19)

Other eastern alpine areas:

1 - Green Mountains, VT
2 - Monadnock, NH
3 - Mount Cardigan, NH
4 - Mount Chocorua, NH
5 - Franconia Ridge and Bond Range, NH
6 - Carter and Mahoosuc Ranges, NH/ME
7 - Western Maine Mountains, ME
8 - Mount Desert, ME
9 - Cape Breton Highlands, NS
10 - Mont Logan, QC
11 - Table Mountain (Wreckhouse), NL
12 - Annieopsquotch Mountains, NL
13 - Southern Limestone Barrens, NL
14 - Table Mountain (Stephenville), NL
15 - Buchans Plateau, NL
16 - Blue Mountain, NL
17 - Cloud Mountain, NL
18 - White Hills, NL

19 - Northern Limestone Barrens, NL
20 - Collines de Brador, QC
21 - Forteau Tablelands, QC/NL
22 - Coastal Barrens, NL
23 - Red Wine Mountains, NL
24 - Benedict, Mountains, NL
25 - Kiglapait Mountains, NL
26 - Kaumajet Mountains, NL
27 - Torngat Mountains, NL
28 - Harp Lake Highlands, NL
29 - Irony Mountain/Shefferville Highlands, QC
30 - Monts du Lac Belmont, QC
31 - Monts du Lacs Manitou and Magpie, QC
32 - Mont de Babel, QC
33 - Monts du Lac Berté, QC
34 - Montagnes Blanches, QC
35 - Laurentides, QC
36 - Adirondack Mountains, NY

1. Introduction

1-1 Eastern alpine tundra provides habitat for many disjunct, subarctic species. Monts Groulx, Québec. (MJ)

This book focuses on the mountains of northeastern North America, encompassing the ancient and weatherworn summits and plateaus of the Canadian Shield and the Northern Appalachian Mountains (see Ch. 2). We've further narrowed our focus to those peaks, plateaus, ranges, and ridges that for various reasons break through the natural treeline to rise above the surrounding boreal or temperate forest, providing southern refugia for rare, relictual, isolated populations of subarctic and arctic plants and animals (1-1).[1] We follow the current convention of calling these treeless ecosystems "alpine tundra," mostly for the simple reason that we'd like to avoid semantics in favor of a more substantive discussion.[2] However, there is great variety of habitat types among the various eastern alpine areas, so in the main bulk of this book, Chapters 7–21, we focus on both the uniqueness and similarities of twelve major, representative, eastern mountain ranges and thirty or so smaller or outlying alpine

---

1 See Chapters 4, 5, and 6.

2 The term *alpine* itself might be a source of some confusion. Though it originally referred to the Alps themselves, it has become generalized, and we take it to mean those ecosystems, associated with mountains, that are high enough in elevation to rise above the natural treeline. The *alpine zone* or *alpine tundra ecosystem* has been called many other things. Because *tundra* refers to the treeless plains of the Arctic, and because the factors responsible for the formation of treeline at high elevations are different than those driving the location of arctic treeline and arctic tundra (Ch. 4), other terms describing the treeless mountains of the East have been tossed about. The elevation of treeline, and the factors influencing its elevation vary throughout the region (Ch. 4), but whatever their origin, we focus on those peaks that rise above the trees, but occur south of the continental treeline. The great peaks of the far north (e.g., the Torngat Mountains, Ch. 21) that rise from the arctic tundra itself to the higher nival zone (i.e., the zone continuously covered in snow), and treeless peaks of anthropogenic origin are, for the time being, beyond the scope of this book, though we touch on them briefly in Ch. 21.

1-2 *The Colorado Rockies support more alpine tundra than the entire eastern United States combined. Mummy Range, Rocky Mountain National Park, Colorado. (MJ)*

areas (or areas peripheral to our purpose here). The alpine mountains are topographically complex, covering a wide range of slopes and aspects. This provides critical refugia for boreal and subarctic species—important in the context of threatening climate scenarios, and simply as part of a well thought-out plan for the conservation of mountain biodiversity.

In North America, most of the alpine territory occurs in *the West*. In Washington, Idaho, Montana, Oregon, Wyoming, California, Nevada, and Colorado, the total area of alpine habitats is really shocking, especially to someone raised on summits in southern Québec, New England, or New York. This is to say nothing of the alpine expanses in Alaska, Yukon, British Columbia, and Alberta. Consider this fact: In Colorado's Rocky Mountain National Park, let alone all of the Colorado Rockies, there is more than ten times the total area of alpine tundra as on all of the northeastern U.S. peaks combined (1-2). The view west on a clear day from the Rocky Mountain Front Range is sea of Katahdins and Gros Mornes. Many lifetimes will yet be spent exploring the sprawling, treeless wilds of the Sierra Nevada, the Cascades, and the Rockies, and the other dazzling, western alpine ranges, including the sky islands of Arizona and New Mexico. There is even alpine tundra in Mexico, on Pico de Orizaba, Popocatépetl, and other massive strato-volcanoes within a day's drive of Mexico City.

There is also spectacular alpine territory in eastern North America, of course, and those areas are the focus of this book. In the story of eastern alpine tundra, geography is key. In

1-3 *Perhaps the most famous major eastern alpine peaks are, clockwise from top left: Mount Washington, Katahdin, Mont Jacques-Cartier, and Gros Morne. (MJ)*

this chapter, we present a very general introduction to alpine environments and more specifically to the geography of alpine habitats east of the Rocky Mountains. The best-known eastern alpine ecosystems are found south of the Gulf of St. Lawrence, including the Adirondacks in New York (Ch. 21), Mount Washington in New Hampshire (Ch. 7), Katahdin in Maine (Ch. 8), Mont Albert and Mont Jacques-Cartier (Ch. 9 and 10) in the Gaspésie region of Québec, and Gros Morne in Newfoundland (1-3). Undoubtedly, their relative popularity is due to their proximity to Boston, Portland, Montréal, and Québec City. The best-known and most recognized of these is enormous Mount Washington. Our collective understanding of eastern alpine ecology is susbtantially rooted in studies and surveys undertaken there since the mid-1800s.

1-4 *East of the Gulf of St. Lawrence, alpine tundra occurs sporadically along the western coast of Newfoundland. Serpentine Valley, Lewis Hills, Newfoundland. (IATNL)*

Many mountaineers based in New England or southern Québec (whether by choice or chance) will, at the end, be able to measure their commitment to these mountains in decades. But in this book, Mount Washington, Katahdin, Gros Morne, and the Gaspé peaks

1-5 *Throughout interior Québec there are numerous alpine areas of various size and complexity. Monts Groulx, Québec. (MJ)*

play another role: to provide context—and a familiar launching-point—for a broader treatment of low-latitude alpine ecosystems in eastern North America. Beyond Mount Washington, beyond Katahdin, and beyond Gaspésie are many, many other mountains.

Distribution of Alpine Tundra in Eastern North America

The handful of well-known alpine peaks south of the Gulf of St. Lawrence and the St. Lawrence River in New York, Vermont, New Hampshire, Maine, and southern Québec together compose a relatively small fraction of the alpine ecosystems in eastern North America. Alpine-capped mountains, rising above a surrounding sea of spruce-dominated boreal forest, continue north in great arcs along both shores of the Gulf of St. Lawrence. To the east, these mountains emerge from the sea and run the full length of Newfoundland's west coast via the various Appalachian chains (1-4). West and north of the Gulf of St. Lawrence, high alpine peaks run at disjointed intervals through Québec and Labrador along the Laurentian Plateau of the Canadian Shield (1-5). In this way, the Gulf of St. Lawrence forms a rough boundary region between three distinctive mountainous regions: the Northern Appalachians (covered in Ch. 7–11), the Newfoundland Appalachians (Ch. 12–17), and the Canadian Shield (Ch. 18–21).

Generally, eastern alpine areas are clearly demarcated. In fact, one of the reasons that they are worth writing about is simply how starkly different they are from any other place. Alpine

Plate I. Characteristics of eastern alpine ecosystems

Alpine ecosystems are generally distributed on the tops or shoulders of relatively high mountains, above the natural limit of tree growth. The treeline forest is often composed of conifers such as balsam fir, and white and black spruce. Mont de la Passe, Gaspésie, Québec. (MJ)

Mountains in eastern North America support hundreds of species of northern and subarctic plants and animals generally found much farther north, such as woodland caribou (see Ch. 5 and 6). Highlands of St. John, Newfoundland. (MJ)

Eastern alpine areas are often found in association with mean annual temperatures near or below freezing, as well as prominent soil patterns caused by severe freeze-thaw cycles. Permafrost has been documented in many areas (Ch. 3 and 4). Monts Groulx, Québec. (MJ)

Areas of late-lying summer snow are frequent and widespread in eastern North American alpine areas (see Ch. 4 and 5). Mont Jacques-Cartier from Mont Comte, Gaspésie, Québec. (MJ)

Many eastern mountains show evidence of alpine glaciers active during the Pleistocene ice ages, such as glacial cirques, tarns, and moraines (see Ch. 3). Torngat Mountains, Labrador. (LSC)

Most of the eastern alpine areas were covered by the continental Laurentide Ice Sheets during the Pleistocene ice ages. Evidence of continental glaciation such as erratic boulders abound in eastern alpine habitats, where exposed bedrock and mineral soils are common. Monts Groulx, Québec. (MJ)

ecosystems tend to be found on and around the high, bare summits of relatively tall mountains. But in some exposed areas near the sea, including many areas of Newfoundland and coastal Labrador, as well as at high latitude in northern Québec and Labrador, it becomes less clear what constitutes *alpine* landscapes and what exhibits certain alpine characteristics as a result of mechanisms other than those associated with high mountains. The elevation at treeline varies with climate (itself a function of multiple variables discussed in Ch. 4), slope, aspect, substrate, and disturbance regime, and in certain areas the mountain ecosystems coalesce with the low-elevation tundra of the far North. Such is the case in northern Labrador, where alpine ecosystems merge with sea-level tundra of the Ungava Peninsula. These great Labrador ranges, including the Torngat, Kaumajet, and Kiglapait Mountains, are in some ways more similar to the mountains of the Arctic Cordilleran mountain ranges on Baffin and Ellesmere Islands than to the more southern alpine summits. Confusingly, though, they are in some ways more similar to the true European Alps themselves. Put more simply, in the far northern reaches of our eastern region, alpine areas may be contiguous with low-elevation tundra ecosystems, but in the south, the distinction between *alpine* and *non-alpine* is always fairly clear. Farther south than New Hampshire's White Mountains, large alpine ecosystems are entirely absent. Mount Greylock (1064 m / 3491') in Massachusetts, Spruce Knob (1482 m / 4863') in West Virginia, and Mount Mitchell (2037 m / 6684') and Clingmans Dome (2025 m / 6643') in North Carolina, among others, are capped by subalpine spruce-fir forests. But nowhere do these states support the alpine diversity of Mount Washington and the more northern mountain ranges.

Conservation of the Eastern Alpine

The importance and vulnerability of alpine systems are recognized worldwide. Global programs have been established to evaluate and monitor this rare ecosystem type, which constitutes just 3% of the planet.[3] Because of the varied elevation and topography of mountain regions, they function as reservoirs of biodiversity. In addition to their potential wilderness value, intact mountain systems often offer ecosystem services such as erosion control, air and water filtration, groundwater recharge, and wildlife habitat. But these systems are fragile: development, natural resource extraction, invasive species, acid precipitation, and climate change may disrupt them.[4] Regrettably, we live in a time when many people denigrate or belittle these particular threats to biodiversity and wilderness, but most readers of this book are willing, we hope, to consider the complexity and value of natural ecosystems.

Despite widespread global attention, the alpine mountains of eastern North America have been largely ignored in recent global discussions of alpine ecosystems. This may be largely because the eastern mountains are limited in extent and elevation. Nonetheless, the eastern mountains bear strong biological affinities to the arctic regions 1600 km (1000 mi) to the north, supporting hundreds of species of subarctic plants and several species of distinctive arctic fauna. Furthermore, the climatic, edaphic, and biological processes underway in the eastern alpine areas closely resemble those found in the larger alpine ecosystems of western North America and Europe.[5]

---

3   e.g. http://www.gloria.ac.at/; Körner (1995)
4   Lesica and McCune (2004); Pauchard et al. (2009)
5   See Chapters 2, 3, and 4.

Rising above the boreal forest, the topographic diversity found in eastern alpine areas promotes a range of microclimates and ecological niches that support hundreds species adapted to a variety of life zones, from the temperate and boreal (or *Canadian*) forests to the alpine tundra. These topographically diverse areas offered safe haven to species through the changing climates of the past, and they will do so again in the future. For this reason, the eastern alpine mountains are extremely logical (and necessary) conservation targets.

Some of these ranges are very wild—but likely won't remain so for long. Extensive logging creeps ever closer to some; mining claims are distributed near others; high-tension wires may cut across still more. Other wild ranges are just beginning to be developed for recreational hiking. New trails have been cut into several, and many more are planned. All of this begs the question, what is the ideal outcome? There are many models from which to choose. Do we really want heavily developed roads to parking lots, and highly publicized and accessible trailheads that encourage use (and concentrate it), with restricted use elsewhere? Or should we minimize the development of access routes, creating a context for explorers to take more time and find their

1-6 Mount Washington (top) and Katahdin (bottom) are the only mountains in the northeastern United States that support thousands of hectares of alpine ecosystems. (MJ)

own way, dispersing a much lower volume of use throughout a much larger area, offering each hiker a wilder experience? It isn't a simple question, there certainly isn't only one answer, and it certainly isn't even ours to answer. We only hope to engage others in the conversation. The sheer diversity of the eastern mountains demands multiple, complex, and creative management and conservation scenarios.

Northern Appalachians

New England — The New England states claim two alpine tablelands of regional significance: Mount Washington and the Presidential Range in New Hampshire (Ch. 7), and Katahdin in Maine (Ch. 8) (1-6). Numerous smaller ranges support isolated occurrences of alpine vegetation or alpine plants; in New Hampshire, these are largely restricted to the White Mountain National Forest and the individual summits of Mount Moosilauke, Cannon Mountain, Mounts Lafayette and Lincoln (Franconia Ridge), Bond Range in the Pemigewasset Wilderness, and isolated occurrences on Carter Dome just east of the Presidentials. In Maine, isolated alpine communities occur in the Mahoosuc Range, a peripheral massif within the White Mountain region, as well as on Sugarloaf and Saddleback Moun-

1-7 The Appalachians rise from the sea at Table Mountain in southwestern Newfoundland, after ending their 3000-km continuous length in Gaspésie and Nova Scotia. Table Mountain, Newfoundland. (IATNL)

tain, Mount Abraham, and Bigelow Range. In Vermont, the largest occurrence of alpine tundra occurs atop rocky Mount Mansfield, with smaller occurrences on Camel's Hump and Mount Abraham (Ch. 11).

Gaspésie — North of New England, the next major alpine range of the Appalachians occurs on Québec's Gaspé Peninsula. Here, major tundra complexes are found along the spine of the Monts Chic-Chocs, in particular on Mont Albert (Ch. 9), and on Mont Jacques-Cartier and the Monts McGerrigle (Ch. 10).

Isolated arctic-alpine plant species occur on the high barrens and ravine ledges of Nova Scotia's Cape Breton Highlands (Ch. 11). No part of Cape Breton, however, is dominated by arctic or alpine species, and this region—though windswept and boreal—is only briefly explored in this book.

Newfoundland Appalachians

Alpine tundra appears again across the Cabot Strait on the southwest corner of Newfoundland (1-7). Here, various, large, alpine mountain ranges are found along the coast and up to 100 km (60 mi) inland. In the south, these ranges include Table Mountain, the Annieopsquotch Mountains, and numerous other promontories associated with the Southern Long Range Mountains (Ch. 12). North and west of these ranges lie the mountains of the Bay of Islands (Ch. 13), which include four major alpine mountain complexes (Lewis Hills, Blow Me Down, North Arm Hills, and the Tablelands) south and north of the city of Corner Brook and the namesake Bay of Islands. The northernmost of these, Tablelands, lies within Gros Morne National Park, which also encompasses its namesake summit (Gros Morne, Ch. 14) and many thousands of hectares of alpine tundra on such additional mountains as Killdevil, Big Hill, the Rocky Harbour Hills, and Big Level. Diverse and wild alpine tundra continues from Big Level in large discontinuous stretches along the Long Range Mountains to Blue Mountain, and assumes spectacular form on the twin Highlands of St. John (Ch. 16), slightly north of the town of Port aux Choix. At the northern end of Newfoundland's Great Northern Peninsula, the White Hills rise above the city of St. Anthony, and numerous limestone hills culminate in stunning barrens overlooking the Strait of Belle Isle. Although the limestone barrens host world-class displays of arctic and subarctic wildflowers, and several have a distinctly alpine aspect (any trip to Newfoundland is worth a visit to Burnt Cape, Watt's Point, and Cape Norman), they are only briefly discussed in this book (see Ch. 17 for a summary overview and recommendations for additional reading).

Canadian Shield

Labrador — Thus ends the Appalachian chain in North America. This book explores only the western shores of the Atlantic, and does not follow the Appalachians to the Old World.

However, alpine ecosystems resume across the Strait of Belle Isle on the mainland shores of Labrador, where the ancient Canadian Shield (twice as old as the oldest Appalachian peaks) dominates the landscape. Immediately behind the ferry town of Blanc-Sablon, Québec, low hills rise on both sides of the Québec–Labrador border. These are the metamorphic Collines de Brador; nearby are four sedimentary plateaus known as the Forteau Tablelands. Coastal tundra makes intermittent appearances along the Labrador coast—at Red Bay, Battle Harbour, and Cartwright, the low hills are entirely above treeline. Two hundred and sixty kilometers (156 mi) north, on the south shore of Lake Melville, are the Mealy Mountains (Ch. 20), the northernmost and perhaps the most spectacular of the eastern alpine ranges fully explored in this book (1-8).

1-8 The Mealy Mountains of Labrador are one of the more rugged and remote alpine ranges in the east. (MLL)

1-9 The Monts Groulx in Québec rise above a sea of old growth spruce and encompass the largest island of alpine tundra in eastern North America. (MJ)

North of the Mealy Mountains, along the wild coast of subarctic Labrador, are four massive ranges of real consequence. These great ranges offer world-class mountaineering and in many ways are more closely related to the Arctic Cordillera and the mountains of eastern Baffin and Ellesmere Islands and Greenland than to the "islands" of alpine tundra farther south. South of Makkovik, Labrador, the Benedict Mountains hug the ocean. Near Nain, Labrador (which is accessible via the M/V *Northern Ranger*, a passenger and freight supply ship) the Kiglapait Mountains rise over 1000 m (>3280') at Man O'War Peak. One hundred kilometers (60 mi) farther north, the Kaumajet Mountains rise even higher, above 1200 m (>3937') at Brave Peak. Another 150 km (90 mi) farther north, the great and inaccessible Torngats rise to an almost unimaginable (for this latitude and longitude) 1652 m (5420') at Mount Caubvick. Several Torngat summits rise above 1500 m (>4921') around Nachvak Fjord. The mountains are undeniably alpine, and they are surrounded at sea level by tundra of the continuous Low Arctic. Many of these northern mountains do not have names—a signature of their remoteness and wildness. They are not fully treated in this guide, under the slightly disingenuous pretense that they do not rise above a surrounding forest and thus are not islands of alpine tundra.[6]

Interior Québec — Several great, rambling alpine mountains rise above the Laurentian Plateau of interior Québec. The Monts Groulx (Ch. 18) are by far the most extensive and accessible (1-9). Slightly smaller in overall area are the Monts Otish north of Lac Mistassini (Ch. 19) and the Montagnes Blanches west of the Manicouagan Reservoir. Smaller summits, or ranges, near Lac Berté and Lac Manitou rise well above treeline. Small areas of of tundra occur frequently near the Québec-Labrador border, especially in the vicinity of Schefferville and Labrador City (see Ch. 21).

South of the Monts Otish and Monts Groulx, the southern tier of the Canadian Shield is dominated by two enormous massifs, both of which are mostly subalpine or anthropogeni-

6   Islands of alpine tundra are a form of "inselberg" or ecological island.

cally bare. The Laurentian peaks, south of the Sanguenay Fjord and northwest of Québec City, rise well over 1000 m and encompass the famous subalpine summits Mont du Lac des Cygnes and Mont Raoul Blanchard. In New York, the Adirondacks rise to the highest point on the Canadian Shield south of the Torngats at Mount Marcy (1629 m / 5344'), a spectacular, steep, alpine summit. About one dozen Adirondack summits support alpine ecosystems, including Algonquin Peak and Whiteface Mountain (1-10).

1-10 Most of the alpine ecosystems in New York's Adirondack State Park are located on Mount Marcy and Algonquin Peak. Algonquin Peak from Mount Marcy, New York. (JG)

This book covers each region from south to north. In this way, each section is led by accounts of relatively familiar and accessible mountains. The later chapters in each section tend to profile more remote, poorly explored ranges, where a few rough roads come near the mountains and the very first trails are being cut. The book itself closes with several untamed ranges: the Mealy Mountains and the Monts Otish. There is no easy ground access to these two ranges in summer; there are no formal trails, and there is definitely no safety net.

Access

Accessing many of these peaks can be an adventure in itself, although in several instances it's not as difficult or expensive as one might assume. Several of the more remote ranges in this book may be accessed either by the Québec-Labrador Highway (Route 389), which joins the St. Lawrence coast-road (Route 138), at Baie-Comeau, Québec, or the Trans-Labrador Highway (Route 500) (1-11). Accessing the peaks of Newfoundland usually involves a ferry trip from North Sydney, Nova Scotia, to the southern port town of Channel-Port aux Basques (1-12). The coastal roads of western Newfoundland often come within a few kilometers of the various alpine areas, including the Trans-Canada Highway (Route 1) in the south and the Viking Trail (Route 430) in the north, although several ranges profiled are inaccessible by road. We have only included general descriptions in this book; specific directions, trail descriptions, and compass bearings are not included. Many of these ranges

1-11 Several mountain ranges in interior Québec are accessed directly or indirectly via the Québec Labrador Highway (Route 389) (left). Route 389 also provides access to the Trans Labrador highway (Route 500), portions of which remain under construction (right). (MJ and LW)

do not have formal trails, and those that do already have extensive trail guides. Instead, we've attempted to try to tell the whole story of the mountains: the geology, weather, ecology, their conservation outlook, what unifies them, and what makes each unique.

It's important to point out how easy it is to romanticize all of these mountains from the warm, dry comfort of a New England or southern Québec summer. Raw and rotten weather can come to any big, northern mountain, but in the alpine peaks of the Canadian Shield it can inspire its own sort of faltering despair. The flies are sometimes miserably and freakishly thick (1-13)! Luckily, cold rain or snow usually tempers the enthusiasm of the flies, so seldom are both a problem. Because of all this, the shoulder seasons are nicest; there is sometimes a window in June when the flowers have opened and the flies haven't emerged. Then, there is a time in September when the flies are gone, and the weather is tolerable, and the light is long. Bands of red and yellow color the edges of small lakes.

1-12 Many adventures in Newfoundland begin with a six hour ferry from Nova Scotia. Exhaust towers on the Newfoundland Ferry at night. (NC)

1-13 Alpine exploration presents an array of challenges, blackflies being among of the more predictable. Black fly swarms in the Monts Groulx, Québec. (MJ)

Finally, several eastern mountain ranges have yet to receive formal or permanent protection from mining, road-building, and development, and in many ways they are the primary purpose for writing this book. We hope to reach a wider audience with this information, gradually building an informed constituency for the eastern alpine ecoregion and, ultimately, the conservation of its biodiversity. As a step in this process, we conclude this book with a conservation vision (Ch. 22).

Mountain Building 2

Exposed serpentine outcrops in the Lewis Hills, Newfoundland. (IATNL)

Map 2-1. Geological provinces of the mountains of eastern North America.

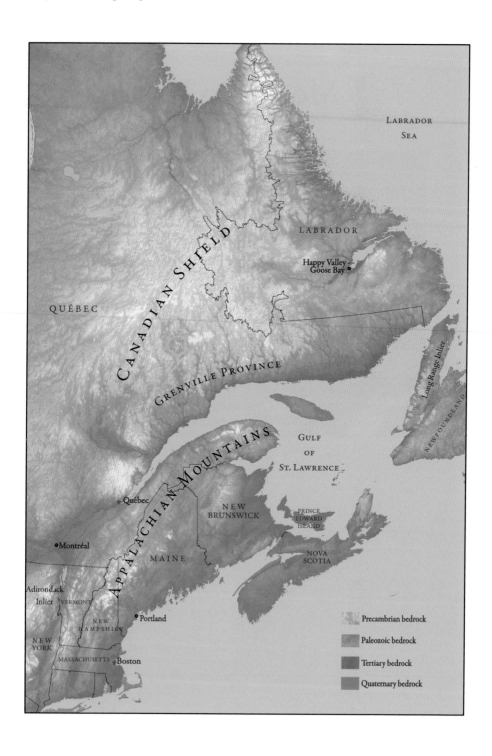

Mountain Building

2-1 The major eastern mountains may be roughly grouped by their location either on the Canadian Shield, a craton of extremely old gneiss and granite (top) in Québec and Labrador, or within the major chains of the Appalachian Mountains in New England, Québec, and Newfoundland (bottom). Mont Provencher, Monts Groulx, Québec (top); Presidential Range, New Hampshire (bottom). (MJ)

Eastern mountains can be relatively neatly divided into two major groups: those that make up the Appalachian Mountains in New England, the Gaspé Peninsula, and Newfoundland, and those in New York, central Québec, and Labrador, which were formed from ancient Canadian Shield rock more than 1 billion years ago (Map 2-1). Of course, the story is not really that simple.

The history of mountain-building events, or **orogenies,** in eastern North America is complex, and the major events have been obscured by hundreds of millions of years of erosion. In this chapter, we attempt to organize the complicated and interwoven events into a coherent narrative. Almost all of the major orogenies took place more than 100 million years ago. In the case of the mountains of the **Canadian Shield,** the important events occurred more than 1 billion years ago, making these some of the oldest mountains on Earth—far older than such well-known ranges as the Alps, the Andes, or the Himalaya. Because the Canadian Shield mountains are so old, they have a long and complex history of major tectonic events and have been subjected to uplift and folding many times over.

2-2 The volcanic ranges of the southern Canadian Shield are composed variously of anorthosite and gneiss and owe their initial formation one billion years ago to the Grenville Orogeny. The shield ranges include (clockwise from top left): the Adirondack Mountains of northern New York; the Laurentians of southern Québec; the Montagnes Blanches of interior Québec; and the Mealy Mountains of southeastern Labrador. (MJ, GD, and MA)

The origins of every mountain range on Earth—old and young—should be first considered in the context of drifting continents and the modern theory of **plate tectonics**. Plate Tectonics illuminates the origins of any given eastern mountain range. Throughout time, earthen material has been deposited and consolidated across the entire planet, and subsequently these materials have been exposed to different climates and pressures, which has created a wide variety of rocks on different tectonic plates. The collision of these tectonic plates creates **orogenic** zones where material piles up, folds, faults, and erodes to form mountain ranges on the surface. The combination of diverse bedrock, deformed through a series of major tectonic interactions, gives rise to a diverse mosaic of alpine substrates, which contributes to the richness of eastern alpine biodiversity (Ch. 5). In this chapter, we outline the **endogenic** processes—processes within the Earth, such as rock formation and tectonic activity—that are the beginning of the story of the physical landscape of eastern mountain ranges. Today's landscape is the combination of these past endogenic processes and ongoing **exogenic** processes, which are driven largely by climatic factors, eroding the Earth's surface from without (erosion and climate are the subjects of Chs. 3 and 4, respectively).

This chapter focuses on the ancient events that lifted the mountains above the surrounding landscape; the following chapter addresses the ice-age and modern conditions that formed many prominent and well-known alpine features. In the following paragraphs, we outline the major mountain-building events that gave rise to the eastern mountains, from the oldest to the most recent. We also present a succinct overview of long-term erosional processes, with an emphasis on the resulting topographic differences and their consequences for today's eastern alpine landscapes. Wherever possible, we've tried to illustrate these phenomena with images from the eastern mountains.

Grenville Province: The Earliest Building Blocks

The **Grenville geological province** is a belt of the Canadian Shield composed of billion year-old metamorphic and igneous rocks. It is exposed at the Earth's surface from the coastline of Labrador to the southern Great Lakes, where it dives under various series of younger Paleozoic rocks all the way to the U.S.–Mexico border. Many great, ancient, and alpine ranges adorn the Grenville portion of the Canadian Shield in New York, Québec, and Labrador, including the Adirondacks (Ch. 21), Laurentians (Ch. 21), Monts Groulx (Ch. 18), Montagnes Blanches (Ch. 21), and the Mealy Mountains (Ch. 20). The most common rock found in this region is **granitic gneiss** (Pl. I), although there are also several regions of **anorthosite**, such as the massif that forms the high peaks of the Adirondacks. Generally, the bedrock is highly metamorphosed, and its formation dates to the Proterozoic era. Many ranges in this region are more than 1 billion years old.

Around 1.5 billion years ago, great seas opened between two of the largest continents in existence at that time: **Laurentia** (corresponding roughly to the modern Canadian Shield) and present day South America. Marine deposits accumulated on submerged land along the southern margin of Laurentia. These seafloor sediments appear later in our narrative and are important to the development of the Monts Otish in Québec (Ch. 19), which straddle the northern border of the Grenville region. As the tectonic plates moved apart, pulses of rising magma formed **igneous intrusions**, including the anorthosite of the Adirondacks (Ch. 21) and Mealy Mountains, the **gabbronite** of the Monts Groulx, and many dikes (igneous intrusions) visible today throughout the region (2-2).

The **Grenville Orogeny** began as the Laurentian and the South American plates came together again around 1.4 billion years ago, closing the ocean that had formed. The closing of this ocean created **subduction zones** where there was active magmatism (the granitoid rocks of the Mealy Mountains are located near the boundary of this subduction zone). From 1.4 billion to 900 million years ago, the two continents progressively collided, triggering a long series of major mountain-building events. These collisions, and the resulting "thickening" of the continental crust, resulted in the delamination of the upper mantle and the lower crust. This caused magma to enter the shear zone between mantle and crust, forming the anorthosite massifs mentioned earlier. Today, the volcanic massifs of the Grenville province represent a small part of the Canadian Shield, but form key components of several major eastern mountain ranges: the High Peaks of the Adirondacks and, to a lesser extent, the Monts Groulx, Mealy Mountains, and Benedict Mountains.

The Grenville Province basically comprises a classic folded mountain range. The main part of the **Grenville mountain range** is made from sedimentary and igneous rocks, first formed during seafloor spreading, then subsequently folded, faulted, stacked, and moved tectonically. It's estimated that at the time of its origin, this mountain range was of Himalayan proportions, rising some 10,000 m (more than 30,000') above sea level. The rocks that were buried under this material were subjected to such pressure and temperature that they were highly metamorphosed. Today, these basement rocks are exposed at the surface, but upon formation these rocks constituted the internal foundation of the Grenville mountain range. The surviving mountains of the Grenville province represent the roots of these ancient mountains, long since eroded, and in some cases uplifted again.

Plate I. Selected bedrock types representative of eastern alpine areas

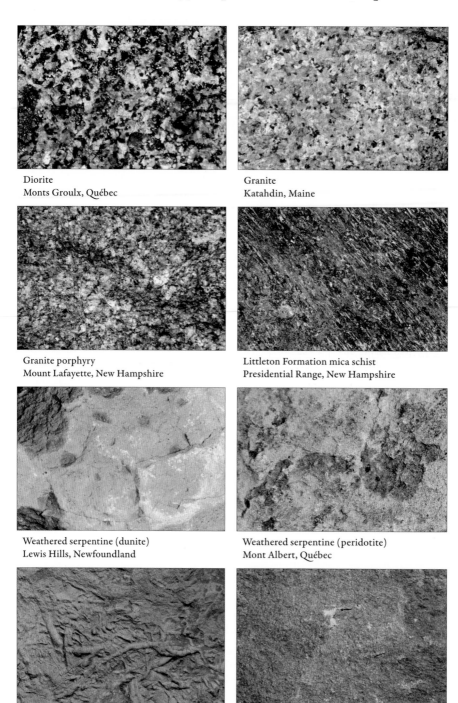

Diorite
Monts Groulx, Québec

Granite
Katahdin, Maine

Granite porphyry
Mount Lafayette, New Hampshire

Littleton Formation mica schist
Presidential Range, New Hampshire

Weathered serpentine (dunite)
Lewis Hills, Newfoundland

Weathered serpentine (peridotite)
Mont Albert, Québec

Limestone
L'Anse Amour, Labrador

Green shale
Mont Logan, Gaspésie, Québec

As an illustration of the complexity of the Canadian Shield orogenies, consider the Adirondacks, the famous mountains of northern New York. The Adirondacks were originally part of the Himalayan-scale Grenville range, but all the mountainous material eroded to become a flat lowland (called a **peneplain**) by the time the region began uplifting again 147 million years ago. This recent uplifting event was probably due to pulses from a hot spot. Because the Adirondacks are still rising at the rate of 2 to 3 mm/year, they are considered young mountains, despite their association with the ancient Canadian Shield (2-3).

By contrast, the other massifs of the Grenville Province, originally uplifted more than 1 billion years ago, continue to diminish through erosion, including Québec's Monts Groulx, which are made mostly of gabbronite, but also of anorthosite, and diorite, all of which are plutonic rocks. Similarly, the Mealy Mountains of southern Labrador witnessed many folding events and are composed both from metamorphosed Precambrian rocks of Laurentia, mostly gneiss and orthogneiss, as well as plutonic rocks from the Grenville Province (2-4).[1]

The impact of the Grenville tectonic collisions created a cascading effect throughout the Canadian Shield that reworked the extremely old belt of gneiss and paragneiss to the north that includes the continent's oldest exposed rocks—some of which have been dated to 3.7 billion years. The collisions gave rise to some of the mountain ranges that are located north of the Grenville region, on the Superior Province, where we find the Monts Otish of central Québec. With the addition of the

2-3 The Adirondack Mountains are part of the Grenville Province of the Canadian Shield. Once part of the Himalayan-scale Grenville mountain range, the region eroded to a peneplain and subsequently began uplifting 147 million years ago. By eastern standards, they are young mountains. Whiteface Mountain, New York. (MJ)

2-4 The Mealy Mountains of Labrador witnessed many folding events and are composed both of metamorphosed Precambrian rocks of Laurentia and plutonic rocks from the Grenville Province. (MLL)

Grenville Province, the Canadian Shield took on its modern form. The Grenville geologic province was mostly completed in the Precambrian era (i.e., before 542 million years ago), before the diversification of the major biological lineages extant today.

Hundreds of millions of years after the Grenville Orogeny, a series of continental collisions resulted in the formation of the Appalachian Mountains (discussed later). These, and ongo-

---

1   James et al. (2002)

2-5 *The Appalachians are a huge mountain range extending from Newfoundland to Alabama. Prominent alpine Appalachian summits include Katahdin in Maine (left) and Mont Xalibu in the Monts Chic-Chocs in Québec (right). (MJ and LJ)*

ing erosion processes and a number of isolated (but significant) events (such as magmatic pulses along a hot spot), continued to shape the landforms of the Canadian Shield to create today's distinctive mountain landscapes.

The Appalachians: Uplift and Differential Erosion

In general terms, the Appalachian Mountains were created by a series of three major tectonic events during the Paleozoic era (542–251 million years ago), during which great supercontinents broke apart and reassembled. The Appalachians are a huge mountain range extending from Newfoundland to Alabama, and in a strict sense, they include many of the great alpine ranges of eastern North America: Katahdin (Ch. 8), the Monts Chic-Chocs (Ch. 9), the Monts McGerrigle (Ch. 10), the Annieopsquotch Mountains (Ch. 12), Gros Morne (Ch. 14), and others (2-5). In a more broad sense, the Appalachians include the higher peaks of the White Mountains (Ch, 7).

Around 650 million years ago, a heat and pressure build-up underneath the aging Grenville Mountains lifted the entire region, leading to extensive continental rift. Multiple landmasses separated and became isolated, including Laurentia (the modern Canadian Shield, discussed earlier) and Gondwana, the forerunner of today's South America and Africa.[2] An early version of the Atlantic Ocean, called **Iapetus**, opened between the northern (Laurentia) and southern (Gondwana) continental masses. On the margin of Laurentia, shallow-water sediments and deepwater sediments were deposited in a warm, tropical sea, evidence of which is visible today as numerous fossils of shells, fish, and corals throughout our area (these sediments will later fold and form the constitutive material of the Appalachian Mountains). Around 500 million years ago, the cycle was inverted, and the Iapetus Ocean began to close. As was the case during the Grenville orogeny, this created a subduction zone between the continental margins and the ocean, and a volcanic arc developed along the continental margin, something like Japan and the Pacific "Ring of Fire" today.

Appalachian orogenies: Taconic, Acadian, and Alleghenian — Around 450 million years ago, the oceanic plate (under the Iapetus Ocean) continued to collide with the Laurentian plate, and the volcanic island arc finally collided into the margin of the continent. This was the first of the three Appalachian orogenies, known as the **Taconic orogeny**. The Taconic

2   Bourque (2010)

collision progressed from south to north, finishing in Newfoundland and is responsible for the creation of the Monts Chic-Chocs in the Gaspé Peninsula, dominated by Mont Albert (2-6).[3] Fifty million years later, the Iapetus Ocean had completely closed. Avalonia, a small landmass derived from Gondwana, collided into Laurentia, causing the second Appalachian orogeny, known as the **Acadian orogeny**. During this phase, volcanic rocks that had been deposited in the Iapetus Ocean were folded and lifted, to superimpose themselves over the older Taconic mountain range. The extensive folding was accompanied by the

2-6 The Taconic orogeny progressed from south to north, finishing in Newfoundland, and is responsible for the creation of the Monts Chic-Chocs on Québec's Gaspé Peninsula. The Chic-Chocs are dominated by Mont Albert. (MJ)

development of granitic **plutons** and intense metamorphism.[4] Following the Acadian orogeny, a period of intense erosion took place, possibly as much as to 0.6 m (2') every 10,000 years.[5] Finally, 380 million years ago, the ocean between the West African component of Gondwana and Laurentia began to close. The two continents collided, and the oceanic crust subducted underneath Laurentia. This was the third and final Appalachian orogeny, the **Alleghenian**, which marked the completion of the supercontinent **Pangaea**, when most of the landmasses on earth were contiguous. The Pangaean landmass remained relatively stable for approximately 100 million years, during which time the Appalachian ranges were a central and prominent feature. After that time, Pangaea began to fragment and give rise to the Atlantic Ocean and the modern continents.

In short, the Appalachians were built by a series of rifting and collision phases as the supercontinent Pangaea formed, but this is an extreme oversimplification. Further understanding of the eastern mountains may be found by examining the rifting and collision processes as they occur today; for example, along the marine gradients from the submerged continental margin to the bottom of the ocean, numerous rock types were created as the Proto-Atlantic Oceans opened. Limestones, sandstones, and shales were formed in this marine environment.[6] Additionally, during the extension periods, **ophiolite** (an intact sequence of crust and mantle) was formed at the oceanic ridge.[7] Ophiolites, and associated **serpentines**, can be found today on Mont Albert in Gaspésie, Québec (Ch. 9), and the Bay of Islands (e.g., the Tablelands) and White Hills regions in Newfoundland (Ch. 13 and Ch. 17) (2-7). During the continent collisions, subduction zones between the oceanic crust and the continental margin supported volcanism and plutonism, which explains the presence of volcanite and granitoid rocks.

To summarize several hundred million years of chaotic geologic processes: the Appalachian orogeny consisted of three distinct tectonic events, interspersed with extensive development of oceanic depositional environments and sedimentation.

3	Brisebois and Brun (1994); Minarik et al. (2003); Bourque (2010)
4	Brisebois and Brun (1994)
5	AMC (2009)
6	Brisebois and Brun (1994)
7	The contact between crust and upper mantle is known as the Mohorovicic Discontinuity, visible at several locations in the Bay of Islands.

2-7 Serpentines, or metal-rich rock derived from mantle under the seafloor, can be found today on Mont Albert in Gaspésie, Québec and the Lewis Hills and the tablelands in the Bay of Islands in Newfoundland (pictured). (IATNL)

2-8 Tabular summits, such as Mont de la Passe in the Monts McGerrigle represent resistant lithology and a former peneplain surface. (LJ)

Differential erosion of the Appalachian landscape — From a modern, topographic point of view, it is important to remember that today's Appalachian landscape bears no resemblance to the original relief from these three major orogenies. After their original uplift, more than 400 million years ago, the Appalachians were reduced to a **peneplain** by erosion through fluvial and glacial processes. Eventually, the peneplain was uplifted and dissected by further erosion, forming the prominent summits on the landscape today. However, the events that led to the final uplift of this peneplain are still under debate among the scientific community. As it was uplifted, this peneplain became a plateau that was furrowed by differential erosion working more quickly on friable rocks, allowing the harder rock nucleus to emerge. The tabular summits, such as the Chic-Chocs, the Monts McGerrigle, Table Mountain, and the Bay of Islands Mountains represent this more resistant lithology (2-8). It may be counterintuitive, but although many Appalachian summits are high enough to support an alpine climate and ecosystems, most of these summits are the result of impressive differential erosion rather than orogenic relief.

Hot Spots: Uplifts from the St. Lawrence to the White Mountains

As described, the fragmentation of the supercontinent Pangaea began approximately 200 million years ago. It took 40 million years before the modern continental landmasses began to become independent. At this time, the North American plate was moving northwest, while the African plate was pulling away in the opposite direction, creating the opening of the Atlantic Ocean. As it drifted, North America passed over a prominent volcanic hot spot, called the **New England hot spot**, which periodically ejected magma into the overlying sedimentary layers. As this magma rose, it slowly cooled and crystalized into granite. As the granite upwelled into bedrock,

2-9 *The White Mountains of New Hampshire were partly formed as North America passed over a volcanic hot spot following the breakup of Pangaea. Carter Notch, New Hampshire. (MJ)*

it deformed the overlying rock layers, accelerating the erosion of the sedimentary rock. As sedimentary layers fractured and eroded, the granite plutons expanded in response to decreased pressure, which created fractures between layers. The earliest visible result of these magmatism episodes is a linear series of mountains in the Saint Lawrence valley called the Monteregian Hills. Although these mountains are beyond the scope of this book, they provide proof that the plate drifted along a nearly straight line. As the North American plate continued to drift, the Appalachians passed over the New England hot spot, where it caused the uplift of Mont Megantic in Québec, and subsequently contributed to the uplifting of the White Mountains in New Hampshire (2-9). This episode created the Conway Granite of the eastern White Mountains as well as the plutonic structure of Cannon Mountain in the Franconia Notch region (Ch. 11).[8]

Although the geological history of North America from the Precambrian to the breakup of Pangaea allows some understanding as to how the mountains rose, this only tells a part of the story. As mentioned earlier, the Grenville and Appalachian tectonic events created mountain ranges of Himalayan proportions. Erosion processes are responsible for diminishing the mountains' height and shaping them into their current form.

Refining Today's Alpine Landscapes

Erosion explains the difference between younger, steeper mountains, and older, flatter ones (2-10). Folding and uplifting of material during orogenies creates steep mountains that are not necessarily stable. It is easier on steep slopes for gravitational processes to transport material downslope. As material is transported and deposited over time, the slopes become gentler. The mountains of eastern North America are all considered to be old mountains, as evidenced by their relatively low altitude and gentle slopes. However, as will be discussed in the following chapter, some eastern mountain ranges (such as the Chic-Chocs in Québec), because of their lithology, have been rejuvenated relatively recently by the glacial episodes of the Pleistocene ice ages.

8   Reidy (2004); AMC (2009)

2-10 Erosion explains the difference between younger, steeper mountains, and older, flatter, more stable mountains. Rope Cove Canyon, Lewis Hills, Newfoundland (left); Mont Jauffret, Monts Groulx, Québec (right). (IATNL and MJ)

2-11 The eastern mountains can't be easily explained by a series of orogenic uplifts. Many of today's mountain landscapes result from impressive differential erosion, which left behind nuclei of only the most resistant rocks, forming the diverse mountains and landforms we see today. Northern Long Range Mountains, Newfoundland. (IATNL)

Summary

Although the mountains of the Northeast do not rise to the impressive heights of the younger mountain ranges of the Earth—the Himalaya, the Andes, or the Alps—their long and complex history is responsible for a varied lithological mosaic, which is associated with a high diversity of environments. Eastern mountain ranges are variably composed of sedimentary rocks deposited in deep and shallow environments, igneous rocks resulting from magmatic intrusion and volcanism associated with the orogenic activity, and metamorphic rocks created under extreme temperature and pressure. The topography associated with the different tectonic events of eastern North America was somewhat ephemeral on a geologic timescale. Rather than being associated with simple orogenic uplifts, many of today's mountain landscapes result from impressive differential erosion, which left behind nuclei composed of only the most resistant rocks and created the topographic diversity we see today (2-11). Whether the result of tectonic or erosional processes, these mountains still undergo the effects of climate and time through a variety of processes over multiple temporal and spatial scales, such as weathering, mass movements or fluvial erosion (next chapter).

Ice Age and After 3

Frost-sorted soil polygons, Burnt Cape, Newfoundland. (MJ)

Ice Age and After

3-1 Relatively recent and ongoing erosion has shaped the fine contours of the eastern mountains. North Arm Hills, Bay of Islands region, Newfoundland. (IATNL)

The past 2.6 million years have witnessed significant cooling of the Earth and a series of glacial cycles, in which continental ice sheets have covered as much as 30% of the Earth's surface. At the height of the last glacial event, the **Laurentide Ice Sheet** covered almost all of northeastern North America. Most of the glacial advances occurred within the **Pleistocene epoch**, informally known as the "ice age," which technically ended 12,000 years ago with the most recent melting of the Laurentide ice, ushering in the **Holocene epoch**. Together, the Pleistocene and Holocene compose the **Quaternary Period**, which is the more inclusive and objective term for the most recent ice age. The Pleistocene designation ignores the apparent fact that we are simply experiencing a temporary glacial recession, or **interglacial** period.[1]

The eastern mountains discussed in this book had essentially taken their current form by the beginning of the Quaternary Period. However, the glacial advances of the Laurentide ice, as well as fine-scale erosional processes during the ice age and during the past 12,000 years, have dramatically shaped the fine contours of the eastern mountain ranges. In fact, recent and ongoing erosion has resulted in some of the most distinctive and important features of eastern mountains (3-1). In this chapter, we focus on the more visible and significant of these processes, which may be roughly grouped into three categories: **glaciation** (continental and alpine), **mechanical weathering**, and geomorphic process such as **mass wasting** (avalanches and landslides) and **fluvial dynamics** (sedimentation and floods).

---

1   Increasingly, scientists recognize that we are entering a new geologic epoch, the Anthropocene, in which the major climatic regimes and new geologic strata reflect the influence of humans.

Glaciers

3-2 Erratic boulders deposited by receding ice sheets are common, and sometimes strikingly visible, on most of the eastern mountain ranges. These may influence the surrounding vegetation by interrupting the prevailing winds or by influencing soil chemistry. Monts Groulx, Québec. (LW and MJ)

Most of the eastern mountains are unglaciated today, although the Laurentide ice sheet once covered most of northeastern North America north of New Jersey during the height of its most recent expansion (approximately 21,000–18,000 years ago).[2] As the Earth's temperature rose, the vast ice sheet began to retreat toward northern Québec. However, the colder climate of the eastern mountain environments allowed the development and persistence of local mountain glaciers and ice caps. For example, on the Gaspé Peninsula, local glaciers did not disappear from the Monts Chic-Chocs until after the **Younger Dryas** glacial recurrence, which ended 11,500 years ago.[3]

Glaciers are created when the annual accumulation of snow exceeds the annual melt. Decade after decade, snow piles up in towering layers that may reach hundreds or thousands of meters thick, transforming under their own weight into hard ice. The immense weight of a glacier puts tremendous pressure on the bottom layers, deforming them. This deformation (and associated melting), along with gravity, allows the glacier to move away from the center of mass. As a glacier moves, it abrades the floor and sides of a valley.

Both the continental-scale Laurentide ice sheet and smaller mountain-based glaciers influenced the eastern alpine summits in clear ways. Glacial relics such as **erratic boulders**, deposited by the ice sheets as they melted, and **bedrock striations**, which formed as glaciers dragged rocky debris across exposed bedrock, are still visible throughout the eastern alpine areas (3-2).

Glacial valley formation — A glacier possesses fluid characteristics and will occupy all available space. Before glaciation, mountain valleys typically have a V-shape, produced by the erosion caused by water flowing downward through preferred channels. When a glacier occupies a valley, because of its high erosional capacity, it will widen and deepen the valley, changing its V-shape profile to a U-shape profile (3-3). After the glacier retreats, the headwalls, over-steepened by glacial erosion, are in disequilibrium. This results in a high rate of

---

2   However, perennial snowfields and small cirque glaciers exist in the Torngat Mountains of northern Labrador (Ch. 21).

3   Hétu and Gray (2000)

3-3 *Valleys cut primarily by streams develop a distinctive "V" shape (left), which widens to a "U"shape when occupied by a glacier (right). Lewis Hills, Bay of Islands region, Newfoundland. (MJ)*

cliff ablation as large amounts of rocky debris are sloughed onto the valley floor. The debris accumulates in the bottom of the valley, creating scree slopes along the walls (3-4). These slopes usually have a linear profile, which can be modified by fluvial mass wasting processes. Scree slopes may be stabilized by vegetation. On many scree slopes, geomorphologic processes and vegetation are in a dynamic and precarious equilibrium. With climate change, biologic and geomorphologic processes evolve, making the treeline move upward or downward accordingly. Through these processes, glaciation in northeastern North America rejuvenated the topography, by scouring the old, stable slopes and revealing geomorphologically active slopes.

Fjords — Where glaciers met the sea, such as at Gros Morne National Park, great fjords were carved. The more famous of these include Western Brook Pond and Ten Mile Pond near Gros Morne (Ch. 14 and 15). Other fjords may be seen in northern Labrador (Ch. 21)

Glacial Cirques — **Cirques** are mountain amphitheatres, surrounded by steep cliffs or slopes, with the open side facing downstream (3-5). Cirques may form when small glaciers

3-4 *Scree, or rocky debris, accumulated at the foot of unstable slopes exposed by glacial abrasion. Molly Ann Canyon, Lewis Hills, Bay of Islands region, Newfoundland. (IATNL)*

3-5 *Glacial cirques such as Tuckerman Ravine (at left in photo) are amphitheatre-shaped hollows in the mountainside, formed when snow and ice accumulated in an existing depression. The shallow basin at center is Raymond Cataract, which is a younger cirque than Tuckerman Ravine. Mount Washington from Wildcat Mountain, Presidential Range, New Hampshire. (MJ)*

form in depressions on the sides of valleys, but glacial cirque formation may also be encouraged by preexisting cirques, whether glacial or not.[4] The glacier accumulates in the depression, and then erodes the surrounding headwall and the floor of the cirque through rotational plucking[5] and abrasion. As these processes continue, the size of the cirque increases. The

---

4   Nelson and Jackson (2003)

5   *Plucking* is the process through which a glacier erodes the bedrock where it is already fractured.

3-6 *Tarns are pools or ponds formed among glacial debris and scree in formerly glaciated cirques and valleys. The North Basin Ponds on Katahdin in Maine are typical examples. (MJ)*

3-7 *When water trapped in seams in the rock expands and contracts during severe freeze-thaw oscillations, the rock shatters in a process known as cryoclasty. Tablelands, Bay of Islands region, Newfoundland. (MJ)*

downstream, open side of the glacial cirque is called the *lip*. The presence of this topographic feature is either created by the rotational erosional movement or by the presence of a **terminal moraine**.[6]

When the glacier ultimately melts, it creates amphitheater-shaped valley-heads. These landforms, often called "gulfs" or "ravines" in New England, "canyons" or "gulches" in Newfoundland, and "*ravins*" or "*bassins*" in Québec, are visible in many mountain ranges in eastern North America. Backcountry skiers and mountaineers are familiar with Tuckerman and Huntington Ravines on Mount Washington, two of the most famous cirques in the eastern region. These are the two steepest, and most accessible, glacial cirques among the many others that surround Mount Washington: Oakes Gulf, Gulf of Slides, Ammonoosuc Ravine, Great Gulf, Jefferson Ravine, and King Ravine. In many ways, it is not surprising that so many cirques occur on the Presidential Range. As the highest point of northeastern North America, Mount Washington was covered by mountain glaciers for longer than smaller mountains were, allowing the formation of numerous archetypal glacial cirques. Moreover, the lithology surrounding the summit of Mount Washington is primarily of coarse-grained mica schist interbedded with quartzite, which is prone to plucking and abrasion, therefore facilitating the creation of cirques.[7] The debris at the floor of a cirque, which may be a combination of scree from the valley walls and glacial debris from the terminal moraine, support rocky pools known as **tarns**. The Lakes of the Clouds on Mount Washington are tarns, though they don't lie on the floor of a cirque. The ponds on the floor of Katahdin's North Basin are also classic tarns (3-6).

---

6   A *terminal moraine* is an accumulation of mineral debris of different sizes that is transported by the glacier and rejected at the deposition zone, at the downstream extremity of the glacier.

7   Thompson (1999)

Mechanical Weathering

Weathering can involve physical (mechanical), chemical or, to a lesser extent, biological processes. In cold (and alpine) environments, mechanical weathering is the most prevalent of these. **Mechanical weathering** may be defined as the gradual destruction of surface bedrock caused by external environmental conditions. Weathering creates earthen materials that can be transported, by breaking down bedrock into substrates of different sizes, from large boulders to clay. Erosion is essentially the combination of weathering and the transportation of the weathered material. In this section, we focus primarily on three mechanisms common in alpine environments: **frost shattering, freeze-thaw disturbance,** and **exfoliation**.

Frost-shattering — Frost shattering can occur when fluctuating air temperature causes rocks to contract (at low temperatures) and expand (at warmer temperatures), a process known as **thermoclasty**. When water trapped in seams in the rock expands and contracts during freeze-thaw cycles, the rock shatters in a process known as **cryoclasty** (3-7). In both cases, it is not the temperature itself as much as the temperature swings and the number of cycles that cause frost shattering. In the eastern alpine ranges, it is quite common for daily temperature cycles to swing above and below the freezing point during the spring and fall.

Cryoturbation — Freeze-thaw cycles—usually involving water—can lead to a variety of patterns in the rock and soil, many of which are visible throughout the eastern mountains. These prominent "sorted soil" formations are hallmarks of the alpine environment and include **soil boils, soil stripes,** and **rock polygons** (3-8). These patterns are a direct result of mechanical weathering and can eventually produce a greater diversity of micro-habitats for lichens, bryophytes, and plants than can those offered by sheer bedrock. In some cases, these frost features are relicts from a colder pe-

3-8 Sorted soils, such as polygons and stripes, are frost-derived features common on eastern mountains with exposed mineral soils. Lewis Hills, Bay of Islands region, Newfoundland. (NC)

3-9 Severe frost action may result in "seas" of broken bedrock, known as felsenmeer. *Felsenmeer is especially abundant on Mount Washington. Presidential Range, New Hampshire. (MJ)*

3-10 When soils slump downslope over underlying permafrost, solifluction pools may result. Lewis Hills, Bay of Islands region, Newfoundland. (MJ)

3-11 Cannon Mountain in Franconia Notch, New Hampshire, is one of the best examples of an active exfoliation dome. White Mountains, New Hampshire. (MJ)

3-12 Avalanches are frequent on the steeper and higher mountains of our region, such as the Monts Chic-Chocs in Québec. (DG)

riod in the past but in many cases the freeze-thaw cycles may still be active. A severe form of frost action manifests as a "sea" of broken bedrock, known as **felsenmeer** (German for "sea of rock"), which is especially common on Mount Washington, Katahdin, Mont Jacques-Cartier, Gros Morne, the Highlands of St. John, and other mountains in our region (3-9). In cases where damp soils have slumped over frozen soils, bedrock, or permafrost, a process known as **solifluction**, deep pools may form. These are especially common on the mountains of Québec and Newfoundland (3-10).

Exfoliation — The third mechanical weathering process commonly encountered in eastern alpine areas is **exfoliation**. Exfoliation occurs when the deterioration of a granitic cliff-face creates a pressure decrease in the bedrock, allowing the underlying rocks to expand. Cliffs that undergo exfoliation will look like they have been peeled like an onion. The best example of an exfoliation dome in the eastern mountains is perhaps found the east face of Cannon Mountain in Franconia Notch in New Hampshire (3-11), but many other exfoliation domes are covered by soil and vegetation in this region. Although not as important or generalized as frost shattering, this process is particularly important in the White Mountains, several of its summits have been created by magmatic (granitic) intrusions.

Mass Movements of Snow and Soil

Steep mountain slopes are not always in equilibrium, which can result in the mass movement of snow and debris downslope. These processes change the landscape, creating niches for different alpine plant communities (see Ch. 5).

Avalanches — Avalanches occur often on the steeper and higher mountains of our region, when the gravitational downslope force exceeds the tensile force holding the snowpack in place (3-12). The primary conditions for an avalanche to be triggered are the accumulation of a sufficient amount of snow and a weak layer in the snowpack (which decreases the tensile force) (discussed further in Ch. 4). The weak layer is usually the result of some climatic scenario characteristic of a particular mountain range. For example, in the Monts Chic-Chocs

(Ch. 9 and 10), extreme avalanches occur under one of several scenarios: 1) above-average total snowfalls; (2) high frequency of snowstorms; (3) major rain events and facet-crust development; (4) sequences of freezing rain and strong winds, and (5) early-season weak layers of faceted crystals known as **depth hoar**.[8] The frequency of these scenarios varies among eastern alpine ranges. For example, although scenario #5 is not rare on the Gaspé Peninsula, the formation of depth hoar in the White Mountains is almost unheard of. From a safety standpoint, it's important to recognize that avalanches occur frequently in eastern North America, but skiers and alpinists often venture into the backcountry in avalanche-prone zones without the knowledge and the proper gear to deal with this hazard.

3-13 The steep slopes of the Adirondacks provide the perfect environment for landslides. Numerous ribbon-like exposures of bedrock on Basin Mountain are the result of landslides during large precipitation events. Basin Mountain from Saddleback Mountain, Adirondacks, New York. (JG)

Landslides — The steep slopes and the high amount of liquid precipitation in the Adirondacks and the White Mountains create the perfect environment for landslides (3-13). Landslides can be defined as a downward translation of material following a shear surface. One of the classic types of landslides found in this region is the **skin slide**, also referred to as a **debris slide**, **debris avalanche**,

3-14 Tropical Storm Irene triggered numerous new landslides in the White Mountains of New Hampshire, such as this one at Crawford Notch. (JPM)

or **skin flow**. However, the term *flow* is inappropriate because it refers to viscous transport of material instead of solid transport. Skin slides are characterized by surficial debris sliding over steep rocky slopes (usually over 35°), bringing the overlying vegetation along with it. They usually leave a ribbon-like stripe in the vegetation, where often the bedrock will be exposed. Landslides will usually take place during high precipitation events. The passage of Tropical Storm Irene over New England in August 2011 triggered many landslides in the White Mountains (3-14).

Debris flows — Debris flows are characteristic of torrential precipitation events. When such an event occurs, water flowing on the surface will concentrate in streams. As the stormwater converges and increases in velocity on steep slopes, it mobilizes surficial materials, from loam and vegetation to pebbles to boulders. As the debris flows downstream, it may dig deep channels, and as transported material loses velocity on the sides of the stream, levees develop on the borders of the channels. The material will then be deposited on a lobe further down where the slope is gentler (approximately 10–15°), with coarser material deposited before finer grained material. On the Gaspé Peninsula in 2009, the authors witnessed an event of huge debris flows on Mont Saint-Pierre, after a storm that dropped 66 mm of rain in one hour. The channels created on the scree slope were a few meters deep (1–5 meters) and

3-15 Debris flows are characteristic of torrential precipitation events. A slurry of loam, vegetation, and boulders will concentrate in narrow channels and leave distinctive, entrenched troughs in the mountainside. Mont St. Pierre, Gaspésie, Québec (left) and Lewis Hills, Bay of Islands region, Newfoundland (right). (DG and MJ)

between 3 to 8 m wide; approximately 10,000 m³ of material had been removed from the upper part of the talus and transported further down in the lower forested part of the slope. This anecdotal event illustrates the powerful impact that water along with gravity can have on the landscape (3-15).

Summary

Glaciers and the quaternary environment have had a major influence on the topography of the mountains of eastern North America, on a variety of scales. The resulting topography influences both the meteorological (Ch. 4) and edaphic (or soil) conditions, which in turn influence the associated flora and fauna (Ch. 5 and 6) (3-16). These interactions also happen at different scales: a large boulder on a summit plateau may shelter alpine plant communities from prevailing winds. An active avalanche path will create a disturbance regime that favors early-successional species. Different aspects of a single summit are exposed differently to wind and sunlight, which, along with the lithology, create a vast variety of ecological micro-habitats. These concepts are explored further in Chapters 4 and 5.

3-16 Changes to the mountains' topography from glacial action and ongoing erosion in turn influence both the meteorological and edaphic environment of the mountain. Lewis Hills, Bay of Islands region, Newfoundland. (MJ)

The eastern alpine areas are the synergistic sum of the interactions between countless, constantly changing physical and ecological variables, and it is therefore necessary to consider the alpine environment of the northeast as a dynamic system. The diversity and relative significance of any given alpine system, as well as its sensitivity to climate change and anthropogenic activity, is linked to many physical variables, and to the interactions between these variables.

Climate and Weather 4

Mount Washington from Jackson, New Hampshire. (MJ)

Climate and Weather

4-1 *Bewildering fog is a common occurrence in of all of the eastern alpine areas. Crawford Path, Mount Washington, New Hampshire. (MJ)*

A region's climate is defined by long-term weather, not any particular weather event itself. Climate is determined by four major factors: latitude, altitude, proximity to large bodies of water, and exposure to regional air mass circulation patterns. In this chapter, we'll examine how weather and climate influence eastern mountains.

Eastern mountains lack some of the typical human health hazards associated with high altitude, such as low atmospheric pressure and extreme UV exposure, but they overcompensate with their raw, powerful, wet, and volatile weather. Anyone who has spent much time in the eastern mountains can muster up a reasonably interesting story about high-elevation weather. Sooner or later, ridgeline after ridgeline, there comes a day when impenetrable fog suddenly catches up to you, obscuring the next landmark or blaze or cairn (4-1). Or perhaps instead, the wind roaring and humming over a barren ridge felt capable of blasting you into the cirque below. Or perhaps biting cold numbed your wet hands and clouded your judgment. Strong mountain winds can also take on a dark and life-threatening quality, likely playing a factor in at least a quarter of the deaths above treeline on Mount Washington. Those who have failed to take mountain weather seriously have paid dearly for it, and the "story" of eastern mountain weather has been told many times over through the lens of their lethal experiences.[1] But the weather, as it turns out, is more than a feature of eastern alpine landscapes. It is also a driving agent in their persistence (4-2).

The alpine and subalpine ecosystems explored in this book are encompassed and bounded by the pronounced tree limit on the shoulders of high mountains. This transitional boundary between high-elevation conifer forests and alpine tundra is known as the

4-2 *Severe mountain weather is a driving factor in the persistence of eastern alpine tundra. Table Mountain, Newfoundland. (MJ)*

1   Of the weather-related fatalities on eastern mountains, only those on the Presidential Range of New Hampshire (Ch. 7) have been exhaustively profiled. See Howe (2000).

4-3 The alpine-treeline ecotone in our region ranges from sea level in parts of Labrador and Newfoundland to 1500 m in places on the shoulder of Mount Washington, in response to climatic conditions. Red Bay, Labrador (top) and Ridge of the Caps, Presidential Range, New Hampshire (bottom). (MJ)

alpine-treeline ecotone, or simply "treeline." In the eastern mountains the terms **arctic-alpine tundra** and **alpine tundra** are frequently used interchangeably; regardless of the term used, the important characteristic is a low-stature plant community with minimal to no understory that includes species associated with northern, low elevation tundra (Ch. 5). Globally, the factors limiting the growth of trees at high elevations and on mountain ridgelines have been well-studied,[2] and the alpine-treeline ecotone generally correlates well with average growing season temperatures below 6.7°C (44°F), a temperature at which coniferous trees experience limited tissue formation.[3] However, the primary drivers of treeline worldwide have not been unified into a single global theory,[4] and local or regional factors are often of major importance. Treelines can be depressed (i.e., occur at lower elevations or warmer temperatures) as a result of several temperature-independent factors, including soil disturbance processes, air and soil humidity, seasonal weather patterns, soil depth and chemistry, herbivory, and seed availability. Though the 6.7°C growing season isotherm generically describes treeline at the global scale, treeline in our region is rather anomalous, occurring at much warmer temperatures than the global average[5] and at comparatively lower elevations for corresponding latitudes, suggesting than factors other than temperature are important in our region.

The elevation of the treeline on eastern mountains varies dramatically throughout the region, from near sea level on the exposed capes of northern Newfoundland and coastal regions of Québec and Labrador to at least 1,687 m (5535') on the slopes of Mount Washington, New Hampshire (4-3). Rather than being driven solely by low temperatures, the actual elevational position and general structure of the treeline on most eastern alpine mountains appears to result from an interaction between winter icing, cold climatic thresholds, and soil (edaphic) processes. Treeline on larger eastern mountains can vary in elevation by 500 meters or more in response to topographic changes in slope, concavity-convexity, and aspect (4-4).

Global warming and regional climate change dominate today's conversations about the future of eastern alpine tundra, but clouds and wind are likely more important than temperature is in shaping the past and future fate of the subalpine and alpine ecosystems in much of our region. The common explanation that hiking up Mount Washington or Katahdin (or Mont Jacques-Cartier or Gros Morne) is climatically equivalent to driving from the low-

---

2   e.g., Hoch and Körner (2003); Körner et al. (2003); Körner and Paulson (2004)
3   Körner and Paulsen (2004)
4   Körner (2003)
5   Cogbill et al. (1997)

elevation forests of New England (or Gaspésie or Newfoundland) to the tundra of the Low Arctic may ignore many of the actual causal agents for the pronounced elevational gradients in our mountain vegetation. The graded vegetation patterns on eastern mountains—from northern hardwoods such as yellow birch (*Betula alleghaniensis*) and American beech (*Fagus grandifolia*), to spruce-fir forests, to stunted krummholz subalpine forest, to alpine tundra—may mimic some of the changing plant communities one would experience by driving north (certainly for the dominant and most visible plant species), but on most eastern mountains, reduced growing seasons at various elevations are less responsible than commonly suggested in this "hike upward" analogy (4-5). Rather, the principal driving factors limiting tree growth on eastern summits (at least on the more southerly peaks that have been well-studied, e.g., the Presidential Range, Katahdin, and Mont Jacques-Cartier, as well as the coastal peaks of Newfoundland) are wind combined with frequent cloud exposure, and associated ice accretion and abrasion, which result in the mechanical degradation of the forest.

4-4 Treeline on larger eastern mountains can vary in elevation by 500 m or more in response to topographic changes in slope and aspect. Mont de la Passe, Gaspésie, Québec. (LJ)

4-5 Graded vegetation patterns on eastern mountains—from lowland hardwood forests to mid-elevation spruce and fir, to tundra—superficially mimic the vegetation changes observed along a latitudinal gradient from New England to northern Québec. Carter Range from Mount Madison, New Hampshire. (MJ)

Wind, Clouds, and Ice

On a global scale, the eastern mountains aren't particularly high, but their winds can be more memorable. These mountains lie at the entrance to the strong Atlantic storm track, where frequently changing weather fronts give rise to tempestuous winds that rake the higher elevations of the northeastern United States and eastern Canada. During these storms, the airflow up and over the mountain summits is accelerated in a manner similar to the effect of placing your thumb over the open end of a garden hose. The result can be quite impressive, as seen often in the wreckhouse region of Newfoundland (Ch. 12). In fact, until recently, New Hampshire's Mount Washington (Ch. 7) held the world-record wind speed at 327 kph (231 mph), which remains the highest mountain wind ever recorded, and the summit endures long stretches with winds in excess of 120 kph (74 mph), near the lower bound of a category 1 hurricane. This is all the more impressive considering that the forces delivered by wind increase quadratically, not linearly, with wind speed. The force that the air exerts, say on a small balsam fir (*Abies balsamea*), is fourfold greater when the wind speed doubles. As a result, winds combined with frequent winter clouds and icing events play a pivotal role in the ecology of the eastern mountains, providing a mechanistic stress agent that contributes to the long-term persistence of remnant alpine ecosystems stranded atop the highest peaks. Without the inhibitory effect of strong winds and icing events on the development

4-6 The orographic uplift of a moving airmass over a mountain can result in the formation of a lens-shaped lenticular cloud if cooling is sufficient. Carter Dome, New Hampshire. (MJ)

of treeline forests, certain of these alpine ecosystems would likely have been overgrown by subalpine forest thousands of years ago.

Despite the strong effects of mountain winds and clouds, it's important also to acknowledge the role that cold summer temperatures play in defining the environment of alpine zones. Temperatures at higher elevations tend to be predictably lower than are those in the valley due to **adiabatic cooling**, which occurs as prevailing surface winds move across the landscape and, upon hitting an obstruction such as a large mountain, are displaced upward, an effect known as **orographic ascent**. As the surface airmass is lifted over the mountain, it encounters progressively lower pressures (recall that pressure decreases with height in the atmosphere), and as a result expands into its new surroundings.[6] This expansion of the airmass also requires a reduction of its temperature (so energy can be conserved) and under ideal conditions, the rate of cooling for ascending dry air, known as the **dry adiabatic lapse rate**, is 9.8°C/km (5.5°F/1000'). For a mountain summit rising 1500 m (5000') above the valley floor, adiabatic cooling would result in a nearly 15°C (27.5°F) disparity in summit to valley daytime temperatures. It is worth noting that the record high temperature on Mount Washington (1916 m / 6288') is 22°C (72°C) while that at nearby Berlin, New Hampshire (283 m / 930'), is 37°C (98°F),[7] a difference very close to that predicted by the adiabatic lapse rate.[8] As a result of this systematic cooling mechanism, organisms in the alpine zone typically experience much colder conditions than do those that live in the valley.

4-7 Orographic clouds are commonly composed of supercooled water droplets, which freeze on contact with cold surfaces, forming rime ice. Feathers of rime may grow 10 cm per hour, in the direction of the oncoming clouds. Lakes of the Clouds, Mount Washington, New Hampshire. (MJ)

---

6   Even the casual observer can notice this effect by bringing along an unopened bag of potato chips on their next trip to higher elevations; provided you can resist consuming the chips before reaching the mountain summit you will find that the bag has expanded during your voyage upward.

7   National Climatic Data Center (NCDC)

8   Although it should be noted that these record temperatures were not recorded on the same day and thus not necessarily comparable.

4-8 Spindly conifer branches are efficient at collecting rime ice and frequently break under the weight of accumulated ice, especially during windstorms. Mont Boissinot, Monts Groulx, Québec, and Mount Adams, Presidential Range, New Hampshire. (MJ and KK)

Although the dry adiabatic lapse rate may be useful during benign weather days, the actual vertical variations in temperature, the **environmental lapse rate**, typically is considerably more complex. For example, when the air reaches 100% humidity the lapse rate is reduced to 3°F/1000' elevation owing to the heat released during the process of condensation. Other factors may come into play as well, such as the comparative roles of stratification and turbulent mixing within the atmosphere. In fact, at times temperatures in the alpine zone may even exceed those at lower altitudes, a scenario known as a temperature inversion. Inversions tend to form during clear, calm nights, which favor rapid radiational cooling of the ground surface. The air immediately adjacent to the ground is also cooled, and as a result becomes denser, drains down the mountain slopes, and accumulates in the valleys below. The result is a "cold air pool" at low elevations and comparatively warm temperatures aloft. For this reason, record low temperature for the valleys and summits are more comparable than high temperatures. For example, the record low temperature on Mount Washington (–46°F; –43°C) is just slightly colder than that at nearby Berlin, NH (–41°F; –40°C).[9] Notably, the coldest temperatures in valley locations tend to occur one or two days after the coldest temperatures at the mountain summits. When strong winds and associated turbulence usher in cold air following the passage of a storm, the temperature profile will initially be nearly adiabatic, with the coldest temperatures at the mountains summits. As the storm departs, the atmosphere becomes less well mixed, and the cold air is allowed to settle into the mountain basins, providing the coldest valley readings.

In addition to experiencing strong winds and numbing cold, the eastern mountains are prone to extremes of cloudiness and precipitation, which in turn play a profound role in modulating the alpine ecosystem and the local structure of treeline. As discussed earlier, air forced up and over a mountain cools dramatically. If the cooling is sufficient, moisture in the air condenses to form orographic (mountain) clouds, which sometimes manifest as lenticular (lens-shaped) clouds draped prominently over a summit (4-6). As a result of this process, the higher summits throughout our region frequently hide in the clouds; Mount Washington, for example, is one of the cloudiest places in the United States. In the winter,

9   National Climatic Data Center (NCDC)

4-9 *Forests composed of spruce and fir thrive in damp, cool conditions across Canada. In New England, they occur primarily above 800 m in elevation. Monts McGerrigle, Gaspésie, Québec. (LJ)*

these orographic clouds are commonly composed of supercooled liquid cloud droplets, meaning water droplets that remain liquid at temperatures well below the freezing point. When these supercooled droplets impact a cold surface—for example, the exposed limbs of a black spruce—they instantly freeze forming **rime ice**, or frozen fog (4-7). Substantial rime ice can accumulate quickly on any exposed surface (as much as 10 cm or several inches in an hour), with feathers of ice extending into the direction from which the winds originated. The taller and spindlier the surface area—such as conifer branches—the better the collecting efficiency (4-8). This ice can build up so thickly that limbs break, and ice-encased branches swaying in high winds are often chafed or damaged.

Regionally, the best measurements of rime ice accretion rates with elevation come from Mount Washington and, in Vermont, Mount Mansfield. In the 1980s, researchers placed rime ice detectors on the ski area lift towers on Mansfield and a nearby peak, Madonna. Rime ice accumulated more sharply above 822 m (2700'). This is the same elevation at which spruce-fir forest—which prefers cool, damp conditions—begins in central New England (4-9). Rime ice formed more frequently on Mount Washington than on Mount Mansfield because Washington is higher, as well as because of its proximity to the moisture-bearing Atlantic Ocean, a factor enhanced when the weather to the east involves a low-pressure system. Similar phenomena are likely at work in the mountains of Newfoundland (the highest mountains abut the western coast) and coastal Labrador, but perhaps less so in interior Québec, which is several hundred kilometers north of the Gulf of St. Lawrence and equidistant from the shores of James Bay.

In addition to rime ice accumulation, the eastern alpine areas receive copious precipitation in summer and winter alike. The same processes that favor orographic clouds provide enhanced water availability for precipitation over mountain summits and ridges. As a result, during strong storms, which affect the region as a whole, precipitation is systematically enhanced over elevated terrain. Even on days when large-scale weather disturbances are absent, orographic ascent may still be sufficient to generate modest precipitation over the highest

4-10 *Orographic ascent of regional airmasses is often sufficient to cap the summits in clouds, resulting in severe ice accumulation when surfaces are sufficiently cold. Devil's Bite, Northern Long Range, Newfoundland (left) and Mount Adams, Presidential Range, New Hampshire (right). (IATNL and KK)*

peaks (4-10). To examine this relationship, we return to our now-familiar comparison of Mount Washington and nearby Berlin, New Hampshire. The Mount Washington Observatory reports an annual average precipitation in excess of 2.54 m (100 inches) of liquid water (i.e., rain and melted snow) and records a trace or more of precipitation on a staggering 215 days a year! In contrast, the valley site receives less than half that, 1.03 m (40.5 inches), deposited over just 120 days. Clearly, elevation plays a significant role. But it isn't just a numbers game: the abundance of precipitation alters the alpine landscape by adding acidifying nitrogen and leaching calcareous nutrients from nascent soils, as well as increasing erosion rates and, in extreme cases, the lubrication for landslides.

4-11 Strong mountain winds remove snow from exposed ridgelines and deposit it in the lee of ridges and cliffs. Bay of Islands, Newfoundland. (IATNL)

4-12 Glacial cirques such as Oakes Gulf, on New Hampshire's Presidential Range, accumulate the largest amounts of windblown winter snows. (MJ)

During the winter, the interaction of heavy snow and strong winds is of particular importance to alpine vegetation and the treeline. Following a fresh snowfall, snow is scoured from the windward slopes, transported across the landscape, and deposited in the lee, or downwind, terrain. The small fractured particles of snow suspended in the wind act like a sandblaster on any tall, unsheltered vegetation. The result can be lethal, especially when combined with frequent rime icing. Break off the tallest leading buds and these trees will grow horizontally and short instead of vertically, forming **krummholz** (crooked or twisted dwarf forests; see Ch. 5), which hikers may find nearly impenetrable. On the most wind-exposed ridges and summits, neither trees nor shrubs can survive, and only tiny alpine plants persist because they offer minimal collecting surface for rime ice and they hunker down where the air moving across the ground is slowed dramatically due to friction.

Leeward slopes and depressions, however, may provide a slightly less hostile alpine environment. On these down-wind slopes, the net deposition of snow, rather than scouring, builds a deeper snowpack. In some locations, the deep snow protects a unique subset of alpine vegetation, known as snowbed communities, from the extremes of wind, temperature, and icing (4-11). The deeper snow in these locales may also persist longer into the spring and early summer, providing a continual supply of water at the start of the growing season. Of course, there is considerable variability from storm to storm and year to year on which aspects of a mountain receive the most deposition and scouring, though some locations, such as glacial cirques, often develop the deepest snow pack (4-12).

The loading of snow onto steep mountain slopes may affect the alpine and subalpine ecosystems in yet another way: avalanches. As discussed in Chapter 3, snow avalanches occur when the gravitational forces acting on the snow overwhelm the interior bonding within the

snowpack, resulting in a cascade of snow and ice. Avalanches are most common during periods of heavy snowfall, rapid wind deposition, and during the spring when warm temperatures weaken the internal bonding. Avalanches are sufficiently common on some slopes that only small resilient trees can grow there. On other slopes, avalanches have long return intervals and may drastically alter the landscape in catastrophic events, which fell mature stands within the subalpine forest (4-13).

4-13 Avalanches are most common during heavy snowfall and rapid wind deposition of snow, and are sufficiently common on some slopes that only small trees will persist. On other slopes, avalanches have long return intervals and may drastically alter the landscape in catastrophic events, which can fell mature stands within the subalpine forest, as shown here in Ammonoosuc Ravine on Mount Washington, New Hampshire. (MJ)

Counter intuitively, the alpine zone is also prone to extreme dryness. During episodes of strong high pressure, dry air descends from the upper troposphere, warms by compression during its descent, and as a result, becomes dryer still. When these descending layers of dry air, known as **subsidence inversions**, intersect the higher elevations of the eastern mountains, the relative humidity may drop into the low single digits, a condition similar to the driest deserts on earth. Although these conditions rarely persist for more than a few days, the extreme dryness favors rapid sublimation and evaporation of snow and water and may place additional stresses (e.g., desiccation) on plants not buried beneath the snow (4-14).

The conifer species that compose the krummholz subalpine forest (i.e., black spruce and balsam fir) provide a clear indication that weather and climate factors other than growing season temperatures prevent forests from invading the alpine zone. Though stunted, they do grow almost to the top of the highest mountains in the region. But they only survive at the highest elevations in the shelter of depressions and large rock formations (4-15). By contrast, elsewhere in the world, where growing season temperature controls the treeline, the line tends to be cleaner, and full-sized trees stop rather abruptly, without dwarfed individuals of the same species occurring higher up.

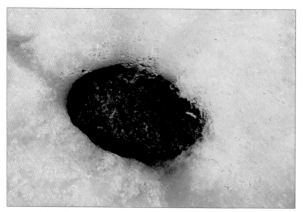

4-14 Alpine plants such as diapensia, that are not protected by accumulated snow, may be damaged by severe winter cold and wind. Presidential Range, New Hampshire. (MJ)

4-15 Krummholz conifer species such as black spruce and balsam fir (left) persist at high elevations in the lee of boulders and other obstacles to wind. Similar phenomena allowed this old tamarack (right) to persist above treeline on the Tablelands in Newfoundland. (MJ)

The wide range of elevations of treeline in eastern mountains correlates with comparisons of valleys to ridges. In valleys or gulfs, or glacial cirques, treeline may extend higher (although the steep headwalls may be free of large woody species). The clouds are often present, but winds that enhance rime ice deposition are weaker. On ridges, treeline is lower. The frequency of clouds may be less, but the accelerating winds over the ridges increase the efficacy of deposition of those clouds.

Another facet of alpine zone climate is the daily, or diurnal, variations in the depth of the **planetary boundary layer** (PBL). The PBL is the lowest layer of the atmosphere within which the air physically interacts with the earth's surface via friction and exchanges of heat and moisture. During the daytime, the depth of the boundary layer is defined by the height to which buoyant plumes of air, heated at the surface, rise. Although the depth of the PBL tends to grow throughout the day, at night the boundary layer depth collapses as turbulent mixing is supplanted by increasing stratification driven by cooling at the earth's surface (for example, the nocturnal inversions discussed earlier). Above the planetary boundary layer, a separate, more laminar and faster airflow typically occurs. More recent research indicates that many eastern alpine summits reside within the planetary boundary layer during midday, but are above it by evening. This nighttime "uncoupling" from the boundary layer results in enhanced evening winds that, in many instances, transport polluted air masses from afar to these otherwise pristine alpine landscapes. The diurnal variation in the PBL depth helps explain why pollutant concentrations atop Mount Washington, especially in the evening, often exceed those in the valleys of the White Mountains where boundary layer winds are weaker and aerosol scavenging more efficient.

4-16 Eastern alpine summits are fairly well "insulated" when in the clouds. Baxter Peak, Katahdin, Maine. (MJ)

Of interest in the northeastern United States, the top of the planetary boundary layer—where fair weather and orographic clouds typically form—coincides with the same elevational range in which subalpine forest and alpine ecosystems survive. This alpine and cloud relationship provides some clues to why the regional global warming trend appears to be weaker at higher elevations and why we have alpine ecosystems surviving at such low elevations in the eastern mountains. Clouds, as it were, play a major, albeit complex, role in regulating earth's climate. In some instances, the presence of clouds may warm the climate system while in others they cool it. In the White Mountains, for example, observations indicate that abundant clouds act to suppress the broader warming trend observed at lower elevations.[10] Regionally, the warming trend is presumed to be driven by greenhouse gas emissions of carbon dioxide, methane, and nitrous oxides in the lower troposphere, which act as a thermal blanket that reduces the radiation of energy back into space during clear nights. However, because clouds and water vapor also act as an efficient thermal blanket, nighttime cooling on the frequently cloud-enshrouded mountaintops is less influenced by the addition of anthropogenic greenhouse gases. In other words, the mountain summits are already well "insulated" when in the clouds, which occurs frequently (4-16). All told, the amelioration of warming trends by the combination of wind, frequent rime ice events, and cloud immersion may help explain why eastern alpine ecosystems have not succumbed to forest invasion, even several thousand years ago when this region's climate was warmer than today. During that warm period, the lower elevation forest transformed to more southern forest species, until a cooler climate returned, building the northern forest we know today.

As the regional climate warms in the coming decades and centuries, the subalpine forests of the eastern alpine will serve as refuges for the subarctic species profiled in the next two chapters and for low-elevation spruce-fir communities, whose cooler habitat is expected to decline along its southern fringe. The subalpine forest served this role during the postglacial hypsithermal period, which lasted from about 9,000 to 5,000 years ago. Then, deciduous forests are believed to have moved northward through New England in response to warming conditions, restricting the spruce-fir forest to the cooler coasts and higher elevations. Gradually, the Earth cooled again, and about 1,000 years before present (YBP) the spruce-fir forest species were able to recolonize the lower elevations of the Northern Forest. It is essential that our activities in eastern mountains not compromise the ability of present-day, upper-elevation spruce-fir forests to serve again as refugia in a warming climate (see Ch. 22).

---

10   Seidel et al. (2009)

Vegetation 5

June alpine flower display, Mount Monroe, New Hampshire. (JS)

Vegetation

5-1 Low, stunted growth forms are hallmarks of alpine vegetation. Here, Lapland rosebay casts its delicate flowers above crumbling serpentine. Lewis Hills, Newfoundland. (MJ)

The isolated subarctic vegetation atop the higher summits is perhaps the most visible and unifying characteristic of the eastern North American alpine region. For many plant species, the eastern alpine areas profiled in this book represent disjunct and fragile outliers at the southern periphery of an enormous range. Often, their main territory extends across expansive treeless areas of subarctic Canada (and in many prominent cases, such as moss campion [*Silene acaulis*] and sibbaldia [*Sibbaldia procumbens*], Europe and Asia). Arctic-alpine vegetation classically exhibits brilliant flowers among low, tangled mats and cushions whether you're in Oregon, Colorado, Iceland, Scotland, or France, and this is certainly the case in eastern North America (5-1). For our part, we often can't help but wonder how these tiny plants survive the harsh alpine winters. In fact, the plant species that do survive above treeline share numerous adaptations that increase their odds of survival despite ice and brutal winds. These plants tend to occur within pronounced vegetation communities composed of species exploiting similar survival strategies. For example, the cushion plant, diapensia (*Diapensia lapponica*), typically grows alongside similarly dwarfed or matted plants. Diapensia and its associated species compose a community that typically flourishes on exposed and windy areas that lack protective snow cover because these plants' prostrate, huddled forms protect them from wind exposure better than would a taller growth form. Vegetation isn't uniform across eastern alpine regions, however—in fact, it is complex and variable. Individual plant species vary greatly in their distributions throughout the region, and the flora of each major mountain range profiled in this book is fairly distinct. But physical processes have arranged these plants into a spectrum of definable communities, and in this chapter, we have done our best to summarize the most frequent and wide-ranging alpine communities found on eastern mountains. Because many of the individual species important to our under-

5-2 *An example of convergence, cushion-type growth allows many species of alpine plants to sustain internal temperatures as much as 20°C higher than ambient temperatures. Here, moss campion (left) and diapensia (right) provide classic examples of alpine cushion formation. Rope Cove, Lewis Hills, Newfoundland. (CE and MJ)*

5-3 *Mat-forming plants are common in eastern alpine ecosystems. On Newfoundland's Burnt Cape, an entire-leaved mountain avens clings to a substrate of shattered limestone. (MJ)*

standing of alpine communities may be unfamiliar to the reader, we've profiled several of the most distinctive plants. We anticipate that many readers will be familiar with the more common alpine plants, but we have made an effort to broadly characterize the full range of variation seen in our area.

Alpine Plant Adaptations

In general, plants (and other organisms) become able to tolerate harsh environments in one of three major ways: through adaptation (or evolution via natural selection), through non-reversible modifications incurred during life, and through acclimation (reversible changes incurred during life).[1] Through any of these three processes, a plant population can locally adapt to certain environments (for instance, severe alpine conditions) though they may still be considered the same species that occurs at low elevations. These locally adapted populations are known as **ecotypes**, and they may eventually become so distinct that a new species evolves. Many examples have been recognized among the clubmosses (Lycopodiaceae) and conifers (e.g., Pinaceae). Regardless of whether they are considered alpine ecotypes or alpine species, plants in alpine environments exhibit several similar adaptations that help them survive conditions that are generally cold, windy, nutrient poor, and often heavily disturbed by frost, avalanches, and herbivores.

When one sets foot in the alpine zone, the most evident adaptations are the growth forms that many alpine plants assume. For one, alpine plants are typically very small and low to the ground. Plants stay out of the wind so they lose less heat and moisture and suffer less damage and desiccation.

The **cushion plant** is one classic growth form found in alpine areas throughout the northern hemisphere (5-2). Cushion arrangements are ideal in windy, low moisture, and low nutrient environments for a number of reasons. Because cushions are low to the ground, wind moves over them without disturbing the leaves, allowing a large surface area for photosynthe-

---

1   Körner (2003)

sis. The inner part of the cushion is protected from the wind and can collect debris and soil (and therefore nutrients), retain moisture, and remain several degrees warmer than the surrounding air. This slight thermal increase can be just enough to allow photosynthesis to continue after air temperatures have dropped below freezing.[2] **Mats** are another common growth form (5-3). They spread more than cushions, but like cushions, the mats remain strategically close to the ground and out of the wind (5-4). Often mats will root where branches touch the ground, which allows them to expand while also staying firmly attached to the frequently disturbed soil. **Rosettes** are a third common growth form in the alpine zone. They are symmetrical, like cushions, and allow sunlight from all sides. Rosette leaves are flat to the ground and achieve direct sunlight as well as radiation from the ground, and minimal wind. Rosettes are limited in how large they can become, however, and cannot expand like a cushion or mat. Flowering stalks grow high above the rosette, allowing them to be readily seen by pollinators and facilitating wind dissemination of the seeds (5-5).

Many of the traits exhibited by alpine plants allow them to tolerate late season (in many cases, summer) frosts, and enable them to survive despite a growing season truncated at both ends. Many alpine plants turn from the deep-greens of summer to rich purples and reds during the shoulders of the growing season, from late-summer until late-spring (5-6). These stunning fall colors, especially notable in common alpine species such as diapensia and alpine bearberry (*Arctous alpina*), are produced by an abundance of **anthocyanins**, a common class of pigment in plants. Anthocyanins allow plants to absorb heat from light, warming plant tissues, increasing frost tolerance, and improving flowering rates and survival in cold alpine climates. Alpine plants tend to grow faster at colder temperatures than do their lowland counterparts, demon-

5-4 Often found together on wind-blistered alpine ridges, diapensia and alpine azalea exhibit cushion and mat growth patterns, respectively. Blow Me Down Mountains, Newfoundland. (MJ)

5-5 Rosettes are a common growth form in the mustards (Brassicaceae) and composites (Asteraceae), as shown here in the arctic bladderpod (Burnt Cape, Newfoundland) and horned dandelion (Monts Groulx, Québec). (MJ)

2   Cushion plants may attain extremely old ages, see McCarthy (1992)

5-6 *From September to May, diapensia cushions are a deep purple-red from an abundance of the pigment anthocyanin. Mount Lafayette, New Hampshire. (MJ)*

5-7 *The eyebrights are among relatively few alpine plant species with an annual life-history. Monroe Flats (left) and Alpine Garden (right), Mount Washington, New Hampshire. (MJ)*

5-8 *Alpine bistort is capable of asexual reproduction through viviparous propagules. Monts Groulx, Québec (left) and Forteau Tablelands, Labrador (right). (MJ)*

strating that, with an early start, they are able to maximize growth during a limited growing season. In fact, most alpine plants, including most of the familiar eastern species pictured in this chapter, are perennial; they don't start from seed every year. True alpine plants with annual life-cycles, such as the eyebrights (*Euphrasia* spp.), are relatively rare (5-7). Perennial species are able to store up excess carbohydrates in their root systems from one year to the next, allowing for quick growth early in the spring. Further, many alpine species are able to reproduce vegetatively (without seed) via runners, rhizomes, or vivipary. Examples include black spruce (*Picea mariana*), an abundant species in the krummholz community, and alpine bistort (*Polygonum viviparum*), a widespread alpine-subalpine herb (5-8).

The leaves of alpine plants are often succulent, waxy, downy, or evergreen. These traits minimize water loss via **transpiration**, increase frost tolerance, provide insulation, or allow a jumpstart on photosynthesis in the spring. These attributes are especially prominent in the saxifrages (Saxifragaceae) and heaths (Ericaceae), which are prominent families in eastern alpine areas (5-9).

Many alpine plants have deep taproots that allow them to take advantage of minimal water in well-drained, low-moisture environments, while preventing ejection from the soil during frost-thaw cycles (5-10). Soil and bedrock characteristics largely determine how much root structure alpine plants will require. These same characteristics don't only influence root structure—they also largely determine chemistry and nutrient availability, and therefore play a dominant role in determining which plants can occupy a site.

Bedrock and Soil

Bedrock composition and soil characteristics are important factors in determining plant distributions, but their significance is amplified in alpine areas where trees are rare and are thus not a factor in the abiotic environment of the understory. There is little to shield the wind or keep soil or moisture in place, making bedrock composition an extremely important factor.

Bedrock composition — Most of the eastern mountains are composed of igneous or metamorphic rock such as granite, gneiss, gabbro, anorthosite, and related rocks, with relatively few composed of calcium-rich sedimentary rock such as limestone. These common rock types support and encompass the major functional groups of alpine plant communities, essentially composing a graded suite of "typical," or default, alpine plant communities throughout much of the region. However, two major rock types present an unusual balance of magnesium/iron and calcium, which may dramatically influence the overlying vegetation. In general terms, respectively, these rocks are **serpentine** and **limestone**, which each support distinctive plant species wherever they constitute all or part of a given high mountain anywhere in our region because of the chemical composition of these rock types (especially the ratio of calcium to other elements) and their different weathering patterns (5-11) (Ch. 2 and 3).

5-9 The waxy leaves of yellow mountain saxifrage prevent water loss through transpiration. Forteau Tablelands, Labrador. (MJ)

5-10 On exposed ridges, frost action can eject shallow-rooted plants from the soil. Deterioration of this diapensia cushion was caused by trail disturbance and concurrent erosion. The open cross-section shows the internal structure of the plant and the deep-reaching roots. Katahdin, Maine. (MJ)

Permafrost — Permafrost has been documented in the Mealy Mountains (Ch. 20), Mont Jacques-Cartier (Ch. 10), and other mountains in the region. Permafrost influences plant community structure by facilitating frost disturbance of the soil (see Ch. 3).

Typical Eastern Alpine Vegetation

What we've referred to as the "typical" eastern alpine vegetation occurs to some extent throughout every one of the major mountain complexes profiled and discussed in this book, and is worthy of note for many reasons outlined later. This particular suite of plants has been profiled elsewhere, particularly in two excellent books published by the Appalachian Mountain Club 30 years apart: Dr. Nancy Slack and Allison Bell's *AMC Field Guide to New England Alpine Summits* and Dr. Stuart K. "Slim" Harris and colleagues' *AMC Field Guide to Mountain Flowers of New England*. These typical alpine plants and alpine plant communities generally occur on all of the major rock types that are not overly rich in magnesium and iron (ruling out serpentine) or calcium (ruling out limestone). The most common plant communities span all of the major classes of rock, including igneous (e.g., basalt, granite)

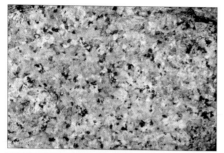

and sedimentary (e.g., sandstone) rocks and their metamorphic (e.g., schist and gneiss) derivatives. Most of the major and familiar mountains throughout the eastern region fit entirely into this broad category, including the Presidential Range (schist and gneiss), Katahdin and Jacques-Cartier (granite), and the top of Gros Morne (quartzite).

Typical eastern alpine vegetation is largely dominated by plants in the heath family (Ericaceae). Prominent among these are many species of cranberry and blueberry (*Vaccinium* spp.), Lapland rosebay (*Rhododendron lapponicum*), alpine azalea (*Loiseleuria procumbens*), and alpine bearberry (*Arctous alpina*). Diapensia and crowberry (*Empetrum* spp.) are also common constituents and are closely related to the heaths. These species are often encountered together on windswept alpine ridges, in frequent association with numerous grasses, sedges, and rushes.

Well-protected **snowbed** or seepage areas (discussed later in this chapter), such as alpine springs and streams and the damp headwalls of glacial cirques, support different vegetation than the exposed ridgelines and summits.

5-11 Many eastern mountains are composed of granite (top), which appears to support similar species of plants as such varied rock types as schist, gneiss, and quartzite. Serpentine (middle; composed of relatively high ratios of iron and magnesium), and limestone (bottom; composed largely of calcium carbonate) support unique and rare vegetation. (MJ)

Some common species in these protected areas are discussed in greater depth in the following profiles of alpine plant communities.

Gradients on the Landscape

Several environmental gradients are often noted as particularly important drivers in alpine environments. These include temperature, growing season length, wind exposure,[3] snow depth, atmospheric moisture, soil moisture (which is influenced by drainage and soil texture), soil chemistry and/or nutrient content. Any given location in the alpine zone (and anywhere on earth, for that matter) would fall somewhere along each of those gradients, and those factors (in addition to others like seed bank and inter/intra species interactions) help determine what species and what plant community occupies that location. These gradients are not independent; they covary and influence one another. Sites with similar locations on multiple gradients can usually be grouped based on topographic features because they are largely influenced by them. For example, ridge tops experience a great deal of exposure, high winds, cold temperatures, little snow cover, and low moisture, all of which limit soil

---

3   Allen and Walsh (1996); Elliot and Kipfmueller (2010)

development and nutrient content. Conversely, sheltered, snow-filled areas tend to occur in lees of topographic features, where wet, deep, relatively rich soil can develop. Kimball and Weihrauch[4] noted that slope, aspect, elevation, and slope shape (convex vs. concave) greatly influenced the type of communities found in New England alpine areas.

Species interactions, in addition to the abiotic environment and the gradients discussed earlier, can play a key role in altering the environment.[5] This is particularly true in the alpine zone where other species can locally influence climate (and therefore spatial pattern, productivity, and species richness) through competition and facilitation.[6] In recent decades, the alpine zone has been a focal experimental ground for the **stress gradient hypothesis.**[7] This hypothesis posits that in low-stress environments, competition is important and species are segregated, whereas in high-stress environments, facilitation is more important, and so species tend to be more aggregated in space.[8] As further evidence of this, if neighboring plants are removed in low-stress environments (e.g., below a species' elevational mean distribution), the remaining species thrive, but if neighbors are removed in high-stress environments (above the elevational mean of a species), remaining neighbors tend to experience higher mortality rates.[9] Results have been inconsistent, however,[10] and the relationship may vary across metrics of severity, scale, and growth form.[11]

The gradients that have been shown to be most important in determining which species occur where vary regionally and with the scale of observation. L.C. Bliss[12] suggested that the most important gradients in determining alpine plant communities in the Presidential Range of New Hampshire are snow accumulation (which is influenced by topography and exposure to prevailing wind direction) and a summer moisture gradient. In examining alpine plant communities from throughout the northeastern United States, researchers suggested that elevation also plays a key role, though elevation could be a surrogate for other abiotic factors: varying temperature, exposure, or moisture.[13] Other alpine researchers around the globe have also evaluated plant communities along gradients. In the west, researchers identified two important gradients that influence plant communities: namely, wind exposure and soil depth, and moisture.[14] In the Swiss Alps, researchers found that communities fell along two slightly different gradients: (1) temperature/wind speed, and (2) soil suction/pH/calcium content.[15]

Regardless of which underlying gradients are most important in determining community distribution, gradients are heavily influenced by topography so it is ultimately more practical to think about alpine plant communities being influenced by topography, albeit indirectly. We might argue that in conjunction with bedrock geology, topography is of primary importance, and we can envision eastern alpine areas as having two primary drivers that account

4   Kimball and Weihrauch (2000)
5   Olofsson (2004)
6   Kikvidze et al. (2005)
7   Bertness and Callaway (1994)
8   Kikvidze et al. (2005)
9   Choler et al. (2001)
10 e.g., Olofsson et al. (1999); Lortie and Callaway (2006); Maestre et al. (2006)
11 Olofsson et al. (1999); Lortie and Callaway (2006); Maestre et al. (2006); Dullinger et al. (2007)
12 L.C. Bliss (1963)
13 Sperduto and Cogbill (1999)
14 Cooper et al. (1997)
15 Vonlanthen et al. (2006)

5-12 *Mountain ranges with stark contrasts in topography, such as Katahdin, Maine, may have rather strikingly different plant communities within a few meters of one another. (MJ)*

5-13 *Krummholz, or stunted conifers, is found near the treeline and in protected areas above treeline. An example of "flagged" white spruce on the Monts Groulx, Québec. (MJ)*

for most of the variation in plant diversity: bedrock geology and topography.

Geomorphic variables such as slope, aspect, and slope shape interact to determine what areas will be blown free of snow in winter, and which areas will lie under many meters of windblown snow (protected from the mountain weather) well into spring. This in turn determines which plants can persist there. Mountain ranges with stark contrasts in topography, such as the Presidential Range and Katahdin, have rather strikingly different plant communities within a few meters of one another (5-12). It's also worth noting that topography is important at multiple scales because similar processes work on both the broad scale and fine scale, and snow can build up in the lees of whole ranges, ridges, single rocks, or patches of trees. Given topography's heavy influence on plant community structure, it's helpful to categorize plant communities into a consistent grouping of species that are often associated with certain topographical conditions near and above treeline.

Alpine Plant Communities

There is a long tradition of classifying plant communities, particularly in alpine areas of the Northeast,[16] and several similar schemes have been developed. In general, these classification systems have primarily focused on the New Hampshire, New England, or northeastern U.S. mountains, and consequently, they work quite well on the potassic summits encountered in that region. The appearance of serpentine and limestone vegetation in Québec and Newfoundland, as well as different plant species found in the larger, more northern ranges causes frameworks that are species-focused to be too specific. For this reason, we have adapted the typical classification schemes developed and used by previous authors. Following their lead, we have identified the following nine alpine plant communities, using the broadest scheme possible, so that they can be applied to most of the alpine ranges in the East. The species vary depending on the bedrock type, so among other factors, we've focused on the abiotic conditions in which the communities are found, rather than the species found in them (though we do make generalities based on observations from across the eastern alpine).

16   e.g., Bliss (1963); LeBlanc (1981); Slack and Bell (2006); Sperduto and Cogbill (1999); Kimball and Weihrauch (2000); Carlson et al. (2011)

Krummholz (Tuckamore) communities are composed of stunted conifers and hardwood trees, often found at the alpine-treeline ecotone.[17] Krummholz also occurs in the protected lee of large boulders and ridgelines that are oriented perpendicular to the prevailing winds (5-13). Krummholz throughout our region is usually dominated by prostrate forms of balsam fir (*Abies balsamea*) and black spruce, especially in the Presidential Range, Katahdin, and other relatively southern mountains. In these areas, red spruce (*Picea rubens*) occurs mostly in the forests below treeline but occasionally is found in the krummholz zone. In the more northerly ranges, from Gaspésie to the interior of the Canadian Shield, as well as on Newfoundland, white spruce (*Picea glauca*) becomes widespread and may be common in the krummholz zone. (In low-elevation forests and bogs, black spruce assumes a characteristic spindly, top-heavy growth form.) Tamarack (*Larix laricina*) and jack pine (*Pinus banksiana*), members of the pine family (Pinaceae) like spruce and fir, are sometimes found in the krummholz regions of eastern mountains, especially in Québec (Pl. I). Throughout the re-

5-14 Extensive birch-alder communities occur throughout the alpine areas of eastern Canada. Monts Groulx, Québec. (MJ)

5-15 Monticello Lawn on Mount Jefferson, in New Hampshire's Presidential Range, is a spectacular example of an alpine meadow dominated by Bigelow's sedge, highland rush, and other graminoids. (MJ)

gion, krummholz provides shelter necessary for other shrubs and herbaceous plants to grow beneath them, resulting in an herbaceous understory similar to that found throughout the boreal (or "Canadian") forests. This protected understory almost always includes a host of common boreal species such as Canada mayflower (*Maianthemum canadense*), starflower (*Trientalis borealis*), bluebead lily (*Clintonia borealis*), and large-leaved goldenrod (*Solidago macrophylla*), as well as occasional species such as northern comandra (*Geocaulon lividum*) and bakeapple or cloudberry (*Rubus chamaemorus*). Although bakeapple is more at home in northern bogs, it is sometimes extremely abundant in poorly-drained krummholz regions in Québec and Newfoundland (Pl. II). Also hidden away in the sheltered subalpine krummholz could be the twayblade orchids, such as heart-leaved twayblade (*Listera cordata*).

Birch-alder is a form of deciduous shrub community, outlined in the classification scheme by Kimball and Weihrauch.[18] We find this classification particularly useful when evaluating the eastern alpine areas holistically. Middle-elevation slopes of many of the northern ranges support extensive thickets of heart-leaved birch (*Betula cordifolia*), glandular birch (*Betula glandulosa*), and green alder (*Alnus viridis*) (5-14). Birch-alder communities can cover hundreds of hectares in the expansive alpine ranges of the Canadian Shield, such as the Monts Otish (Ch. 19) and Monts Groulx (Ch. 18), where they may be augmented by Bartram's

---

17 These "elven forests" are known as *krummholz* in the United States and often referred to as *tuckamore* in eastern Canada.

18 Kimball and Weihrauch (2000)

Plate I. Krummholz (Tuckamore)
Representative tree species of the krummholz-tuckamore communities

White spruce (*Picea glauca*)
Twig. Monts Groulx, Québec

Black spruce (*Picea mariana*)
Twig. Monts Groulx, Québec

Red spruce (*Picea rubens*)
Twig. Smugglers Notch, Vermont

Red spruce (*Picea rubens*)
Cones. Carter Notch, New Hampshire

White spruce (*Picea glauca*)
Cones. Mont Comte, Québec

Balsam fir (*Abies balsamea*)
Cones. Great Gulf, New Hampshire

Tamarack (*Larix laricina*)
Cone. Monts Groulx, Québec

Jack pine (*Pinus banksiana*)
Cone. Mont du Lac des Cygnes, Québec

Plate II. Krummholz (Tuckamore) Understory
Representative plants of the krummholz-tuckamore understory

Creeping snowberry (*Gaultheria hispidula*)
Mont Logan, Québec

Starflower (*Trientalis borealis*)
Flowers. Mount Monroe, New Hampshire

Bluebead lily (*Clintonia borealis*)
Flowers. Mount Monroe, New Hampshire

Red-stemmed feathermoss (*Pleurozium schreberi*, left) and stairstep moss (*Hylocomium splendens*, right) Monts Groulx, Québec

Bunchberry (*Cornus canadensis*)
Flower. Highlands of St. John, Newfoundland

Bunchberry (*Cornus canadensis*)
Drupes. Highlands of St. John, Newfoundland

Northern comandra (*Geocaulon lividum*)
Drupe. Battle Harbour, Labrador

Plate III. Birch–Alder
Representative plants of birch-alder communities

Green alder (*Alnus viridis*)
Monts Groulx, Québec

Glandular birch (*Betula glandulosa*)
Monts Groulx, Québec

Squashberry (*Viburnum edule*)
Monts Groulx, Québec

Skunk currant (*Ribes grandulosum*)
Monts Groulx, Québec

Tea-leaved willow (*Salix planifolia*)
Mount Washington, New Hampshire

Labrador willow (*Salix argyrocarpa*)
Mount Washington, New Hampshire

Clustered lady's mantle (*Alchemilla glomerulans*)
Monts Groulx, Québec

Arctic sweet coltsfoot (*Petasites frigidus*)
Monts Groulx, Québec

shadbush (or serviceberry) (*Amelanchier bartramiana*) and squashberry (*Viburnum edule*) (Pl. III).

Alpine meadows occur on slightly less exposed, less dry, and relatively flat sites (5-15). Alpine meadows are often dominated by Bigelow's sedge (*Carex bigelowii*) and other sedges such as brownish sedge (*Carex brunnescens*), as well as various species of familiar heath such as alpine bilberry and mountain cranberry (*Vaccinium vitis-idaea*). Monticello Lawn and Bigelow Lawn in the Presidential Range (Ch. 7) are famous examples of alpine meadows. Alpine meadows also cover large areas of Katahdin's tableland and may be found in some areas of the Bay of Islands, Newfoundland (Ch. 13) (Pl. IV).

Cushion-tussock communities occur on exposed ridge-top sites that accumulate little snow throughout the winter. Also called **diapensia communities** by some authors, these sites are in fact dominated by diapensia along with highland rush (*Juncus trifidus*), Lapland rosebay, alpine azalea, and alpine bilberry (*Vaccinium uliginosum*). They are widespread on Katahdin (5-16) and other New England mountains where the largest continuous extent of cushion-tussock alpine tundra in New England is found near the Lakes of the Clouds on Mount Washington. This community is relatively rare on some of the plateau-like massifs of the Canadian Shield (Part IV), which are poorly drained and less exposed, such as the Monts Groulx (Ch.18).

Heath-shrub-rush is a community-level designation that tends to be a catch-all for

5-16 *Cushion-tussock communities dominated by diapensia, alpine bilberry, and highland rush are common on Katahdin. Hamlin Ridge from the Saddle, Katahdin, Maine. (MJ)*

5-17 *Vivid red alpine bearberry stands out in this heath-shrub community along the Trans-Labrador Highway north of Red Bay, Labrador. (MJ)*

5-18 *The larger mountain ranges on the Canadian Shield support extensive alpine and subalpine bogs, fens, marshes, and pools. Monts Groulx, Québec. (MJ)*

those exposed and windswept plant communities that don't fall clearly into cushion-tussock (5-17). Heaths are the dominant group in this community. Highland rush, woodrushes (*Luzula* spp.), mountain cranberry, Labrador tea (*Rhododendron groenlandicum*), black crowberry, and alpine bearberry are commonly found in these areas with numerous lichens (Pl. V).

Alpine bogs and fens are high-elevation wetland areas that occur in poorly drained sites with generally low relief (5-18). The northern ranges of Newfoundland and Québec will often have expansive plateaus covered with these communities, because these ranges have

Plate IV. Alpine Meadows
Representative plants of alpine meadows

Mountain firmoss (*Huperzia appressa*)
Mount Monroe, New Hampshire

Stiff clubmoss (*Spinulum canadense*)
Mont Joseph Fortin, Québec

Bigelow's sedge (*Carex bigelowii*)
Mount Washington, New Hampshire

Brownish sedge (*Carex brunnescens*)
Monts Groulx, Québec

Highland rush (*Juncus trifidus*)
Katahdin, Maine

Alpine bilberry (*Vaccinium uliginosum*)
Monts Groulx, Québec

Mountain cranberry (*Vaccinium vitis-idaea*)
Mount Monroe, New Hampshire

Alpine bearberry (*Arctous alpina*)
Monts Groulx, Québec

Plate V(A). Cushion-Tussock and Heath-Shrub-Rush
Representative plants of exposed and windswept ridges

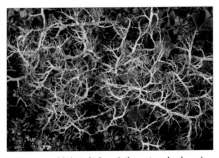

Green witch's hair lichen (*Alectoria ochroleuca*)
Monts Groulx, Québec

Star-tipped reindeer lichen (*Cladonia stellaris*)
Red Wine Mountains, Labrador

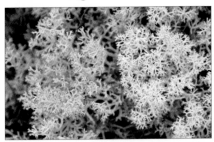

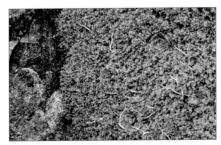

Black-footed reindeer lichen (*Cladonia stygia*) (left)
and Grey reindeer lichen (*C. rangiferina*) (right)
Monts Groulx, Québec

White worm lichen (*Thamnolia vermicularis*) in
hoary rock moss (*Racomitrium lanuginosum*) (right)
Mount Monroe, New Hampshire

Alpine sweetgrass (*Anthoxanthum monticola*)
Monts Groulx, Québec

Bigelow's sedge (*Carex bigelowii*)
Katahdin, Maine

Tufted clubrush (*Trichophorum cespitosum*)
Katahdin, Maine

Highland rush (*Juncus trifidus*)
Gros Morne, Newfoundland

Plate V(B). Cushion-Tussock and Heath-Shrub-Rush
Representative plants of exposed and windswept ridges

Black crowberry (*Empetrum nigrum*)
Monts Groulx, Québec

Diapensia (*Diapensia lapponica*)
Lewis Hills, Newfoundland

Alpine bilberry (*Vaccinium uliginosum*)
Monts Groulx, Québec

Mountain cranberry (*Vaccinium vitis-idaea*)
Mount Washington, New Hampshire

Lapland rosebay (*Rhododendron lapponicum*)
Katahdin, Maine

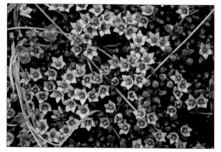

Alpine azalea (*Loiseleuria procumbens*)
Katahdin, Maine

Alpine bearberry (*Arctous alpina*)
Katahdin, Maine

Bearberry willow (*Salix uva-ursi*)
Monts Groulx, Québec

Plate VI. Alpine Bogs, Fens, and Pools
Representative plants of alpine wetlands

Magellan's peatmoss (*Sphagnum magellanicum*)
Monts Groulx, Québec

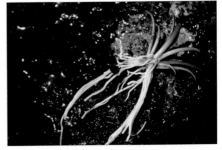

Spiny-spored quillwort (*Isoëtes echinospora*)
Blow Me Down, Newfoundland

Cotton grass (*Eriophorum vaginatum*)
Monts Groulx, Québec

Tufted clubrush (*Trichophorum cespitosum*)
Annieopsquotch Mountains, Newfoundland

Bog rosemary (*Andromeda polifolia*)
Mont Albert, Québec

Round-leaved sundew (*Drosera* rotundifolia)
Blow Me Down, Newfoundland

Small cranberry (*Vaccinium oxycoccos*)
Mount Washington, New Hampshire

Bakeapple or cloudberry (*Rubus chamaemorus*)
Red Bay, Labrador

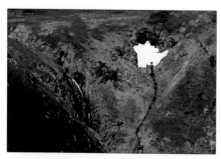

5-19 *Areas of late-lying snow shelter many regionally rare alpine plant species. Here, the last remaining snow in Molly-Ann Canyon in the western Lewis Hills, Newfoundland. (MJ)*

much more subtle topography than the lower latitude peaks of New England and New York. Alpine bogs and fens are dominated by *Sphagnum* spp., mosses and lichen, and common vascular species might include cotton grass (*Eriophorum* spp.), bakeapple (cloudberry), glandular birch (*Betula glandulosa*), and various boreal heaths such as leatherleaf (*Chamaedaphne calyculata*), bog rosemary (*Andromeda polifolia*), Labrador tea (*Rhododendron groenlandicum*) and blueberries and cranberries (*Vaccinium* spp.).

Often found as part of larger mosaics of wetlands, **alpine pools** occur throughout our region at all elevations above treeline. These small ponds are more numerous and diverse on broad, poorly drained, alpine tablelands such as the Lewis Hills (Ch. 13) and Monts Groulx, and are rare on steep, high-relief mountains such as the Presidential Range. Alpine pools often support relatively few vascular plant species. Among the more common are spiny-spored quillwort (*Isoëtes echinospora*) and northern whiterush (*Juncus triglumis* ssp. *albescens*) (Pl. VI).

Snowbeds, also called **snowbanks** or "zabois," occur in sheltered areas where winter snow accumulates to great depths and persists late into the summer growing season (5-19). These communities occur at a range of scales; extremely large snowbeds are found in the bottoms of cirques and ravines and in the lee of prominent topographic features. Smaller snowbeds, ranging upward in size from a few square meters, occur in the lee of rocks or patches of krummholz, and in depressions on the alpine landscape. Mechanical weathering of trees during ice storms is not as prominent a factor in these systems because branches are shielded from the elements by thick snow. For this reason, it is possible for relatively deep soil horizons to build up in these areas, a rare feature in most eastern alpine areas. Boreal species such as bluebead lily, Canada mayflower, starflower, and large-leaved goldenrod (*Solidago macrophylla*) are abundant in snowbeds. Snowbeds that are relatively stable tend to harbor of the greatest diversity of any alpine community type, and locally rare species are found here, including the heath family members moss heather (*Harrimanella hypnoides*) and purple mountain heather (*Phyllodoce caerulea*), as well as alpine clubmoss (*Diphasiastrum alpinum*), sibbaldia (*Sibbaldia procumbens*), and snowbed willow (*Salix herbacea*). Local rarities include alpine and Norwegian cudweed (*Omalotheca supina* and *O. norvegica*). The snowbeds of the Presidential Range in New Hampshire provide habitat for some of the rarest plants in New England, including sibbaldia and alpine cudweed. Snowbed habitats are much more extensive in the northern ranges, including the Monts Groulx, Monts Otish, and Mealy Mountains, and otherwise rare snowbed species can be found growing in abundance there. In the Presidential Range, snowbed communities tend to be on east and southeast oriented slopes, which results from prevailing wind patterns (Pl. VII).[19]

Felsenmeer (literally "sea of rocks"), also referred to as **fellfields**, occur at high elevation sites on many eastern alpine ranges (5-20). The Presidential Range and Katahdin (Ch. 8)

19   Bliss (1963); Kimball and Weihrauch (2000)

Plate VII. Snowbeds
Representative plants of late-lying snow areas

Alpine clubmoss (*Diphasiastrum alpinum*)
Highlands of St. John, Newfoundland

Sibbaldia (*Sibbaldia procumbens*)
Lewis Hills, Newfoundland

Snowbed willow (*Salix herbacea*)
Monts Groulx, Québec

Dwarf bilberry (*Vaccinium caespitosum*)
Mount Adams, New Hampshire

Moss heather (*Harrimanella hypnoides*)
Monts Groulx, Québec

Purple mountain heather (*Phyllodoce caerulea*)
Monts Groulx, Québec

Alpine cudweed (*Omalotheca supina*)
Mount Washington, New Hampshire

Norwegian cudweed (*Omalotheca norvegica*)
Mont Richardson, Québec

5-20 *Expanses of shattered rock thought to be a periglacial phenomenon are known as* felsenmeer. *The term is German for "sea of rocks." North Summit of the Highlands of St. John, Newfoundland. (MJ)*

5-21 *Cliffs in alpine areas often owe their existence to regional or local glaciation. Such is the case with these promontories near Rope Cove in the Lewis Hills, Newfoundland. (MJ)*

encompass extensive felsenmeer barrens. The flat, siliceous summits in particular, including Gros Morne (Ch. 14) and the Highlands of St. John (Ch. 16) in Newfoundland, tend to support expansive fellfields. Serpentine mountains at upper elevations often develop expansive felsenmeer communities. Patterned ground caused by frost action is common in these areas, with rocks sorting into size classes in polygons (in flat areas) or stripes (on slopes). With the exception of serpentine areas, most fell field rock surfaces will be covered with various lichen species including map lichen (*Rhizocarpon geographicum*), target lichen, (*Arctoparmelia centrifuga*), and rock tripe (*Umbilicaria hyperborea*). Various species of heath plants can survive when nestled in the lee areas of the larger rocks, and they often give way to the cushion-tussock or heath-shrub-rush community types where soil has collected, or where finer rock fragments have settled (Pl. VIII).

Cliff communities are highly variable throughout the eastern alpine, occurring on the steepest slopes, and often including high-angle **seepages** and **talus** slopes (5-21). Any number of species might be found clinging to alpine cliff faces, depending on exposure and moisture conditions, and soil and bedrock type. Some of the more broadly distributed species include the sharply sweet-smelling lance-leaved arnica (*Arnica lanceolata*), alpine bittercress (*Cardamine bellidifolia*), and mountain sorrel (*Oxyria digyna*). Wormskjold's alpine speedwell (*Veronica wormskjoldii*) and tall white bog orchid (*Platanthera dilatata*) grow sporadically in seepages and rivulets on cliff walls and surrounding talus in New England. Northeastern paintbrush (*Castilleja septentrionalis*) often grows on stabilized talus near the base of high cliffs (Pl. IX).

By loosely grouping communities into these categories, while being mindful of the underlying gradients that influence community structure, it's possible to visit any eastern alpine range and notice both similarities and differences in the plant species that occur in the various communities. The potassic, limestone, and serpentine ranges of the eastern alpine are strikingly different in many respects, but they share widespread features and processes that unite them.

Vegetation on Serpentine

What we refer to loosely as "serpentine" summits are a handful of both famous and little-known eastern mountains formed from rock that once lay at the contact between the earth's crust

Plate VIII. Felsenmeer
Representative lichens and plants of felsenmeer

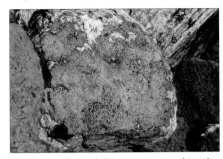

Yellow map lichen (*Rhizocarpon geographicum*)
Tablelands, Newfoundland

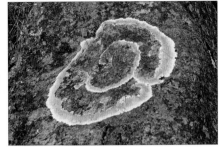

Concentric ring lichen (*Arctoparmelia centrifuga*)
Annieopsquotch Mountains, Newfoundland

Rock tripe lichen (*Umbilicaria* sp.)
Tablelands, Newfoundland

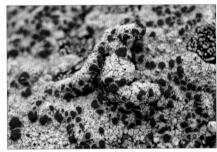

Alpine bloodspot (*Ophioparma* sp.)
Monts Groulx, Québec

Hoary rock moss (*Racomitrium lanuginosum*)
Mont Albert, Québec

Mountain firmoss (*Huperzia appressa*)
Mount Washington, NH

Purple mountain heather (*Phyllodoce caerulea*)
Blow Me Down, Newfoundland

Bluebell or harebell (*Campanula rotundifolia*)
Mont Albert, Québec

Plate IX. Cliffs, Seepages, and Talus
Representative plants of cliffs and seepages

Elegant sunburst lichen (*Xanthoria elegans*)
Tablelands, Newfoundland

Single-spike sedge (*Carex scirpoidea*)
Tablelands, Newfoundland

Mountain sorrel (*Oxyria digyna*)
Lewis Hills, Newfoundland

Alpine bittercress (*Cardamine bellidifolia*)
Lewis Hills, Newfoundland

Northeastern paintbrush (*Castilleja septentrionalis*)
Monts Groulx, Québec

Wormskjold's alpine speedwell (*Veronica wormskjoldii*)
Mount Washington, New Hampshire

Lance-leaved arnica (*Arnica lanceolata*)
Mount Washington, New Hampshire

and mantle. Famous examples include Mont Albert (Ch. 9) on Québec's Gaspé Peninsula and the Tablelands (Ch. 13) in Gros Morne National Park, Newfoundland (5-22). A full exposure of the crust-mantle boundary is known as an **ophiolite suite** and often comprises vast areas of exposed **serpentine**, a metal-rich rock toxic to many of the more widespread alpine plants. Note that there are two related terms: **ophiolite** (pertaining to the sequence of rocks characteristic of the Earth's crust-mantle boundary) and **serpentine** (describing magnesium-rich and iron-rich bedrock, a part of the ophiolite). These are distinct concepts important to understanding the structure and biodiversity of serpentine summits. Serpentine outcrops occur throughout the Appalachian region, including some very large exposures in southern Québec, but these "outcrops" occur in the form of massive, mountains at only three sites: Mont Albert in Québec (Ch. 9), the Bay of Islands complex in southwestern Newfoundland (Ch. 13) and in the White Hills near St. Anthony, Newfoundland (Ch. 17). In the serpentine areas, a number of physiographic considerations are at play. The krummholz may include stunted

5-22 Vegetation patterns and species composition on serpentine may be different from nearby mountains composed of granite, schist, gneiss, or quartzite. Mont Albert, Québec. (MJ)

5-23 The White Hills of northern Newfoundland are isolated and poorly known examples of alpine serpentine vegetation. The common gray moss Racomitrium *dominates the foreground. (MJ)*

forms of tamarack (*Larix laricina*) and white pine (*Pinus strobus*), as well as more widespread subalpine species such as black spruce, white spruce, and balsam fir. Heath-family plants (Ericaceae) are mostly absent from serpentine tundra (although Lapland rosebay is often abundant), and carnation/pink-family plants (Caryophyllaceae) are the most diverse group of flowering plants. This group includes many noteworthy species, including several species of low sandworts (*Minuartia marcescens, M. rubella,* and *Arenaria humifusa*), the striking alpine catchfly (*Silene suecica*), and the widespread moss campion (*Silene acaulis*) (known from alpine sites throughout the northern hemisphere, including many potassic and limestone sites). Other abundant species on exposed serpentine include sea thrift (*Armeria maritima*), willows (e.g., *Salix glauca*), and common juniper (*Juniperus communis*). Swamp thistle (*Cirsium muticum*) is common on the serpentine of Mont Albert, but not the mountains of Newfoundland. On less-exposed sites, the vegetation varies across mountain ranges and includes species such as Canada burnet (*Sanguisorba canadensis*), pitcher plant (*Sarracenia purpurea*), bluebell (*Campanula rotundifolia*), and goldenrods (e.g., *Solidago multiradiata, S. uliginosa*). Topography remains a critical factor, as it is in the all other alpine areas. Snowbeds occur on serpentine mountains also and support royal fern (*Osmunda regalis*), Canada burnet, butterwort (*Pinguicula vulgaris*), and many other species (Pl. X).

Most of the serpentine summits covered in this book also incorporate smaller areas of basaltic or gabbroic rock, which tend to support a more typical vegetation as described

Plate X. Serpentine
Representative plants of high-elevation serpentine

Common juniper (*Juniperus communis*)
White Hills, Newfoundland

Serpentine sandwort (*Minuartia marcescens*)
Tablelands, Newfoundland

Reddish sandwort (*Minuartia rubella*)
Lewis Hills, Newfoundland

Creeping sandwort (*Arenaria humifusa*)
Burnt Cape, Newfoundland

Alpine catchfly (*Silene suecica*)
Mont Albert, Québec

Moss campion (*Silene acaulis*)
Lewis Hills, Newfoundland (CE)

Sea thrift (*Armeria maritima*)
Blow Me Down, Newfoundland

Thistle (*Cirsium muticum*)
Mont Albert, Québec

earlier in this chapter. Such is the case on Mont Albert, where a prominent ridge of amphibolite along the northern edge of the mountain supports vegetation more similar to Mont Jacques-Cartier than to the adjacent serpentine tableland. This is also the case in the Bay of Islands, where on the Lewis Hills, for example, numerous areas of gabbro and basalt support a vegetation more typical of potassic-siliceous summits than of serpentine. Further discussion of this intriguing system is provided in the chapters on Mont Albert, the Bay of Islands, and the White Hills section of Chapter 17 (5-23).

5-24 Limestone, a calcium-rich sedimentary bedrock, is rare in the eastern mountains. Generally, limestone supports plant species rare on other mountains. One such outcrop occurs at the western end of Killdevil Mountain, Newfoundland. (MJ)

Vegetation on Limestone

Most of the summits in our region fall within the two categories outlined earlier (i.e., typical, or the unusual case of serpentine). More rarely in our area does limestone, or other calcareous (calcium-rich) sedimentary rock, dominate the upper reaches of an alpine ridge or massif (5-24). More often, calcium-rich rock appears as banded occurrences on an otherwise silica-rich mountain, or as isolated occurrences on a range. Limestone is almost entirely absent from the New York and New England mountains, although some calciphile alpine species may be seen at the Hudson River Gorge and Wallface Mountain in New York; Smugglers Notch, Hazen's Notch, and Willoughby Cliffs in Vermont; Cannon Cliffs in Franconia Notch and Huntington Ravine on the Presidential Range in New Hampshire; and the North Basin of Katahdin in Maine. In Gaspésie, there are isolated occurrence of calcareous bedrock at Mont de la Table, north-

5-25 The ravines and gullies of Mont Logan on Québec's Gaspé Peninsula support many otherwise rare calcareous plant species. (ML)

5-26 The expansive limestone barrens at Burnt Cape, Newfoundland, support many arctic species. (MJ)

west of Mont Jacques-Cartier (Ch. 10), and prominent occurrences in the vicinity of Mont Logan near the western edge of the Chic-Choc range, culminating in the spectacular plant diversity of Mont Logan's *Bassin de Pease* and the *Ravin à Neige* (Ch. 9) (5-25). In our region, the majority of the noteworthy limestone sites are on the island of Newfoundland, where the lines are blurred between alpine tundra associated with high mountains, and coastal barrens. Nonetheless, clear and noteworthy occurrences of alpine limestone plant assemblages may be found at Table Mountain near Port au Port, on Killdevil Mountain in Gros Morne National Park (Ch. 14), and on the Highlands of St. John northeast of Port aux Choix (Ch. 16).

5-27 Isolated occurrences of limestone near Forteau, Labrador, provide rare habitat for limestone-loving plants on the Canadian Shield. (MJ)

Several magnificent promontories on Newfoundland's Great Northern Peninsula represent a sort of hybrid between coastal limestone barrens and high alpine, such as the windswept and lonesome barrens at Port au Choix, Cape Norman, and Burnt Cape (Ch. 17) (5-26). Immediately northwest of these limestone capes, across the Strait of Belle Isle in Labrador, alpine limestone is rare as a rule on the gneissic Canadian Shield, but locally present in the sedimentary Forteau Tablelands near the towns of Blanc-Sablon, Québec, and Forteau and L'Anse Amour, Labrador (5-27). Limestone assemblages reportedly occur again in the Monts Otish of central Québec but appear to be absent from most other alpine sites in the Canadian Shield region (e.g., the Monts Groulx, Montagnes Blanches, and others) (Ch. 21).

On alpine limestone, the heaths (Ericaceae) that are so common in typical alpine assemblages, and the carnations/pinks (Caryophyllaceae) so visible on serpentine, give way to saxifrages (Saxifragaceae) and mustards (Brassicaceae), which may be locally diverse at some sites. These include such species as yellow, white, and purple mountain saxifrage (*Saxifraga* spp.) as well as tufted saxifrage (*S. cespitosa*) and several drabas (*Draba* spp.) and alpine rock-cress (*Arabis alpina*). Cliff faces underlain by limestone, or influenced by alkaline groundwater, such as the western walls of Newfoundland's Highlands of St. John or the north-facing ravines of Québec's Mont Logan, support numerous species that are rare on the adjacent felsenmeer and heath-shrub-rush areas, including white, yellow, and purple mountain saxifrage (*Saxifraga* spp.). Another group commonly encountered on alpine limestone are the mountain avens (*Dryas integrifolia* and *D. drummondii*), widespread and familiar members of the rose family. Many arctic rarities, such as velvetbells (*Bartsia alpina*) and redtip louse-wort (*Pedicularis flammea*), reach their southeasternmost occurrence on Newfoundland's limestone barrens (Pl. XI).

Limestone barrens, including the clearly alpine as well as those close to sea level, can range from gravel barrens with scattered vegetation and prominently frost-patterned ground to more continuously vegetated meadow or heath in slightly more sheltered microsites. Calcareous heath may be rich in crowberry and alpine bilberry, but several species of dwarf willows, junipers, and saxifrages obtain greater prominence than on more acidic rock types. Meadows with a variety of grasses, forbs, ferns, mosses, and lichens also occur, and these replace the gravel barrens and calcareous heath as the dominant calcareous plant community in the Forteau Tablelands of southeastern Labrador. In even more sheltered areas with krummholz, the litter layer acidifies the soil. This masks the calcareous influence to some extent, and with the exception of a few species, the understory vegetation resembles the typical boreal species found underneath krummholz on acidic rock types (see Pl. II).

Plate XI. Limestone
Representative plants of high-elevation limestone

Yellow mountain saxifrage (*Saxifraga aizoides*)
Burnt Cape, Newfoundland (MK and DD)

White mountain saxifrage (*Saxifraga paniculata*)
Hazen Notch, Vermont

Purple mountain saxifrage (*Saxifraga oppositifolia*)
Lewis Hills, Newfoundland

Entire-leaved mountain avens (*Dryas integrifolia*)
Burnt Cape, Newfoundland

Drummond's mountain avens (*Dryas drummondii*)
Lac Jeannine Mine, Gagnon, Québec

Alpine hedysarum (*Hedysarum alpinum*)
Forteau Tablelands, Labrador

Net-veined willow (*Salix reticulata*)
Forteau Tablelands, Labrador

Moss campion (*Silene acaulis*)
Forteau Tablelands, Labrador

Summary

Although Dr. M.L. Fernald[20] eloquently described three primary types of alpine floras in our region—those associated with potassium-, magnesium-iron-, and calcium-rich bedrock (and similar classifications have been made around the globe in the years since)[21]—in reality, the various plant species we find distributed on these three main bedrock types are a result of underlying gradients in chemistry, soil structure, and moisture that tend to result from these three bedrock classes. At any given point on the landscape, a site will fall somewhere along various abiotic gradients. Some of the landscapes discussed in this book are particularly complex. The Bay of Islands Mountains (Ch. 13), for example, represent extreme gradients in bedrock geology and morphology, and many of the classifications outlined above may be found in close proximity to one another. As a result, it is difficult to cleanly compartmentalize the alpine world into a classification scheme, whether it is based on bedrock types, geography, topography, or the species that are present. Furthermore, such classification needs some level of site-specific analysis and understanding. In the later chapters of this book, we've tried to describe the known flora of each mountain range at some finer level of resolution.

20   M.L. Fernald, noted Harvard botanist and author of many foundational papers on the flora of Newfoundland published this classification in 1907.

21   Fernald (1926); Gigon (1987); Reisigl and Keller (1987)

Fauna 6

American marten, Carter Range, New Hampshire. (MJ)

Fauna

6-1 Eastern alpine landscapes provide opportunities to see northern animals south of their primary ranges. Here, a bull woodland caribou patrols Mont Boissinot in Québec's Monts Groulx. (MJ)

Northern animal assemblages provide one of the most exhilarating aspects of a visit to eastern alpine areas. A journey to the mountains may be fueled by the thrill of encountering woodland caribou, arctic hares, rock ptarmigans, American pipits, or subarctic butterflies (6-1).

Almost any mammal, bird, amphibian, or invertebrate from the surrounding boreal and subalpine forest may be found in the alpine zone during mild seasons and storms. In addition, a handful of alpine or subalpine specialists depend on the treeless habitats to persist. These include a few mammals, several birds, and many invertebrates. As with eastern alpine plants, the majority of the eastern alpine animals are distributed across arctic-alpine or subarctic-subalpine regions. They occur across the tundra of the Low Arctic and extend south along one or several of the mountain chains in North America, Europe, or Asia. In many cases, alpine distributions may represent relics from the period following the last glacial retreat 8,000 to 12,000 years before present (YBP), when the whole region (mountains and valleys) were covered by tundra. In other cases, arctic species may have colonized alpine areas more recently.

As described in Chapter 5, a diverse and bewildering variety of alpine plants have undergone selection for miniaturization and cushion-formation to conserve heat and moisture. Alpine animals have similarly evolved adaptations to deal with the extreme weather conditions, and the reduced aerial shelter that would elsewhere be provided by trees and shrubs. These adaptations essentially involve a minimized ratio of surface-area to body-mass, shortened appendages, modifications in coloration, and the development of specialized thermoregulation behaviors. Arctic hares, for instance—which are the largest, native North American rabbit—can curl themselves into a nearly spherical shape to conserve heat, and have specialized winter fur that is white and is thicker and more air-filled than their summer fur.[1] The white winter fur of arctic hares and other mammals may be largely the evolutionary "result" of reduced shelter from shrubs and trees on the exposed tundra. Among insects and other arthropods, alpine adaptations include darker pigment for greater solar absorption, more hairs, and reduction of wings, mouthparts, and antennae.[2] Tundra-adapted insects with aquatic larvae also tend to use relatively less energy as adults. This is variously accomplished, and insects may exhibit increased synchrony of emergence, mating closer to the site of emergence, or even more incredibly by avoiding mating through parthenogenesis,[3] or by skipping the adult phase entirely.[4]

Mammals

As found in many species of alpine plants (Ch. 5), numerous northern, subarctic, and alpine mammals reach their most southerly latitudes on the eastern mountains. The Presidential Range and the surrounding White Mountains of New Hampshire currently provide the southernmost boreal and subalpine habitats for American marten (*Martes americana*), a small-bodied member of the weasel family. Although martens are rare at lower elevations in New Hampshire, the boreal forests and krummholz near treeline abound with these agile carnivores. Although the marten is not an alpine species, it is sometimes the most abundant carnivore in eastern alpine habitats. A winter visit to any of the eastern alpine areas in Labrador, Québec, or New England will likely showcase numerous trails of marten in the snow near treeline and above (at present, though, marten are not commonly encountered in Newfoundland alpine areas, where the native subspecies *atrata* has a highly restricted range) (6-2).

6-2 One of the most common carnivores in eastern alpine and subalpine areas is the American marten, primarily a forest-dwelling species. An adult marten from the Carter Range, New Hampshire, is pictured. (MJ)

---

1   Wang et al. (1973); Gray (1993); Russell and Tumlison (1996)
2   Ring and Tesar (1981)
3   Parthenogenesis is the reproduction of offspring without a genetic contribution from a male.
4   Downes (1965); Ring and Tesar (1981); Böcher and Nachman (2010)

Arctic fox (*Vulpes lagopus*) is another tundra specialist that occupies both arctic and alpine habitats in the Canadian Arctic and other regions; however, arctic foxes do not occur in most of our alpine areas, with the possible exception of areas in northern Labrador such as the Torngat Mountains (Ch. 21).[5] Arctic foxes do occasionally wander along sea ice as far south as Newfoundland, but they are not thought to breed there.[6] Gray wolves (*Canis lupus*) occur above treeline in the Monts Groulx (Ch. 18), Monts Otish (Ch. 19), Mealy Mountains (Ch. 20), and other ranges in Québec and Labrador (Ch. 21). However,

6-3 *The arctic hare exhibits many noteworthy adaptations to severe alpine conditions, including its ability to curl into a spherical shape, and its white winter coat. Scat is ubiquitous throughout the alpine ecosystem. Lewis Hills, Newfoundland. (MJ)*

wolves, including both *C. l. lycaon* of New England and Newfoundland's endemic white subspecies (*C. l. beothucus*) have been extirpated from all areas south of the St. Lawrence. Canada lynx (*Lynx canadensis*) occur locally in subalpine areas from northern Maine, Gaspésie and Newfoundland north throughout Québec and Labrador. Their status in New Hampshire and Vermont is currently tenuous, although they were present historically. In an exciting turn of events, lynx were recently documented breeding in northern New Hampshire. The White Mountain National Forest and adjacent regions appear to provide abundant habitat and the large cat's preferred prey, snowshoe hare (*Lepus americanus*).[7]

The arctic hare (*Lepus arcticus*), though, represents the primary tundra specialist among the mammals found above treeline in our region (6-3). The distribution of arctic hare extends from arctic areas, along the northeastern edge of Labrador south to the mountains of Newfoundland, which hold the southernmost populations of arctic hare in the world (see Ch. 12).[8] These Newfoundland populations are protected from hunting due to their low numbers. Any rabbit scat found in alpine zones of Newfoundland's western mountain ranges is probably that of an arctic hare. Snowshoe hares have been introduced to Newfoundland, and may compete with arctic hares near treeline. However, arctic hares might never have

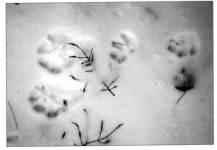

6-4 *Snowshoe hares range commonly to treeline and above in all of the eastern alpine areas. Only in Newfoundland do they co-occur in these areas with another rabbit, the larger arctic hare. Snowshoe hare tracks (pictured at right) are common and distinctive. Pistolet Bay, Newfoundland. (MJ and NC)*

5	Landa et al. (1998); Selås et al. (2010)
6	Northcott (1975)
7	Hoving et al. (2003)
8	Burzynski (1999)

occupied lowland forests even when they were the only rabbit on the island.[9] At lower elevations in Newfoundland or in mainland mountain areas outside of northern Labrador, snowshoe hares are the only rabbit found at treeline (6-4).

Several rodent species, such as eastern heather vole (*Phenacomys ungava*), rock vole (*Microtus chrotorrhinus*) and northern bog lemming (*Synaptomys borealis*) exhibit boreal-subalpine distributions in our region. Rock voles were first described from the headwall of Tuckerman Ravine on Mount Washington (Ch. 7)[10] and are found from Labrador to Ontario and Wisconsin, and along the Appalachians as far south as South Carolina. Rock voles are of conservation concern in most of the states and provinces in which they occur, due to a lack of basic data about their distribution. Northern bog lemmings were also first described from the White Mountains,[11] and are considered imperiled in Maine and New Brunswick, although they (and heather voles) are widely distributed across Canada. Heather voles tend to prefer dry habitats with ericaceous (heath) shrubs,[12] whereas northern bog lemmings occupy a variety of moist habitats such as wet meadows, mossy understory of spruce-fir krummholz or boggy areas.[13] Rock voles live among rocks or talus with moist moss, dense groundcover, and nearby streams or ponds.[14]

6-5 Woodland caribou occur in all of the major alpine areas outside of New York and New England, where they have been extirpated. Clockwise from top left: Adult male on Mont Albert, Québec; single track on the Highlands of St. John, Newfoundland; caribou trail Highlands of St. John, Newfoundland; summer scat in the Lewis Hills, Newfoundland; winter scat in the Monts Groulx, Québec. (MJ)

9	Fitzgerald and Keith (1990)
10	Miller (1895)
11	Preble (1899)
12	Edwards (1955); Foster (1961); Naylor et al. (1985)
13	Clough and Albright (1987)
14	Lansing (2005)

While woodland caribou (*Rangifer tarandus*) occupy forested areas during the winter, open alpine habitats function as important foraging, birthing, and mating grounds in the summer and fall.[15] Caribou are an important and exhilarating part of the northern alpine landscape, and healthy herds will maintain well-worn trails through the krummholz, wide enough to facilitate human access to remote alpine areas. Caribou may play important roles in vegetation dynamics in some alpine areas.[16] In eastern North America, woodland caribou occur primarily on the northern tundra, above the 55th parallel. There, in Québec

6-6 Moose are commonly encountered near treeline in all major eastern alpine areas. In some areas, such as the ranges of interior Québec, moose may be frequently encountered above treeline. Diamond Peaks, NH. (MJ)

and Labrador, three major tundra herds are associated with the *Rivière aux Feuilles*, the George River, and the Torngat Mountains, numbering more than one million animals.[17] Only scattered bands of woodland caribou, numbering as many as 18,000 individuals, occur in the spruce forest between 50°N and 55°N. In this densely forested area, herds are closely associated with elevated tundra massifs such as the Monts Groulx and Mealy Mountains (Ch. 20). South of these mountains, in Québec's Laurentides (Ch. 21), 82 tundra caribou were reintroduced between 1966 and 1972. This herd still survives, albeit reduced in total population size. On the high mountains of the Gaspé Peninsula, a native herd of woodland caribou (*ca.* 150 individuals) is the last herd south of the St. Lawrence.[18] Regrettably, woodland caribou were extirpated from Katahdin and other Maine mountains by about 1908 (see Ch. 8). Today, several mountains in our region, including Mont Jacques-Cartier and

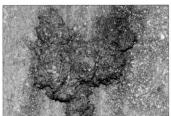

6-7 Black bear regularly feed near treeline in most eastern mountain ranges. In Newfoundland, adult bears are a primary predator of caribou calves. Clockwise from top left: Adult male; berry-based scat (note currant berries in scat); "blood" scat consisting of caribou remains. La Poile Highlands, Newfoundland. (NR)

15   Boonstra and Sinclair (1984); Rayl (2012)
16   Willey and Jones (2012)
17   Courtois et al. (2003)
18   Vandal (1984); Bergerud and Mercer (1989)

Gros Morne, have seasonal hiking restrictions to protect the calving habitat of this sensitive and decidedly iconic species (6-5).

Wolverines (*Gulo gulo*), a large and famously ferocious weasel species, have a range that extends from arctic and boreal areas through both alpine and subalpine areas throughout the northern hemisphere, although they are thought to have been extirpated from most of our focal area.[19] Globally, wolverines are associated with alpine regions, and persistent spring snow cover is essential for successful denning.[20] This suggests that eastern alpine areas, particularly those in Labrador and central Québec may play a critical role in the recovery of wolverines in this region (see Ch. 18).

For the most part, lowland mammals occur sporadically in alpine habitats as a function of large-scale trends in biogeography: for example, moose (*Alces americanus*) (6-6), black bear (*Ursus americanus*) (6-7), red fox (*Vulpes vulpes*), and coyote (*Canis latrans*) occasionally occur above treeline throughout the region. In Newfoundland, coyotes scavenge on caribou calves on open alpine barrens. In some regions of Québec, beavers (*Castor canadensis*) are frequent above treeline (6-8). Although porcupines (*Erethizon dorsatum*) and groundhogs (*Marmota monax*) may sometimes occur above treeline in Labrador, Québec, and New England, they are absent from the entire island of Newfoundland (6-9).

6-8 Beaver occasionally occur in high alpine lakes and ponds, and have even been observed in the Lakes of the Clouds on Mount Washington. Here, a beaver has impounded a high-elevation flowage, creating a small lake. In this case, the dam is constructed of sphagnum peat and muck. Monts Groulx, Québec. (NC)

6-9 Porcupine are common alpine inhabitants in Québec and Labrador, but are absent from Newfoundland. Monts Groulx, Québec. (MJ)

Although most of the larger mammals retreat below treeline for winter, a few small mammals have distributions that extend far north into the arctic tundra and can persist quite well both above and below treeline in the mountains year-round, as evidenced by their presence at our winter cameras stations, or the numerous trails left under the snow in meadows and bogs. This list includes ermine (*Mustela erminea*), southern red-backed vole (*Clethrionomys gapperi*), and masked shrew (*Sorex cinereus*).[21] In Garrit Miller's wanderings on the Presidential Range of New Hampshire,[22] from which he first described rock voles and

---

19   Laliberte and Ripple (2004)
20   Copeland et al. (2010); Johnson et al. (2012)
21   Macpherson (1965)
22   Miller (1895)

a regional subspecies of the red-backed vole (*Myodes gapperi ochraceus*), he also encountered meadow voles (*Microtus pennsylvanicus*) and star-nosed moles (*Condylura cristata*) inhabiting the Alpine Garden on the eastern cone of Mount Washington.[23]

Birds

The birds of the eastern alpine are diverse, and include more than just the many migratory species that pass through the region each fall and spring. Among this list are a few arctic-

6-10 Common ravens are widely distributed throughout all of the major American mountain ranges. (MJ)

alpine specialists, as well as boreal birds, grassland birds, wetland birds, and several cosmopolitan species. The raven (*Corvus corax*) and American robin (*Turdus migratorius*) are among the generalist species common on eastern alpine summits (robins are particularly frequent on the Canadian Shield ranges) (6-10). Other generalist species occur more sporadically (within their known range) above treeline, including golden eagle (*Aquila chrysaetos*), bald eagle (*Haliaeetus leucocephalus*), peregrine falcon (*Falco peregrinus*), and the red-tailed hawk (*Buteo jamaicensis).*

6-11 Ptarmigans are creatures of the far north and high mountains. Pictured, clockwise from top left: willow ptarmigan in the Monts Groulx, Québec; rock ptarmigan on Killdevil Mountain, Newfoundland; rock ptarmigan nest in the Lewis Hills; ptarmigan scat in the Lewis Hills, Newfoundland. (CE and MJ)

---

23   Miller's 1893 sightings are intriguing but the taxonomic validity of these observations warrants further exploration (Miller 1895).

6-12 American pipits are arctic and alpine specialists. Pipits are known to breed on Mount Washington, Katahdin, Mont Jacques-Cartier, and the Monts Groulx, and apparently occur on many other mountain ranges in our area. Pictured (from left): adult, Mount Washington, New Hampshire; nest, Mount Washington, New Hampshire; fledgling, Monts Groulx, Québec. (JW and MJ)

6-13 Horned lark breed throughout the mountains of Newfoundland and in the interior mountains of Québec south to Mont Albert. (BT)

A few birds in our region are restricted to arctic-alpine breeding distributions and are found nesting primarily above treeline. Rock ptarmigans (*Lagopus muta*) are perhaps the quintessential subarctic bird. These chicken-like members of the grouse subfamily reside year-round in arctic conditions, turning white in winter from their summer brown. In the Arctic, rock ptarmigan migrate relatively short distances southward in winter, but in the mountains they move down to lower elevations. The alpine peaks of Newfoundland are the only home to the endemic subspecies (*L. m. welchi*), representing the southernmost populations of rock ptarmigans in North America.[24] Newfoundland's rock ptarmigans aren't well-protected by hunting regulations because the regulations do not distinguish this species from the related and more widespread willow ptarmigan (*L. lagopus*) (6-11). Outside Newfoundland, willow ptarmigans breed from the Arctic regions south to the higher mountains of interior Québec, including the Monts Groulx and Monts Otish.

The suite of arctic-alpine breeding birds found on eastern mountains includes the American pipit (*Anthus rubescens*), a medium-sized migratory songbird that characteristically bobs its tail as it walks along the ground. Pipits breed primarily on the arctic tundra, but are known to breed on the alpine summits of Mount Washington, Katahdin, Mont Albert (Ch. 9),[25] and the Monts Groulx (6-12) and likely breed throughout the other northerly ranges. In addition, pipits breed above treeline on the western mountains. Although horned larks (*Eremophila alpestris*) breed widely in grasslands across North America, the summit of Mont Albert represents the southernmost breeding site for one subspecies, the northern horned lark (*E. a. alpestris*) (6-13).[26]

---

24   Holder et al. (2004)

25   Townsend (1923); Palmer and Taber (1946)

26   Townsend (1923); Demille (1926)

6-14 Bicknell's thrush (left) occurs in coniferous forests near treeline throughout New York, New England, and southern Québec. Gray-cheeked thrush (right) is a similar species found throughout Québec and Labrador. (PJ and CE)

6-15 Of the many sparrows in our area, the two most common near treeline are the white-crowned sparrow (left), which is most common on the Canadian Shield, and the white-throated sparrow (right), which ranges throughout the mountains in our area. (CE)

With further research in these areas, we may yet find several other arctic species nesting on summits far from their known breeding grounds. In May in the Monts Groulx, our field team has observed snow buntings (*Plectrophenax nivalis*), Lapland longspurs (*Calcarius lapponicus*), rough-legged hawks (*Buteo lagopus*), and Barrow's goldeneyes (*Bucephala islandica*; a western duck known to breed on high-elevation lakes along Québec's North Shore). These species are known to breed in alpine areas in other regions, so it is reasonable to believe that some may make similar use of the larger eastern alpine areas. Might other far-northern species—such as common redpoll (*Acanthis flammea*), northern wheatear (*Oenanthe oenanthe*), snowy owl (*Bubo scandiacus*), or gyrfalcons (*Falco rusticolus*)—breed in some of the alpine areas far south of their known nesting range? Recent studies of snowy owls wintering at Boston's Logan Airport have indicated that the Monts Groulx in Québec may be a stopover for some owls tagged in Massachusetts.[27]

In the highest parts of the krummholtz, important breeders include blackpoll warblers (*Setophaga striata*), Bicknell's thrushes (*Catharus bicknelli*) (6-14), and yellow-rumped warblers (*Setophaga coronata*).[28] Dark-eyed juncos (*Junco hyemalis*) and white-throated sparrows (*Zonotrichia albicollis*) breed up into the alpine tundra and are among the most abundant alpine birds. In central Québec, the white-crowned sparrow (*Zonotrichia leucophrys*) is a common species at treeline and its song is heard commonly throughout the alpine tundra of the Monts Groulx and other ranges (6-15).

27   Smith (2012)
28   Sabo (1980); Atwood et al. (1996)

6-16 Many species of boreal bird occur frequently in our alpine areas, although they are not alpine specialists. These include, as pictured clockwise from top-left: gray jay, white-winged crossbills, and spruce grouse (Monts Groulx, Québec); and dark-eyed junco (Mont Richardson, Gaspésie, Québec). (MJ)

6-17 Alpine meadows in Newfoundland provide habitat for grassland species like the Savannah sparrow. Tablelands, Newfoundland. (MJ)

6-18 Marshy meadows above treeline in Newfoundland provide breeding habitat for the wetland-loving greater yellowlegs. (Blow Me Down Mountain, Newfoundland). (MJ)

A number of boreal birds, or their songs, may drift up into alpine areas from the forest below. These include gray jays (*Perisoreus canadensis*), white-winged crossbills (*Loxia leucoptera*), red crossbills (*L. curvirostra*), spruce grouse (*Falcipennis canadensis*), boreal chickadees (*Poecile hudsonicus*), pine siskins (*Spinus pinus*), magnolia warblers (*Setophaga magnolia*), golden-crowned kinglets (*Regulus satrapa*), ruby-crowned kinglets, black-backed woodpeckers (*Picoides arcticus*), winter wrens (*Troglodytes hyemalis*), and swamp sparrows (*Melospiza georgiana*) (6-16).[29]

Eastern alpine meadows, especially in western Newfoundland, also provide breeding grounds for the savannah sparrow (*Passerculus sandwichensis*) (6-17), and the wetland-loving greater yellowlegs (*Tringa melanoleuca*) (6-18).

29   Adams et al. (1920); Palmer and Taber (1946)

6-19 *The most widely distributed amphibian in eastern alpine habitats is the American toad (left; Lewis Hills, Newfoundland), which occurs on almost every mountain range in our area. A colorful northern variety, the Hudson Bay toad (right; Monts Groulx, Québec) may be encountered in Québec and Labrador. (MJ)*

6-20 *Wood frogs are the second most widely distributed amphibian in eastern alpine environments. At left, a dark male rests beside a pinkish female in the White Mountains, New Hampshire. At right, an unusual "northern" variety is pictured near Katahdin, Maine. (MJ)*

6-21 *Mink frog, though widely distributed through boreal Canada, is not a widespread component of alpine wetland faunas (left; Mont Joseph Fortin, Québec). The spring peeper (right; Lakes of the Clouds, Mount Washington, New Hampshire) is widespread in New England and Québec alpine areas. (MJ)*

Reptiles and Amphibians

At least six species of amphibians occur above treeline in the mainland alpine areas.[30] Of these, the American toad (*Anaxyrus americanus*) (6-19), and wood frog (*Lithobates sylvaticus*) (6-20) are certainly the most widespread and abundant. American toads and mink frogs (*Lithobates septentrionalis*) are common on the tundra of southern Newfoundland, where they have been introduced from the mainland, whereas American toads, wood frogs, and spring peepers (*Pseudacris crucifer*) (6-21) breed in a variety of wetlands above treeline in the mountains of Québec. Blue-spotted salamander (*Ambystoma laterale*), a relative of the

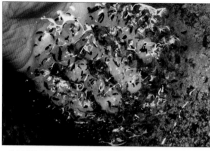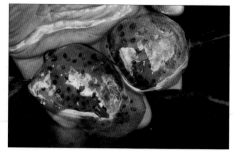

6-22 *The most common and visible egg masses in the eastern alpine areas are those of wood frog (left; Katahdin, Maine) and spotted salamander (right; Monts Groulx, Québec). They are easily told apart with practice by the looser, lumpy appearance of wood frog eggs. (MJ)*

6-23 *The only reptile likely to be encountered in the eastern alpine areas is the garter snake. Mont du Lac des Cygnes, Québec. (MJ)*

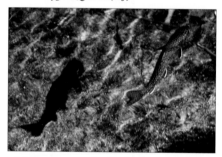

6-24 *Brook trout are locally abundant in the mountains of Québec. Monts Groulx, Québec. (MJ)*

spotted salamander (*Ambystoma maculatum*), occurs north to the Churchill Valley of central Labrador and may be locally present in mountains of the Canadian Shield, including the Monts Otish.[31] Spotted salamander occurs widely throughout the subalpine areas of New England, New York, and Québec, but has not yet been documented in Labrador. The bulbous eggs of spotted salamander are easily told from wood frog eggs, deposited at the same time in early spring (6-22). Green frogs (*Lithobates clamitans*) and red-spotted newts (*Notophthalmus viridescens*) are generally uncommon in alpine areas but have been documented at Lakes of the Clouds on Mount Washington.[32]

The garter snake (*Thamnophis sirtalis*), a versatile and variable species, ranges throughout the mountain regions of New England and Gaspésie and evidently also occurs in Québec as far north as the Mistassini-Monts Otish region (6-23).[33]

Fish

Fish are generally absent from the alpine pools of the White Mountains, Katahdin, and Gaspésie. A few members of the family Salmonidae are found in high elevation lakes of eastern alpine areas, most notably arctic char (*Salvelinus alpinus*), which is often the only fish species found in alpine lakes.[34] Brook trout (*S. fontinalis*) also occur in water bodies above

---

31   Fortin et al. (2011)
32   Jones (2004); Jones and Smyers (2010)
33   Fortin et al. (2011)
34   Klemetsen et al. (2003)

treeline in our region, including the Monts Groulx, Monts Otish, and Mealy Mountains, and may be locally abundant (6-24).

Invertebrates

Although vertebrates always receive the most attention, most of the faunal biodiversity in eastern alpine habitats is found among the insects and other arthropods. This is also true of every other ecosystem on earth. In the alpine zone of Mount Washington alone, one researcher estimated more than 3,000 insect species had been recorded as of 1940,[35] and this only begins to scratch the surface. In fact, on a calm day, the high points of mountains are an excellent place to observe a large diversity of invertebrates. Many species of Diptera (flies), Hymenoptera (ants and bees) and Lepidoptera (butterflies and moths) congregate on high points and summits to find mates,[36] a phenomenon aptly named "hill-topping." Larger, predatory insects and birds will also go there to forage on the congregations.

6-25 The White Mountain arctic is an endemic butterfly found on the tableland of the Presidential Range. Mount Washington, NH. (BG)

6-26 White Mountain fritillaries are related to Arctic fritillaries and are known only from the White Mountains. Presidential Range, New Hampshire. (MR)

In general, springtails, mites, spiders, beetles, stoneflies, moths, butterflies, and flies are the best represented groups in alpine areas, while many other invertebrates are represented by fewer species.[37] Some of these species may make a disproportionate impression on the alpine adventurer—for example, it is known that the greatest densities of mosquitoes occur on the Arctic tundra (despite the greater diversity of insects in the tropics). This abundance of mosquitoes in arctic (and alpine) areas may be related to the poor drainage of recently glaciated landscapes and the sprawling, stagnant, seasonal wetlands that form from snowmelt pooling over underlying permafrost. This phenomenon is replicated in many of the alpine areas in this book, where swarms of mosquitoes (and black flies, which prefer cold, clear, running water in which to deposit eggs) can inspire real despair. Several dozen species of mosquito occur in our alpine areas: the common mosquito *Aedes vexans* reaches well into central Québec. *Culiseta alaskaensis* and *Aedes hexodontus* reach across interior Québec to Newfoundland and Labrador. *Wyeomyia smithii* oviposits in pitcher plants (*Sarracenia purpurea*). It is possible that some subarctic specialists occur as disjunct relicts in our mountains.

Invertebrates found in eastern alpine areas fall into four rough categories: endemic eastern alpine species, arctic-alpine species, generalists, and transients. The fourth category, transient insects, is the least interesting and we will mostly ignore this group here. These transients

---

35   Alexander (1940)
36   Alcock (1987)
37   Mani (1968)

6-27 Several grasshoppers are locally abundant in eastern alpine areas, including two species of spur-throated grasshoppers (top: Melanoplus bruneri; Katahdin, Maine; middle: M. borealis, Lewis Hills, Newfoundland) and the White Mountain wingless locust (bottom, Katahdin, Maine). (MJ)

consist of insects that do not fare well in the alpine zone, and persist only temporarily after having been carried in by the wind.

The most interesting group of species include those that are endemic to the eastern alpine, occurring only at or above treeline, and only in northeastern North America. The best known of these species, and those with the most restricted distributions are the Katahdin arctic butterfly (*Oeneis polixenes katahdin*), the White Mountain arctic butterfly (*Oeneis melissa semidea*) (6-25), and the White Mountain fritillary (*Boloria chariclea montinus*) (6-26).[38] These three butterflies are protected under the Maine and New Hampshire Endangered Species Acts and occur only on the mountain ranges for which they are named. Another very local endemic is the Gaspé grasshopper found only on Mont Albert (*Melanoplus gaspesiensis*). Biologists believe the Gaspé grasshopper was "stranded" when the tundra, once widespread across Gaspésie, shrank away following the last glacial retreat.[39] Several other spur-throated grasshoppers (e.g., *M. bruneri* and *M. fasciatus*), as well as the wingless White Mountain locust, (*Booneacris glacialis glacialis*), are more abundantly encountered in eastern alpine areas (6-27).

Specialization for alpine conditions is not restricted to the butterflies and grasshoppers. Among the less widely loved groups, many species are known only from eastern alpine areas. In some cases, this may reflect true endemism, but in many cases, this may result from a lack of study. Undoubtedly, many endemic species are yet to be discovered. Among the crane flies, *Tipula broweri* is known only from the summit of Katahdin,[40] *T. insignifica* is known only from Mount Washington and Katahdin, *T. subserta* is known from alpine areas in the Presidential Range and from Labrador, *Nephrotoma penumbra* larvae are found in *Diapensia lapponica* cushions on Mount Washington as well as on Katahdin and in New Brunswick, and *Dicranota modesta* is found in subalpine areas from New York to Newfoundland. Alexander[41] described the following species as characteristic of the Hudsonian (high-elevation conifer) zone, and they are apparently known only from alpine areas in our region: a tachinid fly (*Admontia washingtonae*), three scuttle flies (*Trupheoneura vitrinervis, Spiniphora slossonae, Megaselia subobscurata*), a dance fly (*Rhamphomyia expulse*), a long-legged fly (*Dolichopus brevicauda*), two gall midges (*Argyra obscura, Dirhiza papillata*), an anthomyiid fly (*Spilo-*

---

38   McFarland (2003)
39   Chapco and Litzenberger (2002)
40   *T. boweri* may be synonymous with *T. fragilina*
41   Alexander (1940)

6-28 *Many species of Hymenoptera (ants/bees) and Diptera (flies) act as pollinators in the alpine environment, including (clockwise from top left): March fly (Lewis Hills, Newfoundland); rusty-patched bumblebee on bilberry (Monts Groulx, Québec);* Formica *sp. ants on mountain avens (Lewis Hills, Newfoundland); syrphid fly on fireweed. (NC and MJ)*

6-29 *Spiders are diverse in the eastern alpine areas. Some of the more prominent groups are represented by these species, including (clockwise from top left): mountain spider (Katahdin, Maine); wolf spider (Katahdin, Maine); wolf spider tunnel entrance (Monts Groulx, Québec); boreal jumping spider (Lewis Hills, Newfoundland). (NC, CE, and MJ)*

6-30 Many dragonflies are found foraging and breeding in alpine environments, including Aeshna *spp. (left; Mont Richardson, Québec) and* Somatochlora *spp. (right; Monts Groulx, Québec). (MJ)*

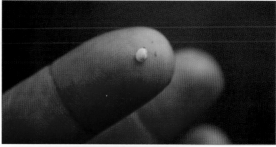

6-31 Insects and spiders are far from the only invertebrates found above treeline. Pictured here, an arctic fingernail clam. Blow Me Down Mountain, Newfoundland and Annieopsquotch Mountains, Newfoundland. (MJ and NC)

gona argenticeps), and a golden looper moth subspecies (*Syngrapha u-aureum "vaccinii"*).[42] In addition, several species characteristic of the Hudsonian zone also occur throughout more southern areas of the Appalachian Mountains. Among others, these include a dance fly (*Rhamphomyia rustica*), a long-legged fly (*Neurigona viridis*), two robber flies (*Laphria altitudinum, Cyrtopogon lyratus*), and the short-legged pygmy grasshopper (*Neotettix femoratus*). Some of these, along with black flies (*Simuliidae*), may be important pollinators of alpine plants (6-28).

While our region represents the northern edge of some Appalachian species' ranges, it also represents the southern edge for many arctic species. The second category of alpine invertebrates includes those that occur throughout the arctic and into other high mountains in western North America, Europe, and Asia. These species often find their southernmost point in the east on the peaks of Mount Washington, Katahdin, or Mont du Lac des Cygnes in Québec's Laurentides (Ch. 21). Many insects, spiders, and at least one mollusk are in this category. This group includes what M.S. Mani[43] listed as the most common ground beetle on Mount Washington, *Amara hyperborea*; the mountain spider (*Aculepeira carbonarioides*) (6-29) found on talus slopes in the Appalachians and Rocky Mountains; the arctic/alpine sweat bee (*Lasioglossum boreale*); parthenogenic plant bugs (*Nysius groenlandicus*) found on *Dryas* mats; a stonefly (*Diura nanseni*) found in eastern North America, Northern Europe, and Northern Asia; and two lesser dung flies (*Rudolfina digitata* and *Aptilotus spatulatus*) found

---

42   Some of Alexander's identifications and taxonomy should be revised because the reference is 70 years old. For example, *Syngrapha u-aureum "vaccinii"* is no longer a valid subspecies

43   Mani (1968)

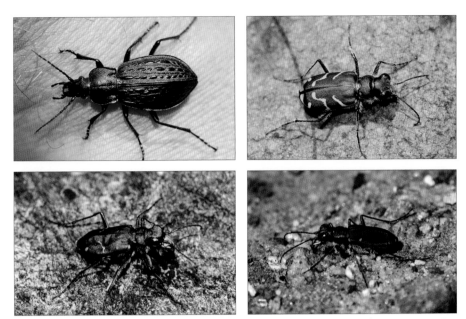

6-32 Some of the many beetles found in alpine areas of the East include ground beetles (top left; Lewis Hills, Newfoundland); and many tiger beetles (clockwise from top right: oblique-lined, sidewalk, and long-lipped; Lewis Hills, Newfoundland, and Mahoosuc Mountains, Maine). (NC, CE, MJ)

in eastern high alpine meadows and in the western United States and northern Canada. Several subarctic dragonflies occur near or above treeline throughout our region, including darners, or hawker dragonflies, (*Aeshna* spp.) and emeralds (*Somatochlora* spp.) (6-30). The arctic fingernail clam, *Sphaerium nitidum*, occurs locally above treeline in Newfoundland and possibly other ranges (6-31).

The third category of invertebrates in the eastern alpine includes species that seem to be very good generalists, persisting in both alpine or subalpine areas and lowland areas. For instance, the wingless White Mountain locust (mentioned earlier), which was first described from the top of Mount Madison, New Hampshire, is characteristic of the subalpine zone in New England; yet it can also found in places such as sea-level bogs on Mount Desert Island, Maine (which bear some resemblance to subalpine wetlands). Consider also the cobweb spider (*Crustulina sticta*), which lives among stones on beaches at sea level, among stones in crowberry barrens, and among stones at the tops of mountains including Mount Washington and Killdevil Mountain in Newfoundland (Ch. 14).[44] Perhaps the extreme microclimate of beach stones and alpine stones are not too different, and having adapted to one habitat, this spider can persist well in the other. This group includes many species of beetle, including several species of tiger beetle (*Cicindela* spp.) (6-32) and the widespread boreal white-spotted sawyer (*Monochamus scutellatus*), which is abundant at all elevations in the Monts Groulx and Monts Otish and throughout the region (6-33).

Many arctic-alpine distributions are driven by the insect's adaptations to these extreme abiotic environments. However, for some insects with very restricted host plants, such as leaf-mining flies, their distributions are governed by the host plant distribution. There is still

6-33 *The white-spotted sawyer is an abundant boreal beetle often found near treeline. Monts Groulx, Québec. (MJ)*

much to learn about these species. Of all the arctic-alpine anthomyiid flies, no larval host plants have yet been identified with two interesting exceptions. In a large genus of flies that specialize in mining plant leaves, *Pegomya ruficeps* is one of a few odd species that feed on bolete mushrooms in the genus *Leccinum*.[45] The arctic-alpine plant, mountain sorrel (*Oxyria digyna*), is the known host for one anthomyiid fly (*Pegomya icterica*). However we do not know if the fly range extends south into the Appalachian alpine areas following the mountain sorrel distribution (a closer look in the ravines of the Lewis Hills, Mont Albert, or Mount Washington might clear this up). Clearly, we need to encourage more people to study these fascinating creatures.

---

45   Ståhls et al. (1989)

Mount Washington *and the* Presidential Range

NEW HAMPSHIRE

7

Mount Washington from the Cog Railway Station, New Hampshire. (MJ)

Map 7-1. Mount Washington and the Presidential Range, New Hampshire.

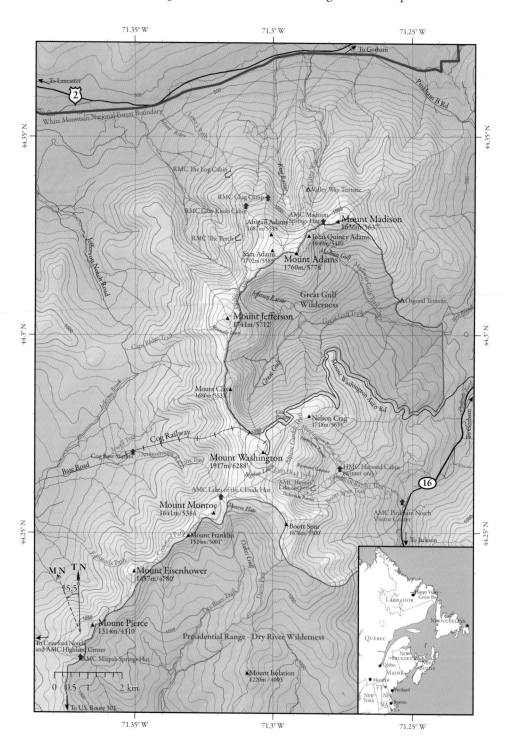

Mount Washington and the Presidential Range, New Hampshire

7-1 *The Presidential Range encompasses the largest expanse of alpine tundra in the eastern United States (MJ)*

It wasn't by chance that we chose to begin this long section of the *Eastern Alpine Guide*, a series of range profiles, in New Hampshire's White Mountains. Mount Washington and the Presidential Range have achieved iconic status among eastern hills.[1] The Presidential Range is likely familiar to literally millions of people, including many who seldom climb mountains. Some of this familiarity probably originates in part from the marketing strategies of the railway, auto road, and weather observatory that laid claim to the mountain between the 1850s and the 1930s, but it is nonetheless entirely deserved.[2] Mount Washington is aesthetically stunning and staggeringly unique east of the Mississippi River, and because of its broad appeal in the northeastern United States and eastern Canada, it has served as an elegant envoy for the eastern alpine tundra (7-1). Mount Washington's value in this regard perhaps exceeds its undeniable scenic value. In addition to the editors, many of the contributors to this book cut their teeth on its flanks, or on its summit, and so it seemed natural and intuitive to open the main portion of this book with an overview of the Presidential Range. Many would agree that for its great height, alpine extent, isolation from other alpine areas, and long history of scientific exploration and artistic portrayal, there's no eastern mountain as ecologically, culturally, or aesthetically significant as Mount Washington.

---

1   Mount Washington lies at the center of New Hampshire's Presidential Range and is the highest mountain in that range.

2   If "This Car Climbed Mount Washington" and "Home of the World's Worst Weather" aren't familiar slogans, you probably aren't from New England. These are trademarks of the Mount Washington Auto Road and Mount Washington Observatory, respectively.

7-2 *Mount Washington's alpine plants have attracted scientists and naturalists since the mid-19th century. Dwarf mountain cinquefoil (left) occurs on only two mountains in the world and was among the oddities that attracted botanists. Several species were collected to near extinction, such as sibbaldia (right). (MJ)*

Mount Washington, or *Agiocochook* in native parlance,[3] harbors diverse and isolated arctic-alpine vegetation—this fact is generally well-known. The arctic-alpine plant diversity on its ridges and in its gullies is greater than that found on Katahdin in Maine or any of the lower alpine summits in New Hampshire, Vermont, or New York. The alpine plants of the Presidential Range have been famously well-studied by some of New England's greatest scientists and naturalists—from Henry Thoreau, Louis Agassiz, and Edward Tuckerman (contemporaries of a sort) to Merritt Lyndon Fernald and other botanists and ecologists of the modern era. Several arctic-alpine plant species, such as moss campion (*Silene acaulis*), sibbaldia (*Sibbaldia procumbens*), snowbed willow (*Salix herbacea*), and nodding saxifrage (*Saxifraga cernua*) reach their southernmost location in eastern North America on Mount Washington's alpine tableland or in its great ravines, and at least one plant—dwarf mountain (or Robbins') cinquefoil (*Potentilla robbinsiana*)—occurs almost nowhere else on the planet (more on that, later). Dwarf mountain cinquefoil and its rose-family relative, sibbaldia, and several other rare species, were collected nearly to extinction by 19th-century botanists (7-2).

Mount Washington and the Presidential Range are famous for heavy hiker and tourist traffic, but much of the Presidential Range's alpine area receives relatively little attention most of the year. Away from the highest summits and major trails such as the Crawford Path

7-3 *By visiting Mount Washington in the off-season, or in the early morning, it's possible to avoid the crowds. Bigelow Lawn from Mount Washington summit. (LW)*

3   "Agiocochook" reportedly means, "place where the concealed one dwells." (Mudge 1992)

and Gulfside Trail (together forming most of the Appalachian Trail as it traverses the Presidential Range), Ammonoosuc Ravine Trail, and Tuckerman Ravine Trail, it's still relatively easy to find solitude with some minor behavioral adjustments (7-3). For example, even the most famous peaks and ridges of the Presidential Range are often silent and empty at odd hours (early morning) and in the off-season, which here runs straight from Labor Day until the middle of June. In early- to mid-June, it's possible to experience the beginning of the alpine flower season before schools let out and the mountain becomes crowded. In many ways this is all beside the point because the rugged biodiversity of Mount Washington's alpine area makes a trip worthwhile regardless of crowds—or perhaps even unpleasant weather, which can occur in any month (see Ch. 4 for more on alpine weather, much of it drawn from Mount Washington) (7-4). Note, though, that more than one hundred people have died on Mount Washington's upper slopes of hypothermia, drowning, falls, and exertion, or some combination of these. Many of these tragic deaths could have been avoided through better anticipation of bone-chilling, sopping wet alpine conditions, a point reiterated by the Mount Washington Observatory and the Appalachian Mountain Club (AMC).[4]

Simply wandering along the Crawford Path or Gulfside Trail will introduce a curious walker to the alpine essentials: the floral "trinity" of diapensia (*Diapensia lapponica*), alpine azalea (*Loiseleuria procumbens*) and Lapland rosebay (*Rhododendron lapponicum*) are in clear evidence throughout much of June (7-5), and most of the alpine plant communities outlined in Chapter 5 are clearly seen from the major hiking paths. But Mount Washington's most unique alpine plants and spectacular landscapes may be found away from the ridgeline trails, in areas of late-lying snow as well as along alpine seepages and creeks, such as those

7-4 An 1800-m (6000') mountain only 100 km from the coast, positioned at the convergence of several storm tracks, Mount Washington is often cloaked in freezing fog. Nelson Crag, Mount Washington. (MJ)

7-5 By mid-June in most years, depending on the springtime conditions, a prominent triumvirate of alpine plants draws wildflower watchers: diapensia (top), alpine azalea (middle), and Lapland rosebay (bottom). (MJ)

---

4   For an extremely thorough treatment of the deaths on the Presidential Range, see Howe (2000). For semiannual analysis of recent accidents, read the Appalachian Mountain Club's *Appalachia* journal.

7-6 *The headwalls of cirques such as Tuckerman Ravine support some of the most interesting alpine plants found on Mount Washington. (MJ)*

7-7 *The Alpine Garden is located above Huntington Ravine and is famous for its floral displays in June and July. Bluebell, pictured here, is late to flower and may be seen well into August. (MJ)*

on (and below) the headwalls of Oakes Gulf, Tuckerman Ravine, Huntington Ravine, and Great Gulf (7-6). Other really noteworthy botanical areas include the Alpine Garden, a sloping mosaic of seepages and meadows above Huntington Ravine, and Monroe Flats, a windswept and frostheaved dome near the Lakes of the Clouds (7-7).

In this chapter, we attempt to view Mount Washington through a fresh lens. So much has been written and published on the Presidential Range that we undoubtedly fall short and, in places, fall back on clichés. Still, Mount Washington is a place where clichés carry great power and significance, and we hope there is some value in reiterating them where they are, in fact, true. In any case, in this book as well as your own explorations, Mount Washington can serve clear purpose as an introduction to and foreshadowing of the complex alpine landscapes to the north. Furthermore it serves as a familiar launching-point for this broad treatment of eastern alpine landscapes.

Lay of the Land

Most of the alpine area of the Presidential Range occurs around the central peak of Mount Washington (1917 m / 6288') and north across the Northern Presidential Range, a series of three major alpine peaks including (from south to north) Mount Jefferson (1741 m /

7-8 *The Northern Presidentials run above treeline for about 10 km from the summit of Mount Washington. South view from Mount Adams. (MJ)*

5712'), Mount Adams (1760 m / 5774'), and Mount Madison (1636 m / 5367') (7-8).[5] Together, Mount Washington and the Northern Presidentials encompass about 10 km (6 mi) of exposed ridgeline. In the other direction, running southwest from Mount Washington, a thin alpine ridgeline extends toward Crawford Notch. These are the Southern Presidentials, which extend above treeline for about 4 km (2.4 mi), over Mount Monroe (1641 m / 5384'), Mount Eisenhower (formerly "Pleasant Dome," 1457 m / 4780') and Mount Pierce (formerly "Mount Clinton," 1314 m / 4310') (7-9).[6] Extending northeast and southeast from the Mount Washington summit cone, respectively, are the great arms of Nelson Crag and Boott Spur, both of which encompass more alpine tundra than all of the Southern Presidentials combined, or all of New Hampshire outside of the Presidential Range (7-10). Other named, noteworthy promontories on the Presidential massif are Mount Clay, a low spur on the col between Mount Washington and Mount Jefferson, and Mount Franklin, a narrow ridgeline between Mount Monroe and Mount Eisenhower.

The summit cone of Mount Washington itself is a mosaic of fellfield (broken rock and talus), interspersed with vast alpine meadows such as Bigelow Lawn, where alpine sedges (*Carex* spp.) dominate (7-11).[7] This sprawling expanse is interrupted by the rocky pools at Lakes of the Clouds, and the sheltered floral displays of the Alpine Garden. Other sites on the Presidential Range noteworthy for their relative rarity on the landscape include Star Lake, a small tundra-tarn between Mounts Adams and Madison (7-12); Monticello Lawn at the foot of Mount Jefferson's summit cone; the "Cow Pasture" sedge meadows near the

7-9 The Southern Presidentials are seen here from the summit of Mount Monroe. The pronounced dome at center right is Mount Eisenhower, and the peak at top left is Mount Willey, located on a separate massif. (MJ)

7-10 Boott Spur is almost a mountain in its own right, extending east from Mount Washington and forming the divide between Tuckerman Ravine and Oakes Gulf. (MJ)

---

5   To Americans, we hope it goes without saying that the highest peaks of the Presidential Range are named for the country's first five presidents (Washington, Adams, Jefferson, Madison, and Monroe), but we note the technicality that George Washington was not yet president when the mountain was named for him. Henry Clay and Benjamin Franklin, of presidential stature but never president, somehow take their place among the others.

6   Mount Pierce is named for Franklin Pierce, the only American president to hail from the Granite State (New Hampshire).

7   Bigelow Lawn and Bigelow's Sedge are both named for Jacob Bigelow (1787–1879), a Massachusetts physician and botanist.

7-11 The summit cone of Mount Washington is covered by a mosaic of felsenmeer (rock fields) and sedge meadows. Bigelow Lawn and Monroe Col from Westside Trail. (MJ)

7-12 Star Lake near Mount Madison is one of few permanent waterbodies above treeline on the Presidential Range. (MJ)

7-13 Most of Monroe Flats, the small bare area at the foot of Mount Monroe, is currently closed to the public, but interesting plant species may be seen from the adjacent Crawford Path. (MJ)

summit of Mount Washington; and perhaps most significantly, the blistered crown of Monroe Flats, which supports the endemic dwarf mountain cinquefoil noted earlier, and many other interesting plants (see Ecology, later) (7-13).

The periphery of the Presidential Range has been deeply eroded by a series of prominent alpine glaciers, forming hanging, U-shaped valleys cut into the mountain during localized mountain glaciation events. Several of these **cirques** adorn the eastern face of the range. Tuckerman and Huntington Ravines (7-14), framed by Boott Spur and Nelson Crag, are the most famous of these. Between these cirques, a shallow saddle named Raymond Cataract holds late-lying snow. About one kilometer southwest of Tuckerman Ravine is the wilderness cirque Oakes Gulf, which forms the dramatic backdrop of federally designated Presidential Range–Dry River Wilderness. Just north of Mount Washington's summit cone lies the largest alpine cirque in New England, Great Gulf, which wraps around the shoulders of Mount Clay and Mount Jefferson, where it meets wild Jefferson Ravine, a trailless cirque (7-15). Like Oakes Gulf, Great Gulf is also protected as federally designated Wilderness (capital "W") by the U.S. Congress. At the northern end of the Presidential

7-14 Mount Washington's great cirques include Tuckerman Ravine (left) and Huntington Ravine (right). View from Carter Dome. (MJ)

7-15 Jefferson Ravine is a less-traveled, trailless cirque on the east face of Mount Jefferson. (MJ)

Range, Madison Gulf and King Ravine face opposite directions from the rocky spurs of Mount Adams.

Access

Mount Washington is to New Englanders what Pikes Peak or Mount Evans has become to Westerners: it is the most accessible major alpine peak in eastern North America, and there are many routes to treeline. These have been extensively profiled elsewhere, so we treat them briefly here. As on Pikes Peak in Colorado, two mechanized options are available: from the west, via the colorful Cog Railway (accessible via U.S. 302) and from the east via the Mount Washington Auto Road (accessible via NH Route 16) (7-16). Both "the Cog" and the Auto Road have numerous worthwhile historical and botanical sights along their routes, but these are best seen on foot.

The most popular foot access routes may

7-16 Mount Washington's colorful Cog Railway was constructed in 1869. It's a unique and eerie experience to watch a train hiss through the alpine fog. (MJ)

7-17 The Crawford Path is one of America's oldest recreational hiking paths, constructed by Abel and Ethan Allen Crawford in 1819. Mount Monroe and Mount Washington from Mount Franklin. (MJ)

be broken into four major categories: (1) the Crawford Path, which ascends Mount Washington from Crawford Notch via the Southern Presidentials (7-17); (2) a series of straightforward trails ascending the Presidential Range from the west, including Edmands Path, Ammonoosuc Ravine Trail, Jewell Trail, and Caps Ridge Trail; (3) major trails ascending Mount Washington from the east, including Boott Spur, Tuckerman Ravine, and Lion Head; (4) a latticework of trails ascending the Northern Presidentials from the vicinity of Randolph, New Hampshire, which are largely maintained by the Randolph Mountain Club as well as the AMC and U.S. Forest Service. There are many other trails to treeline, of course, and these are perhaps even more worth exploring than those

7-18 *The Great Gulf is one of two federally designated wilderness areas on Mount Washington. (MJ)*

7-19 *Lakes of the Clouds, the Gray Lady of the Mountains, flagship of the AMC hut system, has a commanding view of Ammonoosuc Ravine. (MJ)*

listed, and sometimes overrun, trails. The less-traveled paths include the Dry River Trail and Davis Path through Oakes Gulf and the Great Gulf Trail and Madison Gulf Trail in the Great Gulf Wilderness (7-18).

Camping above treeline on the Presidential Range is only allowed on snow, and the only growing-season accommodations above (or near) treeline are the AMC's Mizpah Hut (just below Mount Pierce), Lakes of the Clouds Hut (the AMC's elegant, imposing, flagship alpine hut, between Mount Washington and Mount Monroe) (7-19), and Madison Springs

Hut (between Mount Adams and Mount Madison) and the Randolph Mountain Club's Gray Knob and Crag Camps, located on the western slopes of Mount Adams. It's also reasonable to establish a base of operations at AMC's Hermit Lake Shelters in Tuckerman Ravine or the Harvard Mountaineering Club's cabin in Huntington Ravine (winter only). In fact, Mount Washington and the Presidential Range offer more alpine and treeline accommodations than any other eastern alpine range, although there are several shelters available to hikers in the western Chic-Chocs (Ch. 9) and the Monts McGerrigle (Ch. 10).

7-20 Exposed schist is quickly colonized by lichens, including the ubiquitous map lichen. On this rock, Ethan Allen Crawford carved his initials "EAC 1820" nearly 200 years ago. (MJ)

Unlike most of the other alpine ranges profiled in this book, Mount Washington is surrounded on all sides by towns and civilization.

Geology

The Presidential Range is composed of Devonian-aged Littleton Formation schist and gneiss, which originated from seafloor sediments that were subsequently metamorphosed (7-20). A band of Boott Member (schist) occurs near the foot of Mount Monroe, lending to that area's ecological significance by providing more calcium in the soils and groundwater. Severe freeze-thaw cycles have patterned the ground at Monroe Flats as well as several other sites on the mountain, resulting in shattered erratic boulders (cryoclasty), solifluction

7-21 Severe freeze-thaw cycles and frozen soils have produced (clock-wise from top left) shattered boulders and bedrock, soil stripes, and solifluction terraces. Bigelow Lawn and Monroe Flats. (MJ)

7-22 An extinct rock glacier, formerly a mobile mass of rock, ice, and debris, rests on the floor of King Ravine near Mount Adams. (MJ)

7-23 In 2010, a massive avalanche changed the face of Ammonoosuc Ravine on the southwest shoulder of Mount Washington. (MJ)

terraces where soils have slumped over frozen ground, and soil stripes where frost action has sorted the alpine gravels (7-21). Abundant evidence of alpine glaciation and periglacial phenomena exist; for example, the enormous cirques mentioned earlier. Further evidence of extreme cold and localized glaciation may be found at the bottom of King Ravine, where the remains of an extinct rock glacier are still visible (7-22).

Dramatic erosional forces are still at work on the Presidential Range, as evidenced by the severe avalanche that occurred in Ammonoosuc Ravine in January 2010 (7-23).

Ecology

Vegetation — Adequately describing Mount Washington's ecological significance in narrative form depends on the geographic scope of reference. For example, although Mount Washington's total alpine area is relatively small compared with that of the Rocky Mountains or even the larger mountain ranges of Québec and Labrador, the Presidentials harbor the largest extent of alpine tundra east of the Mississippi River in the United States. Compared with the alpine peaks in New York, Vermont, and Maine, its alpine vegetation is diverse and noteworthy. But compared with some peaks in eastern Canada, such as Mont Logan or Mont Albert, its flora is depauperate. Still, several arctic and alpine plant species occur on Mount Washington and nowhere else in New England, including moss campion, nodding saxifrage, and sibbaldia (7-24). Mount Washington harbors the world's largest populations of dwarf mountain cinquefoil (*Potentilla robbinsiana*), native to the Presidential massif and possibly Franconia Ridge (7-25), populations of which had collapsed by the 1980s, but were recovered through the "headstarting"

7-24 Sibbaldia (left) is an extremely rare arctic plant found at one snowbed site on Mount Washington and nowhere else in the eastern United States. Moss campion (right) is more widespread but also occurs nowhere else in the eastern United States. Both are widespread in alpine ranges of eastern Canada. (MJ)

7-25 *Dwarf mountain cinquefoil is Mount Washington's most famous alpine plant, known from a handful of natural and introduced sites. Red tags (right) mark locations of greenhouse-grown headstarts. (MJ)*

efforts of the Appalachian Mountain Club, New England Wildflower Society, and U.S. Forest Service.

The alpine plants that are common on Mount Washington are essentially the same that are common on Katahdin in Maine (Ch. 8), including the bewildering array of northern and alpine lichens such as worm lichen (*Thamnolia vermicularis*) (7-26).[8] However, the distribution of plant communities is different on these two New England mountains.[9] On exposed sites, in heath-shrub-rush and cushion-tussock communities, it is common to find diapensia and alpine bilberry (*Vaccinium uliginosum*) in association with Bigelow's sedge (*Carex bigelowii*), shrubby five-fingers or three-toothed cinquefoil (*Sibbaldia tridentata*),[10] bearberry willow (*Salix uva-ursi*), alpine azalea, and Lapland rosebay. Diapensia, the hallmark species of the ephemeral alpine wildflowers, occurs as a "double" variety at a single station in Monroe Col (7-27). In addition to these standard components of the eastern alpine flora, several unusual rarities occur

7-26 *Worm lichen is an extremely ancient lineage of lichens, found on alpine summits throughout the world. (MJ)*

7-27 *"Double" diapensia, an uncommon variant of an abundant species, occurs rarely in Monroe Col on Mount Washington. (MJ)*

on the exposed tundra, including Boott's rattlesnake-root (a New England/New York alpine endemic, *Nabalus boottii*) and eastern mountain avens (native only to the White Mountains and Digby Neck, Nova Scotia; *Geum peckii*) (7-28).

Let's consider for a moment the alpine flora of Mount Washington's great cirques, which differentiate Mount Washington from all other New England mountains except Katahdin (which has its own share of glacial basins). Along seepages in these cirques, the vegetation in-

---

8   Pope (2005)

9   Kimball and Weihrauch (2000)

10   *Sibbaldiopsis* has been placed within the genus *Sibbaldia*.

7-28 Boott's rattlesnake-root (left) and eastern mountain avens (right) are widespread on Mount Washington but rare on surrounding alpine peaks in New England and New York, and are endemic to the northeastern alpine regions. (MJ)

7-29 Lance-leaved arnica (top) and northeastern paintbrush (bottom) are common species on wet subalpine cliffs such as those in Tuckerman Ravine and the Great Gulf. (MJ)

cludes green false hellebore (*Veratrum viride*), tall white bog orchid (*Platanthera dilatata*), cow parsnip (*Heracleum maximum*), purple-stemmed angelica (*Angelica atropurpurea*), and numerous other species frequent on wet soils at lower elevations. In addition, a suite of interesting alpine and subalpine species such as the sweet-smelling lance-leaved arnica (*Arnica lanceolata*), and northeastern paintbrush (*Castilleja septentrionalis*) are generally distributed on wet cliffs (7-29), with Labrador and diamond-leaved willows (*Salix argyrocarpa* and *S. planifolia*, both widespread willows in the eastern Canadian arctic). More rare, and generally of extreme interest to botanists, are several saxifrages (white mountain saxifrage, nodding saxifrage, and pygmy saxifrage; *Saxifraga paniculata*, *S. cernua*, and *S. hyperborea*), Oakes' eyebright (*Euphrasia oakesii*), alpine mountain sorrel (*Oxyria digyna*), Wormskjold's alpine speedwell (*Veronica wormsjoldii*), alpine bittercress (*Cardamine bellidifolia*), and three snowbed obligates: sibbaldia, snowbed willow, and alpine or moun-

7-30 *Some of the rare alpine species on Mount Washington include pygmy saxifrage (left), snowbed willow (center), and alpine cudweed (right). (MJ)*

tain cudweed (*Omalotheca supina*) (7-30). An interesting hybrid of snowbed willow with bearberry willow (*S.* x *peasei*) occurs on the headwall of King Ravine.[11]

Regrettably, about a dozen invasive or exotic plant species, common in waste areas of the lowlands, have colonized the areas surrounding the handful of buildings above treeline, including the Sherman Adams building and other summit structures on Mount Washington and Madison and Lakes of the Clouds Huts. These include familiar species like common dandelion (*Taraxacum officinale*) and tall buttercup (*Ranunculus acris*), which are well-established at all three locations, and less ubiquitous lowland species like sheep sorrel (*Rumex acetosella*), common eyebright (*Euphrasia nemorosa*), bird or cow vetch (*Vicia cracca*), shepherd's purse (*Capsella bursa-pastoris*), grass-leaved goldenrod (*Euthamia graminifolia*), disc mayweed or pineappleweed (*Matricaria discoidea*), coltsfoot (*Tussilago farfara*), and orange hawkweed (*Hieracium aurantiacum*), which occur around Lakes of the Clouds (7-31). Several of these species, including dandelion, buttercup, and goldenrod, have spread beyond the footprint of the hut to colonize the marshy shore of the Upper Lake, and dandelion can now be found on the headwall of Great Gulf and other subalpine locations. Several species of native plants join in colonizing these disturbed areas, including arctic yellow rattle (*Rhinanthus minor* ssp. *groenlandicus*), which occurs in alpine cirques such as Oakes Gulf, spiked woodrush (*Luzula spicata*) (7-32), and common yarrow (*Achillea millefolium*), which is especially abundant on the summit of Mount Washington where it grows within a few meters of pygmy saxifrage, one of New England's rarest native plants. Greenland stichwort (*Minuartia groenlandica*) grows everywhere on disturbed sites, preferring the dry gravels of well-trod trails.

Mount Washington's alpine bogs support a number of characteristic species, such as tussock cottongrass (*Eriophorum vaginatum*) and such New England rarities as bakeapple or cloudberry (*Rubus chamaemorus*) (7-33).

The alpine vegetation of Mount Washington is essentially an old-growth ecosystem where natural disturbances continue to reign. As a result, very few (if any) of the plant species native to the alpine zone have been extirpated in historical times. Possible extirpations of native plant species from the Mount Washington alpine areas include the horned dandelion (*Taraxacum ceratophorum*),[12] and erect-fruit wintercress (*Barbarea orthoceras*).[13] Unless decisive

11   Much of this information is presented in greater detail by Harris et al. (1964).

12   Pease (1924)

13   Harris et al. (1964)

7-31 Invasive plant species established at developed sites above treeline on Mount Washington include (clockwise from top-left) common eyebright, orange hawkweed, disc mayweed, grass-leaved goldenrod, bird vetch, and dandelion. Lakes of the Clouds, Mount Washington. (MJ)

action is taken in the immediate future, we are also likely to lose sibbaldia from its single known station.

Fauna — The alpine tundra of Mount Washington was mostly spared the ravages of livestock grazing that is widespread in the alpine areas of Iceland, Wales, Scotland, and mainland Europe: New England sheep farmers kept mostly to the rocky lower hillsides. As a result, the vegetation structure and botanical diversity remain largely intact. The same is not true for the vertebrates, which once formed a more prominent component of New England's alpine fauna.

7-32 Native species frequently found on disturbed sites in the alpine tundra include arctic yellow rattle and spiked woodrush. (MJ)

The past two centuries have likely witnessed the extirpation of woodland caribou (*Rangifer tarandus*) from the alpine areas of Mount Washington. Although caribou presence in the White Mountains is really not well-substantiated, the alpine habitat is suitable, if not ideal, and there are reliable records for Coos County, New Hampshire, immediately north of the Presidential Range. Similarly vague records exist for wolverine (*Gulo gulo*), which was once present in eastern North America, but its extent of occurrence is poorly documented. In any case, both are gone from the New England landscape, and today it is only worth noting that the Presidential Range encompasses the most suitable habitat in the region for both species.

7-33 Bakeapple, or cloudberry, one of the hallmark species of cold bogs throughout Québec, Labrador, and Newfoundland, is highly localized and rare on the Presidential Range, occurring in a handful of alpine bogs. (MJ)

Wolves (*Canis lupus*) have also been extirpated from New England. Although they undoubtedly explored the alpine territory on the Presidential Range, they are a generalist species, not wedded to the alpine tundra. For this reason, Mount Washington will not be as critical a component of their eventual return—but nonetheless it's a magical image, wolves at dusk on Mount Washington.

7-34 Red fox is perhaps the most frequently observed carnivore in the alpine areas of Mount Washington. (MJ)

Today the most visible mammal above treeline on the Presidential Range is probably the red fox (*Vulpes vulpes*) (7-34). In most years, one or multiple individuals may be seen about the high huts and the summit buildings, eager for handouts. Red squirrels (*Tamiasciurus hudsonicus*) are also common near and above treeline; their chattering from sprucetops is heard frequently on calm days (7-35). American marten (*Martes americana*) are now rela-

7-35 Red squirrel occurs in the treeline krummholz and occasionally in the alpine tundra. (MJ)

7-36 Moose and moose sign are infrequently seen above treeline. Star Lake, Mount Madison. (MJ)

7-37 Sphagnum bogs such as this one dominated by tussock cottongrass may provide habitat for the enigmatic northern bog lemming. (MJ)

7-38 The gray jay, a widespread boreal species, is commonly encountered at lookouts and huts near treeline. (WD)

tively frequent in the alpine areas on Mount Washington.[14] Porcupine (*Erethizon dorsatum*) and groundhog (*Marmota monax*) occur occasionally in the alpine areas; snowshoe hare (*Lepus americanus*) are abundant but not often seen. In the summer of 2001, a young beaver (*Castor canadensis*) took up residence in the lower Lake of the Clouds, building small mats of willow stems. At the time, there were vague anecdotes of beavers occasionally frequenting the Lakes of the Clouds. Moose (*Alces americanus*) and bear (*Ursus americanus*) both occur infrequently above treeline in the Presidential Range (7-36). The most interesting and enigmatic mammal to occur above treeline may be northern bog lemming (*Synaptomys borealis*), which has yet to be documented from the alpine tundra, although sufficient habitat exists, especially in the bogs near Mount Eisenhower and at Lakes of the Clouds (7-37).

As one might expect, Mount Washington provides abundant habitat for a variety of uncommon boreal birds, but perhaps the most interesting species to occur there is the American pipit (*Anthus rubescens*), at its southeastern-most station. Perhaps the most commonly encountered birds above treeline on Mount Washington are the yellow-rumped warbler (*Setophaga coronata*, which nests in krummholz at treeline), the white-throated sparrow (*Zonotrichia leucophrys*), the dark-eyed junco (*Junco hyemalis*), and the common raven (*Corvus corax*). Subalpine forests surrounding the alpine areas provide critical habitat for Bicknell's thrush (*Catharus bicknelli*), and gray jays (*Perisoreus canadensis*) are common (7-38).

Open water is rare on the Presidential Range, confined mostly to the Lakes of the Clouds and Star Lake and isolated bogs along the ridgeline, and so the number of breeding amphibian species is relatively low. Wood frogs (*Lithobates sylvaticus*), spring peepers (*Pseud-*

---

14   We have recorded marten on infrared cameras foraging in daylight near Lakes of the Clouds, and we frequently see marten scat along the major trails above treeline.

7-39 Wood frogs (left) breed at Lakes of the Clouds in June and their tadpoles are easily seen in July and August. Spring peepers (center) and American toads (right) arrive in mid-June and may be heard calling from alpine ponds throughout the summer. (MJ)

acris cruicfer*), and American toads (*Anaxyrus americanus*) breed regularly in both locations, and typically they arrive in that order each spring (7-39). In the alpine ponds their breeding season begins two to four weeks after breeding occurs at Tuckerman Ravine, 400 m (1500') lower. There, in Hermit Lake, resides one of the highest known populations of green frogs (*Lithobates clamitans*), which probably fare poorly in the higher ponds for two reasons: their larvae require two years to metamorphose, and the adults require an unfrozen water environment in which to hibernate (7-40). The alpine ponds freeze mostly solid, restricting the breeding amphibian community to those species able to hibernate on land. For the same reason, there are no fish in any of the alpine ponds (7-41).

7-40 Green frogs range nearly to treeline at Tuckerman Ravine, around 1200 m in elevation, but are extremely rare in the alpine ponds such as Lakes of the Clouds. (MJ)

Invertebrates are well-represented on Mount Washington's alpine tundra, and are discussed intensively in Chapter 6. Many thousands of species occupy a variety of environmental niches. In fact, most of the literature on insects and other invertebrates in eastern alpine environments comes from research conducted on Mount Washington.

Syrphid flies and ants are among the most visible insects above treeline; these may be important pollinators of plants that primarily reproduce sexually (7-42).

At least two varieties of Lepidoptera (butterflies) are endemic to the Presidential Range,

7-41 Lakes of the Clouds and other alpine lakes freeze most of the way through and do not support fish. (MJ)

7-42 *Syrphid flies are abundant and diverse in the New England alpine. Here, an adult drone fly rests on Greenland switchwort in Monroe Col. (LW)*

7-43 *Butterflies found in the Presidential Range include the Atlantis fritillary. (MJ)*

7-44 *One of the more distinctive grasshoppers found in the Presidential Range's alpine areas is the wingless White Mountain locust, found commonly on Katahdin and other eastern alpine summits. (MJ)*

these include the White Mountain arctic (*Oeneis melissa semidea*) and the White Mountain fritillary (*Boloria chariclea montinus*). Many other "leps" are seen above treeline on calm and warm days, and following storms, including the green comma (*Polygonia faunus*) and Atlantis fritillary (*Speyeria atlantis*) (7-43).

One distinctive species of grasshopper is encountered with regularity: the wingless White Mountain locust (*Booneacris glacialis glacialis*), a small, bright green, wingless creature that is difficult to confuse with the grasshoppers in the genus *Melanoplus*, which are also abundant (7-44).

The interesting mountain spider (*Aculepeira carbonarioides*), which ranges primarily in the alpine Rockies, occurs here on alpine talus fields.[15]

---

15   The natural history of Mount Washington and the Presidential Range has been detailed in a number of excellent resources including Antevs (1932), Bliss (1963), Harris et al. (1964), and Slack and Bell (2006).

Katahdin 8
MAINE

Hamlin Ridge, Katahdin, Maine. (MJ)

Map 8-1. Katahdin, Maine.

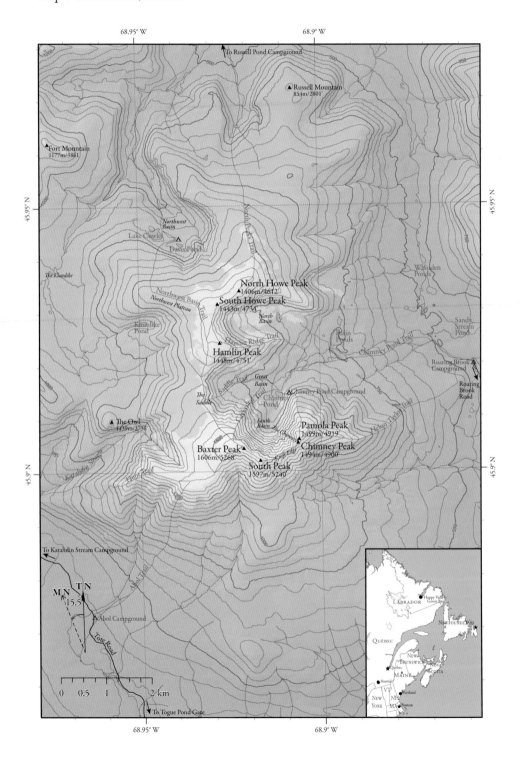

Katahdin, Maine

8-1 Katahdin is surrounded to the north and west by the largest uninhabited reaches of the eastern United States—the North Maine Woods. The Owl and the Klondike from Katahdin. (MJ)

Maine has many mountains, but none compares with Katahdin.[1] Looking back on almost two centuries of devoted reverence for this mountain, it might be easy for the uninitiated to guess that this mountain is overrated. The opposite is true. Great and wild Katahdin awaits rediscovery, celebration, and redefinition by coming generations as long as civilization persists. There are few valid excuses not to see, or explore, this mountain in your lifetime.

Among Katahdin's earliest explorers was Henry David Thoreau, who made the trek from Concord, Massachusetts, in August 1846 (during his second and final summer at Walden Pond). Thoreau's vibrant account of Katahdin in the 1840s survives in his journals, as well as in his clever little essay *Ktaadn*, in which Thoreau redefined his understanding of wilderness based on his frightening experiences at treeline. In the 1840s, the distant hulk of Katahdin hovered at the edge of burgeoning frontier industry. Scattered logging infrastructure in the North Maine Woods—camps and roads—made Thoreau's long journey from Concord feasible and inexpensive.

Today, although Katahdin is many hundreds of kilometers removed from whatever frontier remains, parallels to the Katahdin of Thoreau's day may be found in frontier mountains such as the Monts Groulx-Uapishka (Ch. 18), the Long Range Mountains (Ch. 12 and 15), and the Lewis Hills (Ch. 13), which have been made similarly accessible by haphazard resource extraction.

Katahdin still looms large at the edge of a sprawling wild area, the vast timberlands of northern Maine. Katahdin remains unique in the region because it is surrounded by the most uninhabited reaches of the northeastern United States (8-1). No roads reach to its summit, no colorful trains mar its flank, and no structures whatsoever have been installed above

---

1   *Katahdin* or *Ktaadn* may mean "Greatest Mountain" in Mi'kmaq. For this reason, "Mount Katahdin" is redundant.

8-2 Access to Katahdin's alpine tableland is strictly regulated by Baxter State Park, which encompasses the entire mountain. The strong regulations provide contrast to the Presidential Range. (MJ)

8-3 The Knife Edge (top) connects South Peak to Chimney (bottom) and Pamola Peaks and is one of the more famous arêtes, or thin glacial ridges, in the world. (MJ)

treeline. In this way, Katahdin is a balanced complement to Mount Washington, which by comparison has been exhaustively developed.

Katahdin's unlikely preservation in its primeval state is due partly to its remoteness from Boston and Portland. Even more so, however, Katahdin's preservation is the result of the efforts of former Maine governor Percival Baxter, who worked steadily through the second half of his life in the early 1900s to privately purchase, and subsequently donate to the State of Maine, more than 80,000 hectares (200,000 acres) surrounding Katahdin itself. Baxter's private crusade came only after the Maine legislature rebuffed his efforts to have the area designated a state preserve.

Katahdin's pink and gray granites are of different age and origin than are the mica schists and gneisses of New Hampshire's Presidential Range, but in other ways, the two ranges are ecologically and structurally similar. Katahdin's granitic bedrock supports a suite of largely the same arctic-alpine plant species that also occur on the Presidential Range, including such common species as diapensia (*Diapensia lapponica*) and Lapland rosebay (*Rhododendron lapponicum*) and rarities including alpine cudweed (*Omalotheca supina*). Katahdin's east face is deeply indented by treeless cirques, in much the same fashion as Mount Washington's east face is indented by Tuckerman and Huntington Ravines. In other ways, Katahdin differs strikingly from Mount Washington and the Presidential Range. Besides its more remote and isolated location, the most prominent example is that access to the alpine plateau of Katahdin is more strictly regulated than that of the Presidential Range (8-2).

Lay of the Land

Katahdin is essentially a single mountain with many named summits and ridges, rising steeply on the west side to a broad, alpine tableland below the high pinnacle of Baxter Peak (1606 m / 5268'). East of Baxter Peak, the alpine ridgeline extends sharply to a narrow arête known as the Knife Edge, which culminates at the twin summits of Chimney Peak (1494 m / 4900') and Pamola Peak (1499 m / 4919'), which themselves are separated by a rocky chute known as "The Chimney" (8-3). The mountain sweeps gradually north and west from the high point at Baxter Peak. A broad tableland sweeps southwest toward Thoreau Spring, which lies at the top of Abol Slide, a natural landslide track that is prominently visible on the mountain's shoulder from points along the Penobscot River (8-4). To the north, the alpine tableland drops to the Saddle before rising gradually to Hamlin Peak, and beyond to the Howe Peaks. The Howe Peaks form a northeast-trending arête enclosing the northern wall of a major east-facing alpine cirque, the dramatic North Basin. North Basin is just one of three alpine cirques deeply indenting Katahdin's east face—the more famous South Basin that forms the backdrop to Chimney Pond also lies on the east face. Oth-

8-4 Abol Slide forms a visible scar on Katahdin's southern flank, here viewed from the Penobscot River. (MJ)

8-5 Katahdin's Northwest Basin is the location of Davis Pond, a glacial tarn. (DY)

er cirques indent the western buttresses of the mountains: Northwest Basin, which bottoms out at Davis Pond (8-5), and another cirque supporting remote Klondike Pond. West of the Katahdin massif, the terrain slumps away to the Klondike, a little-known high-elevation tangle of swamp and spruce forest. North of the Klondike, North Brother Mountain rises above treeline and provides memorable views southeast toward Katahdin.

Access

There is sufficient alpine territory on Katahdin to warrant many days, or even weeks, of extended exploration. For this reason, it is worth approaching the mountain from all sides in several seasons. Baxter State Park has two entry points: in the south (Togue Pond, the most popular access point) and in the less-visited north (Matagamon). The Togue Pond Gate is accessible via the Medway/Millinocket exit on I-95 (exit 244) and provides the most direct access to Katahdin. From Togue Pond Gate, the Park's "Tote Road"—an unimproved dirt road—winds north to the various campgrounds that support the main trails up Katahdin: Katahdin Stream and Abol on the west, and Roaring Brook on the east. The most direct route to treeline approximates that of Henry Thoreau, following the Abol Slide from Abol Campground. The Appalachian Trail itself offers a gentler alternative and follows the Hunt Trail from Katahdin Stream Campground. The views from Katahdin's east side are dramatically different than from the west, and it is certainly worth approaching Katahdin from both directions. The east face of Katahdin is accessed easily via the Roaring Brook Trail from Roaring Brook Campground. This provides direct access to the impossibly scenic Chim-

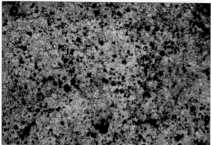

8-6 Most of Katahdin is composed of Devonian-aged granite some 400 million years old. As Katahdin granite weathers, it provides a substrate for mountain sandwort and diapensia to colonize. Katahdin granites are gray near the base of the mountain and rose-colored near the summit. (MJ and LW)

ney Pond and the adjacent North Basin. From Chimney Pond, the Dudley Trail runs directly to Pamola Peak, providing views of Katahdin's cirque and access to the famous Knife Edge.

Geology

Katahdin is composed of 400 million-year-old granite, which formed as part of a larger volcano chain during tectonic collisions in the Devonian period. Katahdin granite, as it is known, becomes more pinkish in color, and more fine-grained, near the summit ridge from Pamola Peak to Baxter Peak (8-6). The pinkish "summit granite" of Katahdin is very distinctive and constitutes a resistant "cap" that protects the underlying weaker rocks.[2] Although most of the mountain is clearly granitic in origin, some evidence indicates that calcareous bedrock influences groundwater near the headwall of North Basin.[3] The entire mountain is strewn with the debris of earlier continental and alpine glaciations. The prominent cirques, now thought to pre-date the Wisconsinan ice advance,[4] are blockaded by extensive moraine systems in the vicinity of the Basin Ponds. An extinct rock glacier, formerly supported by a core of ice, rests on the floor of the North Basin (8-7). Katahdin itself may be a *roche moutonée,* rounded on the north and west from whence came the ice, and steep on the south and east, where moving ice sheared the granite walls.

8-7 The North Basin, a striking cirque on Katahdin's east face, once supported a rock glacier. The extinct remains of the glacier may be seen today on the floor of the cirque. (MJ)

---

2   Rankin and Caldwell (2010)
3   Miller et al. (2005)
4   The Wisconsinan, or Wisconsin, ice advance lasted from approximately 110,000 to 10,000 years before present.

Ecology

Vegetation — Katahdin's granitic fellfields harbor one extremely enigmatic plant. Leafy saxifrage (*Micranthes foliolosa*), a denizen of cold mountains throughout the eastern Canadian Arctic and Greenland, occurs here at its southernmost, and most isolated, colony in North America (8-8). Known from three sites along Katahdin's southern rim, this saxifrage reproduces primarily through **apomixis**, producing **proliferous bulbils** along the stem in place of flowers, which produce live plants without sexual reproduction.[5] The nearest

8-8 The summit fellfields of Baxter Peak support an extremely disjunct population of leafy saxifrage. The nearest populations occur in Québec and Newfoundland. (AH)

populations of this plant are known from Newfoundland and Québec, making it one of the most disjunct alpine plants in the eastern region and perhaps Katahdin's most unique contribution to alpine biogeography. Almost as interesting is a small station of northern willow (*Salix arctophila*), which occurs at only a single site on a slab of Katahdin granite. Apparently only a single individual plant survives, a **staminate** ("male") individual.[6] Numerous other rare alpine plants occur in various gullies, snowfields, crevices, and scree-piles in the South Basin, but these are also represented on the Presidential Range in New Hampshire. Among these are alpine cudweed, alpine bittercress (*Cardamine bellidifolia*), Wormskjold's alpine speedwell (*Veronica workskjoldii*), and white mountain saxifrage (*Saxifraga paniculata*), which occur in the "Chimney" above Chimney Pond and locally in other gullies. The occurrence of white mountain saxifrage may indicate a localized calcareous influence in the Chimney. Mountain sorrel (*Oxyria digyna*) and sibbaldia (*Sibbaldia procumbens*) are notably absent, as both occur in subalpine ravines on the Presidential Range to the southwest and the Monts Chic-Chocs to the north. Katahdin may also harbor a rare southern outpost of milky draba (*Draba lactea*)—but only a damaged specimen has been attributed to the region.[7]

Katahdin's windswept tableland supports alpine vegetation similar in composition to the Presidential Range (Ch. 7) and the McGerrigle Range (Ch. 10), although the actual alpine plant communities are markedly different in their abundance and distribution. Still, familiar alpine plants like Bigelow's sedge (*Carex bigelowii*), highland rush (*Juncus trifidus*), dia-

8-9 Familiar alpine plants common on the Presidential Range in New Hampshire are also abundant on Katahdin. These include diapensia (left), alpine bearberry (center), and Lapland rosebay (right). (LW and MJ)

---

5   Hudson (1987)
6   Willows are dioecious, having male and female plants. MNAP (2004)
7   This specimen, and the likelihood of its rediscovery on Katahdin, are discussed by Dr. Harris and others in the *AMC Field Guide to Mountain Flowers of New England* (1964).

8-10 Some temperate plant species occur in the protected talus of Katahdin's North Basin. Clockwise from top left: white pine, painted trillium, rhodora, and bush honeysuckle. (MJ)

pensia, Lapland rosebay, alpine bearberry (*Arctous alpina*), mountain cranberry (*Vaccinium vitis-idaea*), and black crowberry (*Empetrum nigrum*) are abundant on exposed ridges and frost-disturbed terraces (8-9). Purple mountain heather (*Phyllodoce caerulea*) is found in sheltered rockfields and talus slopes. At middle elevations just below the treeline, balsam fir (*Abies balsamea*), black spruce (*Picea mariana*), Labrador tea (*Rhododendron groenlandicum*), sheep laurel (*Kalmia angustifolia*), and other northern species are abundant. In particular, the North Basin is thickly vegetated with an unusual juxtaposition of alpine-boreal and southern species. These include temperate species such as white pine (*Pinus strobus*), common juniper (*Juniperus communis*), painted trillium (*Trillium undulatum*), rhodora (*Rhododendron canadense*), and bush honeysuckle (*Diervilla lonicera*)(8-10), and alpine species such as single-spike sedge (*Carex scirpoidea*) and yellow sedge (*C. flava*)(8-11), northeastern paintbrush (*Castilleja septentrionalis*), bluebell (*Campanula rotundifolia*), alpine bilberry (*Vaccinium uliginosum*), purple mountain heather, and lance-leaved arnica (*Arnica lanceolata*)(8-12). Patches of boreal bedstraw (*Galium kamtschaticum*) and proliferous fescue (*Festuca prolif-*

8-11 Alpine sedges such as yellow sedge and single-spike sedge occur along the base of the North Basin headwall and in similar habitats elsewhere on Katahdin. (LW and MJ)

era) occur in the shade of krummholz along the higher gullies (8-13). The morainic debris field on the floor of North Basin is dominated by krummholz black spruce (*Picea mariana*) with large patches of alpine bearberry, Labrador tea, and paper birch (*Betula papyrifera*).

A calcareous-loving bryophyte and lichen assemblage was recently discovered near a seepage area in the North Basin,[8] suggesting that additional vascular species, such as saxifrages, may eventually be found in this area. This may also be a good place to search for milky draba. Other species that indicate circumneutral bedrock occur along the bottom of the North Basin headwall, including shrubby cinquefoil (*Dasiphora fruticosa*), which grows commonly alongside lance-leaved arnica and yellow sedge.

The krummholz zone of Katahdin is composed primarily of balsam fir and black spruce, which are extremely common treeline species throughout the eastern region. White spruce is also present, which is noteworthy for several reasons. For starters, white spruce here reaches the southern extent of its main range, which stretches from Newfoundland and Labrador to central Alaska, with outposts or tendrils reaching into the Montana Rockies and South Dakota's Black Hills. The addition of white spruce is noteworthy for several other reasons. The three eastern spruce (red, black, and white) rarely co-occur at a single site, as they do on Katahdin. For example, this is not the case in the White Mountains, where red and black spruce co-occur (but white spruce is absent), or in Gaspésie and points north (where black and white spruce co-occur but red spruce is absent). In some parts of the Adirondacks, northern New Hampshire, and possibly on Vermont's Camel's Hump (Ch.

8-12 Bluebell or harebell (top) and lance-leaved arnica (bottom) are subalpine species common in Katahdin's ravines. (MJ)

8-13 Proliferous fescue is an uncommon alpine species in rills and seepages. (MJ)

11), it is possible to see all three species of spruce. Furthermore, white spruce in Baxter State Park may constitute a part of the "fir waves," visible across the face of the Brothers, which are common phenomena in balsam fir–red spruce forests in the White Mountains (8-14). [9]

8   Miller et al. (2005)
9   Sprugel (1976)

8-14 *"Waves" in spruce and fir on North Brother, photographed from the Appalachian Trail. (MJ)*

8-15 *High-elevation springs and wetlands on Katahdin's tableland provide habitat for northern bog lemming. (MJ)*

Fauna — Katahdin's alpine tableland and surrounding forest supported an apparently large caribou herd until the early 1900s. This herd is better documented than any other in the northeastern United States, because it persisted into the 20th century. As many as 64 caribou were observed on Katahdin at one point in 1894.[10] B.H. Dutcher, a mammalogist, wrote of the Maine caribou in 1903, "The caribou is an animal of the past in the Katahdin region. Today all that remains is its bones in the porcupine dens (sic). From accounts received, there have been two migrations of caribou from northern Maine, within the memory of inhabitants now living. The last of these occurred about six years ago (i.e., 1897). Unfortunately the awakening of public sentiment in regard to the importance of game preservation did not take place while the animals were still abundant, and their absence now can in part at least be attributed to wanton destruction." Subsequent to Dutcher's account, caribou were seen on Katahdin as late

8-16 *Chimney Pond and other subalpine pools on Katahdin provide breeding habitat for (clock-wise from top-right) American toads, spotted salamanders, and wood frogs. (MJ, LW, and CE)*

as 1908.[11] The Katahdin caribou were also the southernmost herd to be well-documented in New England, although some evidence suggests that caribou ranged through New Hampshire's Connecticut Lakes and northern Vermont until the 1830s and 1860, respectively.[12] In December 1963, 17 caribou from Newfoundland were released on Katahdin as part of a larger reintroduction effort.[13] Sightings of caribou were subsequently reported in northern Maine until 1966. A second, state-run effort in the 1980s also failed. Although caribou are gone for the foreseeable future, moose are common throughout Baxter State Park as high as the timberline.

8-17 Northern two-lined salamanders occur in streams and cold ponds to at least 1000 m, and probably higher. Larva from Chimney Pond. (LW)

Black bear are common below the timberline and occasionally explore Katahdin's alpine areas. American marten are more widely distributed in alpine and subalpine habitats.

8-18 Spiders specializing in alpine habitats include several wolf spiders. (LW).

Wet sphagnum bogs on the alpine portion of the tableland probably support a small population of New England's least-known mammal, the northern bog lemming (*Synaptomys borealis*). Several individuals have been documented in the vicinity of Caribou Spring over the past century, one of only two occurrences of this rare taiga-bog species in Maine (the other is at Bigelow Mountain, Ch. 11) (8-15).[14] Rock vole (*Microtus chrotorrhinus*) and southern red-backed vole (*Clethrionomys gapperi*) are also present in the Katahdin alpine areas. American pipits (*Anthus rubes-*

8-19 The wingless White Mountain locust is abundant throughout Katahdin's alpine tableland. (MJ)

cens), arctic-alpine specialists, may also be seen near Caribou Spring and Thoreau Spring. High-elevation wetlands on Katahdin such as Chimney Pond also provide breeding habitat for amphibians, including American toad (*Anaxyrus americanus*), wood frog (*Lithobates sylvaticus*), and spotted salamander (*Ambystoma maculatum*) (8-16). Northern two-lined salamanders (*Eurycea bislineata*) (8-17) are abundant in Chimney Pond and probably occur in streams higher up on the mountain.

Unusual, alpine-adapted invertebrates include the endemic Katahdin arctic butterfly (*Oeneis polixenes katahdin*), the mountain spider, which resides in Katahdin's subalpine talus fields, and wolf spiders (8-18), which occur throughout the alpine tundra. The wingless

---

11   Ludgate (1938); Palmer (1938)
12   Allen (1904); Bergerud and Mercer (1989)
13   Bergerud and Mercer (1989)
14   MDIFW (2003)

8-20 On warm and calm days, hundred of species of low-elevation insects may congregate on Katahdin's ridges and summits, a behavior known as "hilltopping" (top-left; see Ch. 6). Clockwise from top-right, a green click beetle investigates a black click beetle; a cicada rests between boulders on Hamlin Ridge; a Hudsonian white-face dragonfly clings to mosses in the Chimney. (MJ)

White Mountain locust is also abundant above treeline (8-19). Low-elevation insects and other invertebrates may find their way to Katahdin's alpine summits and ridgelines either on fair-weather days, through hilltopping (seeking the highest point on the landscape) or during storm events, which may blow insects off course (8-20).

Mont Albert
and the Monts Chic-Chocs 9
QUÉBEC

Mont Albert, Gaspésie, Québec. (MJ)

Map 9-1. Mont Albert, Gaspésie, Québec.

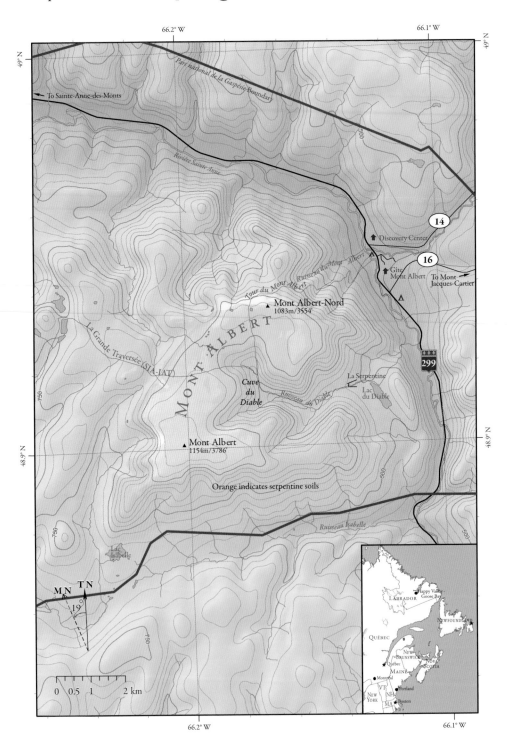

Mont Albert and the Monts Chic-Chocs, Québec

9-1 The serpentine hulk of Mont Albert is visible from most points on the neighboring McGerrigle range. (MJ)

Mont Albert is among the more famous eastern mountains, and is probably the most famous in Gaspésie, the wide peninsula in southern Québec that encloses the southern edge of the Gulf of St. Lawrence. The great, red, serpentine massif rises above the *Rivière Sainte-Anne* immediately west of the granitic summits of Mont Jacques-Cartier and the Monts McGerrigle (Ch. 10) (9-1). The highest reaches of its umber plateau do not quite reach 4000', but Mont Albert is one of the most visually striking and ecologically unique mountains in the region. Although there are larger exposures of serpentine surrounding Newfoundland's Bay of Islands, most of them are much more difficult to access.

From Mont Albert, the thin spine of the Monts Chic-Chocs (in English, sometimes referred to as the "Shickshocks") runs west to the mostly forested region surrounding Mont Logan (9-2). This segment represents the extremes of the Monts Chic-Chocs: serpentine in the east, and sedimentary rocks in the west, with granite surrounding and between. Granite, serpentine and limestone: the alpine mountains of Gaspésie have been roughly divided along these lines since Dr. Fernald of Harvard University first explored them, and proposed this type of bedrock classification, in the early 20th century.[1] Each of the major rock types supports unique alpine vegetation so that generally, a granite mountain in Gaspé is floristically similar to a granite mountain in New England, and so on (although there are some strong local differences between the Gaspé Peninsula and New England, or the Gaspé and Newfoundland). In Gaspésie, three distinct groups of alpine plants, corresponding to the

---

1   Fernald's 1907 paper, "The soil preferences of certain alpine and subalpine plants" was one of the first to explore the relationships between bedrock composition and alpine vegetation. His thorough analysis explored the floristic differences between the granitic and gneissic summits of New England and Gaspésie with the flora of Mont Albert and other areas. His final conclusion, painstakingly supported, that there are three major alpine floras in the region corresponding to serpentine (magnesium-rich), limestone (calcium-rich), and granite/gneiss (potassium-rich) bedrock, is more or less valid today.

9-2 Diverse alpine habitats extend at least as far west as Mont Matawees in the western Monts Chic-Chocs. Mont Matawees from Mont Logan, Québec(MJ)

9-3 Permafrost underlies the highest reaches of Mont Albert's alpine plateau. Expanding and thawing winter ice arranges the serpentine soils into rough polygons. (MJ)

different bedrock, may be viewed within a distance of only 60 km (37 mi.), or about three days in the field.

The high mountains of Québec's Gaspé Peninsula would be the final word on Appalachian alpine diversity if not for the mountains of western Newfoundland and their endless alpine wonders. In any case, the alpine biodiversity of the region is stunning, given that the highest peaks lie only at latitude 49°N (well within the temperate latitudes, and merely 368 km / 230 miles north of Katahdin). Nearly every arctic-alpine plant known from the two great alpine peaks in New England (the Presidentials and Katahdin, Ch. 7 and 8) occur on either Mont Albert or Mont Logan,[2] or on neighboring Mont Jacques-Cartier, as do dozens of other species. Gaspésie is the southernmost station for many of these alpine and arctic plants. Other arctic tendrils reach south to influence this seaswept peninsula: lenses of permafrost apparently underlie the highest elevations of Mont Albert, Mont Jacques-Cartier, and portions of the Monts McGerrigle (9-3).[3] Holocene remains of wolverine (*Gulo gulo*), arctic hare (*Lepus arcticus*), and grizzly bear (*Ursus arctos*) have been unearthed from the Gaspé.[4]

Lay of the Land

The Monts Chic-Chocs are themselves a portion of the more extensive Notre Dame Mountains, which run the full length of the Gaspé Peninsula and are geologically related to Vermont's Green Mountains and the Berkshires of Massachusetts. For our purposes in this book, the Monts Chic-Chocs do not include the enormous, granitic mountains that lie just east of Mont Albert. As noted earlier, these are the Monts McGerrigle, which are

---

2   The alpine plants known from the Presidential Range but not Gaspésie include Robbins' dwarf cinquefoil (*Potentilla robbinsiana*), eastern mountain avens (*Geum peckii*), and Boott's rattlesnake-root (*Nabalus boottii*) (see Ch. 7).

3   Gray and Brown (1979)

4   Miller (2010)

treated thoroughly in the following chapter as a separate mountain range. The alpine portion of the Monts Chic-Chocs, in this strict geologic sense (excluding Mont Jacques-Cartier and the Monts McGerrigle), extends from the serpentine barrens of Mont Albert in the east to the calcareous cliffs of Mont Logan, Mont Matawees, and Mont Collins in the west.

9-4 Mont Albert's north face culminates at Mont Albert Nord, a subsidiary summit composed of amphibolite. (MJ)

Mont Albert (1154 m / 3786') is certainly the most well-known of the Monts Chic-Chocs, and is also the best-understood serpentine mountain in the eastern region. Mont Albert consists of a single plateau, albeit a heavily eroded one, so its geography is straightforward. Albert's northeastern façade presents itself to visitors of the *Parc national de la Gaspésie* arriving from the ocean town of Sainte-Anne-des-Monts, Québec, and culminates at the spur-summit Mont Albert Nord (1083 m / 3554') (9-4). The entire mountain encompasses an exposed, level, plateau supporting about 1600 hectares (~2940 acres) of ultramafic rocks (9-5).[5] Mont Albert's expansive plateau is deeply dissected by the glacier-carved *Cuve du Diable* (Devil's Gulch),[6] a tripartite cirque-complex in serpentine (9-6). Mont Albert seems entirely out of place in the larger context of the western Monts Chic-Chocs and Monts McGerrigle, and could not be more different from the surrounding mountains.

9-5 Mont Albert's alpine tableland encompasses about 1600 ha of alpine and serpentine vegetation. (MJ)

9-6 The Cuve du Diable (Devil's Gulch) is a tripartite glacial cirque that dissects the interior of Mont Albert's plateau. (MJ)

From Mont Albert, the Monts Chic-Chocs continue westward as a linear, east-west chain of mostly forested summits from the Cap-Chat River to *Lac Cascapedia* (Map 9-2). These summits are generally rounded and forested with steep, north-facing slopes. At the western end of this section is Mont Logan (1136 m / 3727'),[7] botanically one of the most interesting mountains in North America and the second tallest of the Monts Chic-Chocs (after Mont Albert) (9-7). The floristic diversity of Mont Logan partly results from the calcareous, sedimentary composition of its flanks and

---

5   In total, 6400 hectares of ultramafic rocks are exposed (Sirois et al. 1988)

6   *La Cuve du Diable* (Devil's Gulch) is a multi-pronged, east-facing cirque complex on Albert's interior tableland.

7   Mont Logan is named for the pioneering Canadian geologist William Edmond Logan (1798–1875), who first climbed the peak in 1844 (Logan 1846). Yukon's Mount Logan, the highest peak in Canada and second highest in North America, is also named for him.

Map 9-2. Monts Chic-Chocs, Gaspésie, Québec.

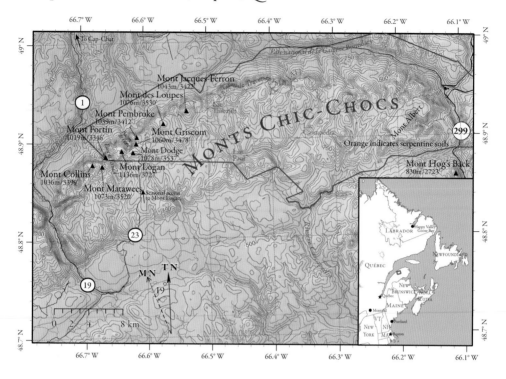

surrounding peaks, which contrast sharply with the ultramafic, igneous Mont Albert and the granitic Mont Jacques-Cartier.

At the far western end of the Monts Chic-Chocs, Mont Logan's pronounced summit cone is buttressed by Mont Fortin (>1019m / 3346') on the west and by Mont Dodge (>1078 m / >3537'), Mont Pembroke (>1039 m / >3412'), and Mont Griscom (>1060 m / >3478') on the east. The Mont Logan summit complex is separated from Mont Matawees[8] (>1072 m / >3520') and Mont Collins (>1036 m / >3399') by the *Passe de Fernald* (Fernald's Pass). Mont Matawees, in turn, is separated from the alpine summit of Collins by a sprawling alpine ridge called the Saddle. Surrounded by summits covered in thick subalpine scrub, the alpine Saddle is one of the more extensive alpine areas at the western end of the Monts Chic-Chocs.

Numerous small cirques and ravines indent the northern face of the Chic-Chocs in this region. Monts Matawees and Collins are bordered by two large cirques, the unofficially named Collins Bowl to the west and *Bassin de Fernald* (Fernald's Basin) to the north. The most dramatic cirque, visually and botanically, is the *Bassin de Pease* (Pease' Basin), the largest cirque of the Mont Logan region. *Bassin de Pease* is a double-cirque, mostly oriented north, occupying about 110 hectares (277 acres) on the north face of Monts Logan, Dodge, Pembroke, and Griscom. The Bassin is delimited to the north by the *Grand Arête,*[9] a calcareous glacier-

---

8   Mont Matawees was named by Collins and Fernald (1925) for the Mi'kmaq word for porcupine, which they had killed on the upper slopes during an attempt to summit Mont Logan.

9   The *Grand Arête* was the "Razorback Ridge" of Collins and Fernald (1925); see Gervais (1982).

9-7 *Mont Logan, although largely forested, is one of the most interesting botanical sites in Gaspésie, after Mont Albert. (ML)*

carved ridge. South of the *Grand Arête* lies the *Ravin à Neige* (Snow Ravine), which holds snow well into the summer. From the east, the Bassin is fed by feeder streams of the *Ruisseau Ouellet*, which rises in the *Vallée Suspendue* (Hanging Valley), a contributory cirque.

Access

Mont Albert is a very popular climb, accessible from the *Gîte du Mont-Albert* on Route 299. A wide, 4.3 km (2.7 mi) footpath ascends steeply to the amphibolite ridge near the summit of Mont Albert Nord, culminating at a small warming hut and two of the most scenic outhouses on earth. The rest of the broad, marshy, serpentine tableland and the impressive *Cuve du Diable* (Devil's Gulch) can be traversed via the 17-km (11 mi) *Tour du Mont Albert*, which follows a developed footpath. As late as mid-June, large patches of snow remain on the ascent trail, on the tableland, and in the *Cuve du Diable*.

In contrast to popular Mont Albert, accessing the Mont Logan summit area can be somewhat confusing. During the summer (outside of moose hunting season), Mont Logan can be accessed by four-wheel roads through the *Réserve Faunique de Matane* (Matane Wildlife Refuge)[10] that lead ultimately to the communications tower on Logan's rocky summit. Alternatively, the Logan area may be accessed by the 17-km (11 mi) *Sentier International des Appalaches* (International Appalachian Trail; abbreviated "SIA-IAT"), which leaves from the Cap-Chat River, and which originates in the *Réserve Faunique de Matane* but crosses into the *Parc national* near the summit of Mont Logan. Mont Logan can also be reached from the center of *Parc national de la Gaspésie* by long overland traverses via the SIA-IAT from Le Huard or Lac Cascapedia.

Geology

As discussed earlier, the bedrock geology of the Monts Chic-Chocs is complex. Although Mont Albert is primarily a massif of emplaced serpentine, rich in magnesium and iron, with occurrences of asbestos (9-8), the northern façade of Mont Albert is composed primarily of

10   The *Réserve Faunique de Matane* may be accessed from the town of Cap-Chat on the St. Lawrence coast. Portions of the *Réserve* are closed to automobile traffic during the fall hunting season, which extends from September to December.

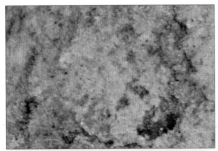

9-8 *Mont Albert is composed of ultramafic rocks derived from the Earth's mantle. Serpentine, or peridotite, weathers to a rich ochre color, but is greenish when freshly broken or polished. Veins of asbestos, such as this one in the Cuve du Diable, are associated with serpentine. (LW and MJ)*

amphibolite, a metamorphic rock composed largely of hornblende. Along the spine of the Monts Chic-Chocs, the bedrock is primarily granitic, although outcrops of calcareous bedrock occur in the vicinity of Mont Logan.

Ecology

Vegetation — The narrow north face of Mont Albert Nord supports alpine vegetation more similar to the granitic McGerrigle massif to the east than to the rest of Albert's adjacent serpentine tableland, although some technical varieties may occur on the amphibolite spur that are unknown on nearby granitic summits (9-9).[11] This amphibolite belt forms a fan-like ridge, which encompasses the summit of Mont Albert Nord and supports pockets of well-developed krummholz. Here, an association of white and black spruce[12] and balsam fir ascend nearly to the plateau around 1000 m (3200'). Several common boreal plants, such as Canada mayflower (*Maianthemum canadense*), creeping snowberry (*Gaultheria hispidula*), and northern comandra (*Geocaulon lividum*), grow on thick mats of the common boreal mosses *Pleurozium schreberi* and *Hylocomium splendens*.[13] Large-leaved goldenrod (*Solidago macrophylla*)—abundant throughout the eastern alpine region in subalpine forests and clearings—is common here also, with Chinese hemlock parsley (*Conioselinum chinense*) (9-10). On exposed parts of the amphibolite ridge, the vegetation is similar to that found on exposed sites on the Presidential Range or Katahdin. For example, heath plants (Ericaceae) dominate the scrubby barrens, including alpine bilberry (*Vaccinium uliginosum*), Labrador tea (*Rhododendron groenlandicum*), and other common acid-loving alpine plants.

The great significance of Mont Albert is not found on its northern ridge, which is similar

---

11   Fernald (1907)

12   Red spruce (*Picea rubens*), which is a dominant component of the subalpine forests in New England and the southern Appalachians, gives way to white spruce on Gaspésie.

13   See Rune (1954).

to many other mountains, but the great serpentine tableland. Serpentine, it should be noted again, is toxic to most plants because it is nutrient-poor and high in metals such as iron, magnesium, nickel, and chromium.

First, alpine species typical of granitic or metamorphic summits like Washington, Katahdin, and Jacques-Cartier manage to survive on Mont Albert's serpentine plateau, such as mountain firmoss (*Huperzia appressa*), several sedges,[14] moss campion (*Silene acaulis*), Labrador tea, alpine azalea (*Loiseleuria procumbens*), purple mountain heather (*Phyllodoce caerulea*), and alpine bilberry.[15] Lapland rosebay (*Rhododendron lapponicum*)—common on both the Presidential Range and Katahdin—is abundant on the serpentine talus, as well as moss campion, which occurs only very sparingly on the Presidentials.

9-9 *The amphibolite of Mont Albert Nord supports many of the alpine species found on the Presidential Range and Katahdin. (MJ)*

9-10 *Chinese hemlock parsley is a common species in the subalpine forests of Mont Albert. (MJ)*

Additionally, a distinctive serpentine-alpine flora is dominated by three species of low, trailing sandworts: serpentine sandwort (*Minuartia marcescens*), reddish switchwort (*Minuartia rubella*), and creeping sandwort (*Arenaria humifusa*). Also abundant are chickweeds (*Cerastium arvense* ssp. *strictum*),[16] alpine catchfly (*Silene suecica*), Siberian sea thrift (*Armeria maritima* ssp. *sibirica*), and field sagewort (*Artemisia campestris*). Swamp thistle (*Cirsium muticum*) is common in protected seepages above treeline (Pl. I). The serpentine talus of Mont Albert provides habitat for at least three plant species currently considered "threatened" in Québec.[17]

In Mont Albert's *Cuve du Diable*, the Aleutian maidenhair fern (*Adiantum aleuticum*) is especially common (although it also occurs throughout the tableland). Two other ferns, *Aspidotis densa* and *Polystichum scopulinum*,[18] occur on some south-facing walls. Other species found commonly on serpentine in the ravines of Devil's Gulch include several grasses,[19]

---

14   Including *Carex canescens* and *C. bigelowii* (Fernald 1907) and *C. scirpoidea* (Rune 1954).

15   Fernald (1907) lists the following species as occurring on both the Mont Albert serpentines and the Washington-Katahdin-Cartier summits: *Huperzia appressa, Carex canescens, Carex bigelowii, Juncus trifidus, Betula glandulosa, Silene acaulis, Rubus arcticus, Empetrum nigrum, Rhododendron groenlandicum, Rhododendron lapponicum, Loiseleuria procumbens, Kalmia angustifolia* and *K. polifolia, Phyllodoce caerulea, Andromeda polifolia, Vaccinium uliginosum, V. myrtilloides* and *V. vitis-idaea, Diapensia lapponica,* and *Nabalus trifoliolatus.* Rune (1954) lists the following plants, also known from granitic or gneissic summits: *Carex scirpoidea, Rhododendron lapponicum, Campanula rotundifolia,* and *Silene acaulis.*

16   Rune (1954) lists *Cerastium arvense* from the Mont Albert tableland; today this encompasses several species, subspecies, and varieties.

17   *Polystichum scopulinum, Salix chlorolepis,* and *Solidago simplex* var. *chlorolepis* (Rajakaruna et al. 2009).

18   Rune (1954).

19   E.g., *Festuca altaica* and *Danthonia intermedia* (Fernald 1907; Rune 1954)

9-11 The western Monts Chic-Chocs, including Mont Fortin and Mont Collins, support many unusual arctic and alpine plants in their north-facing cirques. (ML)

9-12 The Saddle on Mont Collins supports about 13 ha of classic cushion-tussock and heath-shrub-rush plant communities. (MJ)

northeastern paintbrush (*Castilleja septentrionalis*), and at least two unusual goldenrods (Pl. I).[20] Near the floor of the ravines, a thick scrub cover of spruce, tamarack (*Larix laricina*), and scattered red maple (*Acer rubrum*) looks out of place in the Mont Albert moonscape, but it does well also on the serpentine mountains of southwestern Newfoundland (Ch. 13). Most significantly, Mont Albert apparently represents the southernmost location for a number of arctic species, accounting for a large proportion of those that do not reach farther south than the Gaspé Peninsula.[21] Botanically, Mont Albert is a crown jewel of the *Parc national de la Gaspésie* and the Monts Chic-Chocs.

By comparison, many of the major summits in the western Chic-Chocs are largely forested. This is deceptive, and a bit disorienting[22] because it seems contradictory that great alpine diversity would be associated with forested summits (9-11). Indeed, black and white spruce, balsam fir, and associated thickets of Bartram's shadbush (*Amelanchier bartramiana*), green alder (*Alnus viridis* ssp. *crispa*), and Labrador willow (*Salix argyrocarpa*) ascend to the ridgeline at nearly every point. Only Mont Logan rises much above the treeline, and only about 30 m (100'). Another relatively large summit tundra area, the exposed Saddle on Mont Collins, is only about 13 hectares (32 acres) (9-12). Most of the remaining alpine areas on Monts Matawees, Fortin, and Logan are smaller, and exhibit the familiar vegetation typical of the noncalcareous, eastern alpine summits. The dominant plants include mountain firmoss, alpine sweetgrass (*Anthoxanthum monticola*), Bigelow's sedge (*Carex bigelowii*), bearberry willow (*Salix uva-ursi*), alpine bilberry, mountain cranberry (*Vaccinium vitis-idaea*), Labrador tea, alpine bearberry (*Arctous alpina*), black crowberry (*Empetrum nigrum*), diapensia (*Diap-*

---

20 Rune (1954) lists *Solidago chlorolepis*, an eastern representative of the Cordilleran *S. decumbens* now called *S. simplex*, var. *chlorolepis*, and *S. multiradiata*, which is more generally distributed in bogs in northeastern mountains.

21 Among others, Rune (1954) lists *Deschampsia alpina, Carex lachenalii, Arenaria humifusa, Minuartia biflora, Silene suecica, Armeria maritima,* and *Artemisia campestris.*

22 M.L. Fernald's first expedition to Mont Logan in 1922 ended on Mont Matawees, which they mistook for Logan, 3 km (1.9 mi) along the ridge to the east. In 1923, he led a successful, extended expedition to Mont Logan, but was foiled again in 1925, when successive storms held them in *Passe de Fernald* and prevented further exploration.

Plate I. Alpine Plants of Mont Albert
Common plants on serpentine

Harebell (*Campanula rotundifolia*)

Tamarack (*Larix laricina*)

Reddish switchwort (*Minuartia rubella*)

Field chickweed (*Cerastium arvense*)

Alpine catchfly (*Silene suecica*)

Swamp thistle (*Cirsium muticum*)

Multi-rayed goldenrod (*Solidago multiradiata*)

Mont Albert and the Monts Chic-Chocs, Québec— 141

9-13 *The* Grand Arête *and* Ravin à Neige *on Mont Logan support many rare alpine plants. (MJ)*

ensia lapponica), and three-toothed cinquefoil (*Sibbaldia tridentata*)[23] (Pl. II).

The alpine diversity on Mont Logan is found within the numerous cirques that indent the ridge on its north and west faces, and on the north-facing cliffs, ridges, and arêtes, primarily on calcareous sedimentary rock. For example, at Collins Bowl and *Bassin de Fernald*, the summit vegetation is continuous with mosaics of subalpine meadows and clearings associated with late-lying snow and steep topography. Great botanical diversity is found in *Bassin de Pease* on the north face of Monts Logan, Dodge, Pembroke, and Griscom, as well as on the *Grand Arête* and in the *Ravin à Neige*, which holds snow well into the summer (9-13). Several relatively common alpine and subalpine species occur in these subalpine meadows and snowbeds, including moss heather (*Harrimanella hypnoides*), kidney-leaved buttercup (*Ranunculus abortivus*), and glandular birch (*Betula glandulosa*), but a number of rare plants also occur, including several endemics. These plants include, but are not limited to, alpine lady-fern (*Athyrium alpestre*), several sedges,[24] most of the eastern alpine willows,[25] and a suite of snowbed species generally rare throughout the eastern alpine as far south as Mount Washington.[26] Other unusual plants found in the snowbeds and gullies of the *Bassin de Pease* include Oakes' eyebright (*Euphrasia oakesi*),[27] redtipped lousewort (*Pedicularis flammea*), at least two saxifrages,[28] alpine meadowrue (*Thalictrum alpinum*), and Norwegian cudweed (*Omalotheca norvegica*) (Pl II).[29]

A number of calciphiles occur in the exposed calcareous talus of the *Grande Arête* and exposed sites on Monts Matawees, Logan, Fortin, and Griscom, including entire-leaved

---

23 For a description of Mont Logan's alpine flora, see Gervais (1982). For thorough descriptions of these common alpine plants, see Harris et al. (1964).

24 Including *Carex nardina* and *C. rupestris* (Gervais 1982).

25 Including *Salix herbacea, S. arctophila* (abundant on the tundra of the Monts Groulx) and the Canadian Shield but farther south known only from Katahdin), *S. arctica, S. brachycarpa, S. glauca* var. *cordifolia*, and *S. planifolia.*

26 This snowbed/ravine suite occurs in similar habitats throughout Newfoundland, Labrador, Québec and extends south to Mount Washington and includes *Oxyria digyna, Polygonum viviparum, Cardamine bellidifolia, Saxifraga rivularis*, and *Sibbaldia procumbens.*

27 *E. oakesii* occurs sparingly on wet tundra throughout Newfoundland and coastal Québec and Labrador, reaching its southernmost distribution on Washington and Katahdin.

28 At least seven saxifrages (genus *Saxifraga*) occur in the alpine and subalpine areas of Mont Logan, mostly confined to calcareous soils. In addition to *S. rivularis, S. cespitosa, S. cernua*, and *Micranthes nivalis* occur in the *Bassin de Pease* and its gullies. *Saxifraga cernua* was not known farther south until 1968, when it was reported from Huntington Ravine, Mount Washington, by J.A. Churchill.

29 *O. norvegica* is widespread throughout the Chic-Chocs and McGerrigle but is not known from the eastern United States.

Plate II. Alpine Plants of Mont Logan region, Québec
Widespread or noteworthy plants on exposed ridgelines

Mountain firmoss (*Huperzia appressa*)

Bigelow's sedge (*Carex bigelowii*)

White mountain saxifrage (*Saxifraga paniculata*)

Entire-leaved mountain avens (*Dryas integrifolia*)

Moss campion (*Silene acaulis*)

Diapensia (*Diapensia lapponica*)

Bearberry willow (*Salix uva-ursi*)

Hyssop-leaved fleabane (*Erigeron hyssopifolius*)

mountain avens (*Dryas integrifolia*), shrubby cinquefoil (*Dasiphora fruticosa*), a mix of common and unusual saxifrages,[30] and the widespread moss campion (Pl. III).

Fauna — Mont Albert and the Monts Chic-Chocs, together with the Monts McGerrigle, support the last remaining herd of woodland caribou (*Rangifer tarandus caribou*) south of the Gulf of Saint Lawrence. Apparently, the Mont Albert and Mont Jacques-Cartier herds represent different subpopulations and should be managed separately (9-14). A

9-14 The caribou herds on Mont Albert and Mont Logan are largely isolated from the herd on Mont Jacques Cartier in the Monts McGerrigle. (MJ)

five-year telemetry study in the 1980s to 1990s documented only one female (out of 28) to cross between the two mountains.[31] Lichens, a favored food item, are most abundant in the high-elevation spruce-fir forest, but Gaspésie caribou congregate upon and next to the alpine tablelands of Mont Albert and Mont Jacques-Cartier. The caribou may congregate in these open areas to avoid black bear (*Ursus americanus*) and eastern coyote (*Canis latrans*), which may kill and eat caribou calves. In fact, researchers recently found evidence that logging in the vicinity of the Monts Chic-Chocs and Monts McGerrigle could result in localized increases in eastern coyote populations, which could in turn result in "spillover" depredation of caribou.[32] Caribou calve and rest on great spring snowfields, which are known in Québec as *zabois*. The mountains of Gaspésie, therefore, provide the only habitats in eastern North America where moose (*Alces americanus*), whitetailed deer (*Odocoileus virginianus*), and caribou co-occur.

Caribou are not the only northern species thriving, or surviving, on the high plateau of Mont Albert. Northern bog lemming (*Synaptomys borealis*) may occur in high-elevation sphagnum wetlands. Lemmings and voles may be active under the winter snow, leaving trails and scat where summer snow may persist until July. The Gaspé shrew (*Sorex gaspensis*), which resides in mountain talus throughout Gaspésie, northern New Brunswick, and Cape Breton Island, was first described from Mont Albert.[33] The serpentine barrens also support breeding populations of horned lark (*Eremophila alpestris*) and American pipit (*Anthus rubescens*)— noteworthy occurrences in the region. Snowy owls (*Bubo scandiacus*) may use the exposed tableland as a stopover between the wintering grounds on the Atlantic coast and the northern tundra, although Massachusetts-based researchers studying the wintering owls at Logan Airport in Boston did not document frequent or prolonged stops on the high mountains of the Gaspé Peninsula.[34]

Mont Albert, like other alpine mountains south of the St. Lawrence, supports an astonishing variety of insects, spiders, and other invertebrates. A tundra-adapted spur-throated grasshopper, *Melanoplus gaspesiensis*, may be found here and nowhere else, along with more

---

30   *Saxifraga paniculata* and *S. oppositifolia* occur on cliffs and broken talus throughout the Logan complex, especially in the vicinity of the *Bassin de Pease*.

31   Ouellet et al. (1996)

32   Boisjoly et al. (2010)

33   COSEWIC 2006. Gaspé shrew has been synonymized with *S. dispar* according to Rhymer et al. (2004).

34   Smith (2012)

Plate III. Alpine Plants of Mont Logan Region, Québec
Common plants in ravines and snowbeds

Mountain firmoss (*Huperzia appressa*)

Single-spike sedge (*Carex scirpoidea*)

Snowbed willow (*Salix herbacea*)

Sibbaldia (*Sibbaldia procumbens*)

Moss heather (*Harrimanella hypnoides*)

Purple mountain heather (*Phyllodoce caerulea*)

Carolina spring beauty (*Claytonia caroliniana*)

Mountain sorrel (*Oxyria digyna*)

9-15 Mont Albert harbors at least one relict spur-throated grasshopper. Other grasshoppers, such as this huckleberry spur-throated grasshopper, are common. (MJ)

widespread species such as *M. fasciatus* (9-15). Several species of northern fritillaries, such as arctic (*Boloria chariclea*) and Freija fritillary (*B. freija*) are common (9-16). As on the Presidential Range and Katahdin, many boreal insects may be found above treeline on Mont Albert during unusually calm weather or conversely, during windstorms.

9-16 Mont Albert supports a wide variety of butterflies, such as anise swallowtail (top), shown feeding on Chinese hemlock parsley and arctic fritillary (bottom). (MJ)

Mont Jacques-Cartier
and the Monts McGerrigle 10
QUÉBEC

Monts McGerrigle, Gaspésie, Québec. (LJ)

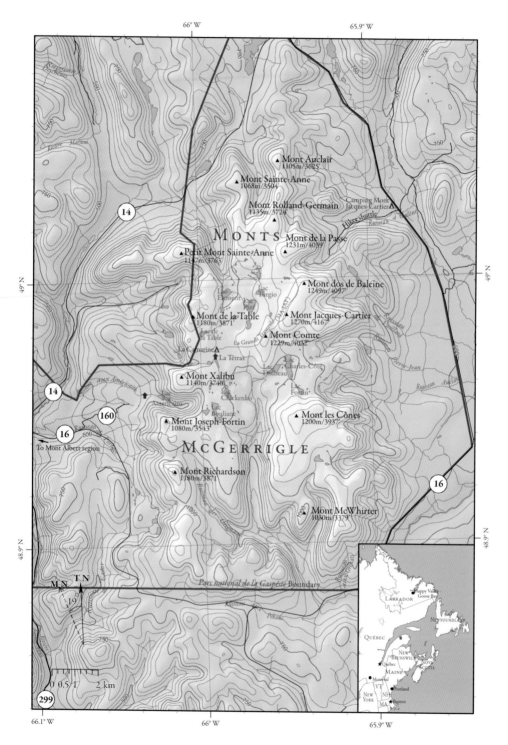

Mont Jacques-Cartier and the Monts McGerrigle, Québec

10-1 Mont Jacques-Cartier and surrounding granitic summits are collectively known as the Monts McGerrigle. (LJ)

Gaspésie's largest alpine ecosystem occurs upon the granitic, enormous tableland surrounding Mont Jacques-Cartier[1] and its secondary summits, known collectively as the Monts McGerrigle (10-1) or McGerrigle Mountains. The Monts McGerrigle, Québec's second-highest mountain range after the towering Torngat Mountains of the Québec–Labrador borderlands, encompass the eastern third of the Gaspésie high peaks region. Mont Jacques-Cartier is altogether unlike Mont Albert and Mont Logan or the other peaks in the Monts Chic-Chocs to the west (Ch. 9). The western Monts Chic-Chocs are essentially a comet-shaped, east-west, linear mountain chain culminating at lumpish Mont Albert, whereas the massif of the Monts McGerrigle undulates between ridges of alpine tundra and densely vegetated, forested saddles. The massif in this way bears superficial resemblance to the mountains of interior Québec (Ch. 18, 19, and 21) or the Northern Long Range Mountains of Newfoundland (Ch. 15) (10-2). A few great cirques indent the tableland of the Monts McGerrigle, but otherwise the gray tableland sweeps gradually upward to culminate at Mont Jacques-Cartier, Mont Richardson, Mont Xalibu, and a few other prominent peaks. Numerous subarctic and alpine plants that are rare on the Presidential Range or Katahdin grow profusely along maintained trails, such as alpine timothy (*Phleum alpinum*), Norwegian cudweed (*Omalotheca norvegica*), and sibbaldia (*Sibbaldia procumbens*) (10-3). Mont Jacques-Cartier and the McGerrigle Mountains, together with neighboring Mont Albert and the western Monts Chic-Chocs, provide habitat for the only remaining woodland caribou (*Rangifer tarandus*) south of the Gulf St. Lawrence.

10-2 Between large expanses of tundra, the high tableland of Mont Jacques-Cartier supports a mosaic of cold bogs and white spruce forest. (MJ)

Lay of the Land

Mont Jacques-Cartier and the Monts McGerrigle are sprawling and somewhat amorphous and are thus difficult to fully absorb from any single vantage point. From most

---

1   Named, of course, for Jacques Cartier (1491–1557), the 16th-century French explorer who explored the Gulf of St. Lawrence and claimed Québec for France.

10-3 *Several uncommon alpine plants, such as alpine timothy, Norwegian cudweed, and sibbaldia (clockwise from top-right) are locally common along hiking trails in the Monts McGerrigle. (MJ)*

angles, the overall impression one gets of the McGerrigle plateau is a steeply rising, flat-topped, unforested grey massif, with about a dozen low, indistinct, peripheral, secondary, alpine peaks. The granitic massif rises to the highest point on the Gaspé Peninsula (and the highest point for more than 300 km (200 mi) in any direction), at Mont Jacques-Cartier (1270 m / 4167') (10-4). Early explorers referred to Jacques-Cartier's plateau by the English name "Tabletop." So it was that when Dr. M.L. Fernald[2] refers to the "northwest shoulder" of "Tabletop Mountain" he is describing the vicinity of a subalpine meadow on nearby Mont de la Table (1180 m / 3871'), a bare summit due west of Jacques-Cartier across a protected spruce valley. Immediately adjacent to the summit of Mont Jacques-Cartier, Mont Comte (1229 m / 4032'), Mont Dos de Baleine (1249 m / 4097'), and Mont de la Passe (1231 m / 4039') are indistinct shoulders projecting off of the larger mountain complex (10-5). Farther to the north and west, near the boundary of *Parc national de la Gaspésie* (Gaspésie National Park), Mont Auclair (1105 m / 3625') and Mont de la Table are partially composed of calcareous sedimentary rock, in contrast with the surrounding granites (10-6). South of Mont Jacques-Cartier, Mont des Cones (1200 m / 3937') stands out in irregular silhouette. Mont des Cones is visible on clear days from most of the McGerrigle summits, and it is the dominant landscape feature at the southern end of the plateau. Classic heath-shrub-rush

2   Fernald (1907)

10-4 *The windswept summit of Mont Jacques-Cartier, once known as "Tabletop," is the highest point in southern Québec and the highest point between Katahdin and the mountains of Labrador. (MJ)*

and cushion-tussock communities, as well as exposed felsenmeer and twisted krummholz (see Ch. 5), continue south toward Mont McWhirter (1030 m / 3379'). Mont Xalibu (1140 m/ 3740') and Mont Richardson (1180 m / 3871') are west of Mont Jacques-Cartier's summit and south of Mont de la Table. These two peaks stand well above the treeline on either side of the *Lac aux Americains,* an east-west elongate cirque named for Dr. Fernald and Dr. A.S. Pease,[3] who catalogued the flora of the Monts McGerrigle in the early 1900s (10-7).

10-5 *The humped summit of Mont de la Passe is one of several "shoulders" of Mont Jacques-Cartier. (MJ)*

A series of smaller, granite-like mountains rise to a low treeline on either side of Route 299 south of the continuous granitic massif surrounding Mont Jacques-Cartier. Mont Hog's Back (830m) and Mont Vallieres-de-Saint-Réal (>914 m / >3000') (and its high point, Pic Sterling) are two prominent mountains in this complex. Both are within the *Réserve faunique des Chic-Chocs.*

10-6 *Mont de la Table is located west of Mont Jacques-Cartier and supports limestone-loving alpine plants. (MJ)*

Access

Most, but not all, of the Monts McGerrigle are protected within Québec's *Parc national de la Gaspésie.* As is true for Mont Albert (Ch. 9), the park's website is a good place to look for information regarding access and seasonal closures. The park is well traveled in all seasons, and has numerous and varied backcountry and front-country accommodations. In many ways, this region feels as developed as the other ranges south of the Saint Lawrence and

---

3   Arthur Stanley Pease spent some years in the shadow of his mentor, Merritt Lyndon Fernald. But Pease was a great botanist and explorer in his own right. His *Vascular Flora of Coos County, New Hampshire* (1924) is a classic New England alpine botany text. Fernald ascribed Pease's name to *Salix peasei,* a hybrid alpine willow known from King Ravine on Mount Adams, New Hampshire.

10-7 Lac aux Americains, *located on the floor of a great cirque on Mont Xalibu, is named for Harvard botanists M.L. Fernald and A.S. Pease. (MJ)*

is as heavily regulated as Katahdin (Ch. 8). Though the trail network is sparser than in the Presidential Range (Ch. 7), trails are heavily traveled, well maintained, and more "improved" than in any of the other major alpine ranges profiled in this book with the exception of those within Gros Morne National Park (Ch. 14). Routes 14 and 16 circumnavigate the entire massif (to the north and south, respectively). Both roads originate at Route 299, the major north-south route between the Monts McGerrigle and Mont Albert. Route 14 provides good access to the handful of alpine peaks that are located outside of the *Parc national de la Gaspésie*. Route 16 provides access to the trailhead for Mont Jacques-Cartier itself. Route 160, a spur-road from Route 16, provides access to the trailheads for Mont Xalibu and Mont Richardson—both popular climbs.

The entire *Parc national de la Gaspésie*, including the alpine portions from Mont Logan (Ch. 9), across Mont Albert, to Mont Jacques-Cartier, may be traversed on foot, following the *Sentier International des Appalaches*—International Appalachian Trail (SIA-IAT). This requires about five days and encompasses about 100 km (60 mi). This route is frequently used by skiers in winter. The traverse may be segmented into western and eastern sections, roughly demarcated at Route 299. The eastern traverse, then, becomes a traverse solely of the Monts McGerrigle massif, from Mont Xalibu to Mont Jacques-Cartier. Halfway between these summits, it is possible to stay an overnight near treeline at *La Camarine* (French for "crowberry," which are abundant) campsite, or at *Le Tétras* hut. Access to certain alpine areas (such as the summit region of Mont Jacques-Cartier), is heavily regulated by the *Parc national de la Gaspésie* in an effort to reduce the stress experienced by the few hundred remaining caribou.

Geology

The Monts McGerrigle are named for a geologist, Harold William McGerrigle.[4] Several of the prominent summits in the western Chic-Chocs, such as Mont Logan,[5] are named for geologists as well. The McGerrigle massif is composed primarily of a plutonic mass of granite and granite-like rock, which intruded into overlying sedimentary crustal rocks during Devo-

---

4   Harold William McGerrigle lived from 1904 to 1970.
5   William Edmond Logan lived from 1798 to 1875 (Commission de toponymie 2011)

nian period (ca. 350 million years ago).[6] The granitic intrusions occurred at the contact between Siluro-Devonian sedimentary rocks to the north and older (Cambrian/Ordovician) rocks to the south. The calcareous bedrock adjacent to the northern end of massif (such as the alkaline outcrops at Mont Auclair and Mont de la Table), apparently originate from seafloor sediments deposited and uplifted during the Silurian and Devonian periods, around 410 mya.[7]

10-8 During the Laurentide glaciation, a local ice cap formed atop Mont Jacques-Cartier. A lense of ice may have protected the summit felsenmeer from the overlying ice cap. (MJ)

Evidence indicates that the Monts Mc-Gerrigle supported a local ice cap during the last glacial maximum, and shallow cirques and glacial valleys adorn the periphery of the massif. Glacial ice from the Monts McGerrigle flowed north and west, toward the Gulf of St. Lawrence, joining ice flows from Mont Albert and the western Chic-Chocs. Most of the glacial features in the Monts McGerrigle are the result of alpine glaciation (10-8). The Laurentide Ice Sheet itself was apparently excluded from the Monts Chic-Chocs and Monts McGerrigle by the presence of a local ice cap.[8] Erratic material and glacial striae are rare on the McGerrigle massif, possibly because the uplands were already protected before glaciation by the development of a "seal" of immobile ice.[9] Lenses of permafrost apparently underlie the highest elevations of Mont Jacques-Cartier as well as surrounding portions of the Monts McGerrigle.[10]

10-9 Alpine cliffs, such as this one on Mont Xalibu, support two alpine species commonly found together: purple mountain heather and moss heather. (MJ)

Ecology

Vegetation — The alpine vegetation of the granitic tableland of the Monts McGerrigle is strongly similar to that of gneissic and granitic Appalachian mountains farther south, such as the Presidential Range and Katahdin.[11] The dominant tableland flora thus consists of familiar alpine species such as mountain firmoss (*Huperzia appressa*), Bigelow's sedge (*Carex bigelowii*), bearberry willow (*Salix uva-ursi*), alpine bilberry (*Vaccinium uliginosum*), al-

---

6   Boucher (2000)
7   Stevens (1989); Samson and Williams-Jones (1991)
8   Olejczyk and Gray (2007)
9   Olejczyk and Gray (2007)
10   Gray and Brown (1979)
11   Fernald (1907)

10-10 Mont Jacques-Cartier's alpine snowfields persist until July or August in some years. These areas provide habitat for many rare alpine plants, including alpine lady fern (bottom). (MJ and LJ)

10-11 Subalpine white spruce forest is dominant near the treeline ecotone on Mont Jacques-Cartier, but is rare farther south. (MJ)

pine bearberry (*Arctous alpina*), alpine azalea (*Loiseleuria procumbens*), and diapensia (*Diapensia lapponica*). In gullies and cliff faces, moss heather (*Harrimanella hypnoides*) and purple mountain heather (*Phyllodoce caerulea*) are common (10-9). In addition, several species occur on Mont Jacques-Cartier that do not occur on either of the southern ranges, including alpine ladyfern (*Athyrium alpestre*) (10-10).[12]

The higher valleys on the alpine tableland—where glacial and fluvial action have eroded the surface to elevations below about 1000 m—now harbor wooded mosaics of bogs and subalpine white spruce forest, a community type not found in New England (10-11), where white spruce is restricted to the U.S.–Canada border region. This assemblage of high-elevation white spruce does not occur farther south and is more typical of the relict forests of the Monts Groulx and Monts Otish (Ch. 18 and 19).

Fauna — The tundra of the Monts McGerrigle and Monts Chic-Chocs is so extensive that until recently the region supported a decidedly subarctic-boreal fauna, including timber wolves (*Canis lupus*), wolverine (*Gulo gulo*), marten (*Martes americana*), lynx (*Lynx canadensis*), and woodland caribou. Of these, only marten, caribou, and lynx persist. Woodland caribou once ranged throughout Atlantic Canada and northern Maine, but the only remnant of the South Shore population resides on the Monts McGerrigle, Mont Albert, and Mont Logan in the *Parc national de la Gaspésie*.[13] Woodland caribou make regular use of the subalpine spruce forest. Numerous caribou trails lead through the forest to treeline, and extensively disturbed wallow pits may be found in the stunted forest (10-12). Caribou calve above treeline on Mont Jacques-Cartier and are fond of late-summer snowfields (10-13). Although the persistence of caribou alone makes the Gaspésie peaks noteworthy, the wolf and wolverine are no longer present anywhere south of the Gulf of St. Lawrence. Instead, the region has given itself over to eastern coyote, which may have contributed to a decline in the caribou herd from as many as 700 to 1500 individuals in the early 1950s to about 140 in 2001.[14] Coyotes occur throughout

---

12   e.g., *Phegopteris alpestris, Asplenium cyclosrum, Cerastium cerastioides, Petasites vitifolia* and *Petasites trigonophylla* (Fernald 1907)

13   Of these areas, woodland caribou are most abundant on the Jacques-Cartier massif.

14   Boisjoly (2006) provides a thorough summary of the caribou collapse and coyote colonization.

the *Parc national* but remain more common in logged, fragmented and suburban areas where prey availability is higher. As more information becomes available about the distribution and habitat requirements of wolverine, it is becoming clear that the McGerrigle massif may have provided the best habitat for this species south of the Gulf of St. Lawrence. The Monts McGerrigle may even have sourced animals south to Katahdin, Washington, and other alpine massifs in New England, which could account for a handful of northeastern wolverine sightings leading up to the mid-19th century.[15]

The high tundra of the Monts McGerrigle, like that of Mont Albert, provides breeding habitat for numerous arctic and alpine birds. American pipit (*Anthus rubescens*) occurs on both mountains—as it does on Mount Washington and Katahdin—in a location uniquely isolated from its primary range on the arctic tundra. Snowy owls (*Bubo scandiacus*) are occasionally reported, but the Monts McGerrigle do not provide habitat for either rock or willow ptarmigan (*Lagopus muta* and *L. lagopus*). White-throated sparrow (*Zonotrichia albicollis*) and slate-colored junco (*Junco hyemalis*) are abundant everywhere at treeline (10-14).

At least five species of frog occur within *Parc national de la Gaspésie*. Of these, the

10-12 Woodland caribou form wallow pits and trails in subalpine forests. (MJ)

10-13 Woodland caribou may be observed throughout the Monts McGerrigle, where they often congregate on late-lying summer snow. (MJ)

10-14 White-throated sparrow (left) and dark-eyed junco (right) are abundant at treeline on Mont Jacques-Cartier, as they are on the Presidentials and Katahdin. (MJ).

15   Wolverine ecology is discussed in greater depth in Chapter 11, but the primary references for this material are Aubry et al. (2007) and Copeland and Yates (2008).

American toad is abundant at all elevations, giving way on the alpine tundra from the typical form to a brightly colored variety that is reminiscent of the Hudson Bay toad, a northern form distributed primarily north of the St. Lawrence. Whether the Hudson Bay toad is a valid form, and whether it occurs south of the Gulf of St. Lawrence, is a mystery waiting for the right researcher. Mink frogs (*Lithobates septentrionalis*) occur in high-elevation ponds to at least 1000 m (3300') in elevation (10-15).

10-15 At least five species of frog occur in the vicinity of Mont Jacques-Cartier, and two are especially notable near and above the treeline: American toads (top) and mink frogs (bottom). This American toad was photographed above treeline on Jacques-Cartier and looks suspiciously like the Hudson Bay toads discussed in Chapter 6.

Other Alpine Sites South
of the St. Lawrence

11

Mount Lincoln from Mount Lafayette, Franconia Ridge, New Hampshire. (MJ)

Map 11-1. Other alpine sites of the Appalachian Mountains south of the Saint Lawrence River.

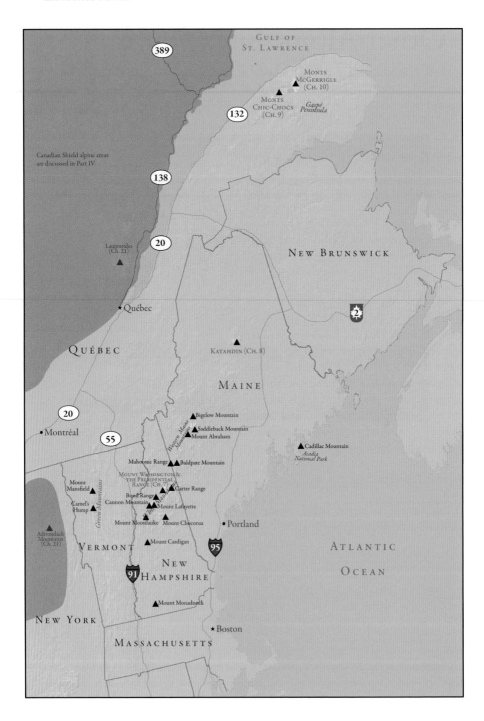

Other Alpine Sites South of the St. Lawrence

Green Mountains, Vermont and Québec

The sublime Green Mountains rise from the eastern shore of Lake Champlain, within clear sight of the High Peaks of the Adirondacks, with which they share almost no geologic connection. Although they are generally lower than the High Peaks, the Green Mountains of Vermont and Québec support a larger combined alpine ecosystem.[1] Most of the Green Mountain alpine, more than 80 hectares (200 acres), occurs atop Mount Mansfield (1340 m / 4395'), the tallest mountain in Vermont (11-1). A much smaller, but starkly impressive, outcrop of tundra survives on Camels Hump, about 24 km (14 mi) south (11-2). An even smaller alpine station survives atop Mount Abraham, just south of the Sugarbush ski resort. Mount Mansfield is best accessed on foot from the north via the Long Trail, which leaves the west side of VT SR 108 near Smugglers Notch, though there is a 4.5-mile toll road that reaches to 3850', just below Mansfield's "nose" (the mountain is often described as profile of a person on his back, the summit being the "chin"). The metamorphosed rock of Mount Mansfield supports a variety of alpine species known from the White Mountains to the east, including Bigelow's sedge (*Carex bigelowii*), alpine bilberry (*Vaccinium uliginosum*), mountain cranberry (*V. vitis-idaea*), black crowberry (*Empetrum nigrum*), diapensia (*Diapensia lapponica*), and Boott's rattlesnake-root (*Nabalus boottii*). Several alpine species have been extirpated through over-collection, including alpine bistort (*Polygonum viviparum*). Camels Hump's four alpine hectares (10 acres) support a smaller subset of these. In addition to these typical mountaintop sites, the Green Mountains encompass several interesting cliffs in Smugglers Notch near the northeast end of Mount Mansfield (11-3), Willoughby Cliffs above Lake Willoughby on Mount Pisgah, and Hazen Notch on Mount Haystack. The calcareous cliffs in the region support yellow, white, and purple mountain saxifrages (*Saxifraga aizoides, S.*

11-1 More than 80 hectares (200 acres) of alpine and subalpine habitats occur atop Mount Mansfield, the tallest mountain in Vermont's Green Mountains. (MJ)

11-2 A small outpost of alpine tundra survives atop Camels Hump, a distinctive Vermont mountain visible from many points in the Adirondacks, Green Mountains, and White Mountains. (MJ)

11-3 The narrow canyon of Smugglers Notch, on the northern end of Mount Mansfield, supports numerous limestone-loving plants, such as saxifrages. (MJ)

---

1   Klyza and Trombulak (1999). The Green Mountains extend into Québec, where the Sutton Range is a conservation target.

paniculata, and *S. oppositifolia*, respectively). In fact, Willoughby Cliffs provides the easiest place in New England to see these unusual saxifrages, which are widespread and locally common in the north.

Mountains in Québec associated with the Green Mountains include Mont Mégantic, Monteregian Hills, and Monts Sutton.

11-4 Famous Mount Monadnock is visible from the Berkshire and White Mountains. Its stony summit was twice burned in the 19th century, and subsequent erosion left it in a quasi-alpine state. (MJ)

Monadnock, Cardigan, and Chocorua, New Hampshire

11-5 Less well known than Mount Monadnock, perhaps, but no less spectacular, Mount Cardigan was similarly left bare by a severe fire in 1855. (MJ)

New England has a tradition of celebrating all of its bare-rock peaks, whether they are naturally so or not. Several famous peaks in New Hampshire became bare following fires and subsequent erosion. Three of these mountains are especially famous and in many ways—aesthetically and ecologically—similar to naturally treeless alpine summits in New England: Mounts Monadnock, Cardigan, and Chocorua.[2] Though not arctic-alpine in the sense described throughout much of this book, these mountains provide examples of a **pyric** (fire-induced) or **anthropogenic** (human-induced) timberlines. The rocky summits are mostly dominated by boreal plants, but have been colonized sparingly by some noteworthy alpine species. Any thorough book covering New England alpine ecosystems should—and for the most part, do[3]—expand upon the unique ecology of these noteworthy mountains.

Mount Monadnock's schist summit[4] was rendered treeless by fires in 1800 and 1820, and lost much of its summit soil during an intervening hurricane in 1815. Today, Monadnock is a world-famous and oft-climbed subalpine peak, and its summit supports the southernmost New England location for several alpine plants which apparently colonized the summit following the disturbance, including mountain cranberry and Greenland swtichwort (*Minuartia groenlandica*).[5] *Monadnock* has since come into general usage to describe an isolated mountain (11-4) in the United States and parts of Canada; elsewhere the term *inselberg* is used.

Mount Cardigan, overlooking Newfound Lake and the southern White Mountains, was similarly stripped of its subalpine summit forest by a fire in 1855 (11-5). Today, Cardigan's

---

2   See Whitney and Moeller (1982)

3   For example, see Harris et al. (1964)—the definitive reference to the flora of New England's alpine mountains.

4   Monadnock's summit is composed of 400 million year-old Devonian schist and is part of the Littleton formation that also appears in the Presidential Range and elsewhere in the White Mountains.

5   Baldwin (1977)

rocky ledges of Kinsman quartz monzonite (an igneous rock related to granite, found throughout the Franconia Notch region) supports several alpine and boreal species, including an unusual population of the capitate sedge *Carex capitata*, which grows below the fire tower at the summit (11-6).

Dignified Mount Chocorua, an isolated monadnock of Conway granite, is the easternmost peak in the Sandwich Range of the White Mountains and an interesting mountain in its own right. Chocorua lies closer to the main ridges and massifs of the White Mountains, and so its subalpine vegetation is less disjunct than that found on either Monadnock or Cardigan (11-7). The mountain is the central location in the 19th-century legend of Chocorua's curse. "Wolves fatten on your bones!" Chocorua screamed as he leapt to his death from the summit cliffs.[6]

Many other mountains, ridgelines, and spurs break through the treeline as a result of past disturbance. Many of these are mentioned in this chapter, but many others are left unnoted. Some additional fire scarred peaks worthy of exploration include Mount Carrigain, Whitewall Mountain, and Kearsarge North (11-8).

Franconia Ridge, New Hampshire

Franconia Ridge—after Mount Washington, the other great alpine ridge of the White Mountains—lies 30 km (16 mi) west-south-west of the Presidential Range. The treeless portion of Franconia runs about 4 km (2.5 miles) from North Lafayette in the north to Little Haystack in the south, passing over Mounts Lafayette and Lincoln along the way (11-9). At 1600 m (5,249') Mount Lafayette is the third highest mountain in New England after the summits of the Presidential Range and Katahdin. The most popular route up Franconia Ridge is the 14.3 km (8.9 mi) Fall-

11-6 *Mount Cardigan's summit ledges support a very unusual population of capitate sedge, a northern species rare even on Mount Washington. (MJ)*

11-7 *Mount Chocorua is the easternmost peak in the Sandwich Range, the southernmost massif in the White Mountain National Forest. (MJ)*

11-8 *The long view from the fire tower on Kearsarge North near Intervale, New Hampshire, takes in many White Mountain ranges. (MJ)*

6   Mudge (1992)

11-9 Franconia Ridge runs above treeline for about 4 km (2.4 mi) and is probably worthy of its own chapter. (MJ)

11-10 The AMC's Greenleaf Hut, pictured in February 2000, on the shoulder of Mount Lafayette above Eagle Lake, has facilitated access to the Franconia alpine areas since the 1920s. (MJ)

11-11 The eastern view from Mount Lafayette is one of the most celebrated wilderness views in the East, taking in all of the Pemigewasset Wilderness from the Twin Range to Mount Carrigain. Fortunately, no trails lead off Franconia's eastern slope — providing welcome balance to the developed western slope. (MJ)

ing Waters-Old Bridle Path Loop Trail, which begins in the Lafayette Place parking lot east of Franconia Notch Parkway / SR-3. The Loop may be completed in either direction, but essentially it follows the Falling Waters Trail to the summit of Little Haystack Mountain and from there continues north, following the Appalachian Trail along the ridge all the way to the summit of Mount Lafayette. The Loop descends via the Greenleaf Trail, which winds past the Appalachian Mountain Club's (AMC) Greenleaf Hut (11-10) above the boggy shores of Eagle Lake where the trail becomes the Old Bridal Path and thence returns to Lafayette Place. On a clear day, the ridge offers sweeping views east into the Pemigewasset Wilderness, including the Twin and Bond Ranges, and across SR-302 to the Presidential Range (11-11). The view to the west takes in Franconia Notch itself and the Cannon Cliffs (11-12). The Franconia ridgeline is exposed and, like any alpine summit, may be dangerous in bad weather. Being equidistant from Boston, Massachusetts, and Sherbrooke, Québec (about 2.5 hr from each), Franconia Ridge is the most accessible large expanse of alpine tundra in the East, and it is often among the most crowded backcountry locations in the White Mountain National Forest. At times, the line of cars spilling out from the parking areas extends several miles in each direction along the Franconia Notch Parkway. Several decades ago, the number of people traversing the ridge caused the bare ground of the trail to expand to a highway almost as wide as the Parkway below. Through the tireless efforts of Laura and Guy Waterman, along with AMC, U.S. Forest Service, and many alpine stewards, the trail was narrowed and off-trail trampling has decreased. Noticeably and fortunately, the alpine vegetation has begun to recover.

Franconia Ridge is granitic, and its 92 hectares (230 acres) of alpine habitats support many of the alpine plants known from the Presidentials, including diapensia, alpine bilberry, bearberry willow (*Salix uva-ursi*), mountain avens (*Geum peckii*), mountain firmoss (*Huperzia appressa*), and purple mountain heather (*Phyllodoce caerulea*). In addition, several plants rare on the Presidentials survive on Franconia, including the White Mountain endemic Robbin's

11-12 Cannon Mountain forms the western slope of Franconia Notch. Alpine azalea, white mountain-saxifrage, common butterwort, and other alpine plants have been documented in recent years. (MJ)

11-13 A few locations on Franconia Ridge support enigmatic populations of Robbins' dwarf cinquefoil, known elsewhere in the world only on Mount Washington. (MJ)

dwarf mountain cinquefoil (*Potentilla robbinsiana*) and alpine bittercress (*Cardamine bellidifolia*) (11-13). Eagle Lake, on the shoulder of Mount Lafayette near Greenleaf Hut, supports one of the region's rare high-elevation amphibian communities. On a warm nights in early June the subalpine forest comes alive with the haunting choruses of wood frogs and spring peepers, up to a month after their breeding season ends in the valley.

11-14 The wild Bond Range as viewed from Franconia Ridge. From left, the summits pictured are Guyot, West Bond and Bond, and Bondcliff. (MJ)

Bond Range, New Hampshire

The wild Bond Range, as part of the larger Twin Range, forms the central part of the Pemigewasset Wilderness, which at 18,540 hectares (45,818 acres) is the largest federally designated roadless area in the White Mountain National Forest (and in New England). Mount Bond, the high point of the range, supports a small, degraded alpine habitat, as does West Bond. The major alpine portion lies on the subsidiary ridge Bondcliff, which also provides some of the best views in the entire White Mountain region. Mount Bond, West

11-15 Carter Notch is a rugged subalpine pass between Wildcat Mountain and Carter Dome. The cliffs and talus fields provide habitat for numerous interesting alpine species. (MJ)

Bond, and Bondcliff are accessed via the Bondcliff Trail, which leaves the Appalachian Trail at Mount Guyot (11-14). The junction at Mount Guyot may be accessed via the Twinway/Appalachian Trail from AMC's Galehead Hut, which itself is best accessed via the Gale River Trail from Gale River Road, which leaves US-3 south of Twin Mountain, New Hampshire. At least 29 alpine vascular plant species occur on the Bondcliff ridgeline, about half that found on the much larger and taller Mount Lafayette.[7]

Carter Range, New Hampshire

11-16 American marten, and their sign, are particularly visible in the Carter Range. Pictured here is an adult marten photographed near Carter Notch in the winter of 2011. (MJ)

Carter Dome (1473 m / 4832') is the highest point in the Carter Range and one of the highest mountains in the White Mountains outside of the Presidential Range and Franconia Ridge. Although dwarfed by Mount Washington, which is located only 10 km (6 mi) due west across NH SR-16, Carter Dome's crags, spurs, flanks and talus fields harbor alpine plants. The actual summit of Carter Dome is covered by scrubby balsam fir (*Abies balsamea*), but Mount Hight (1425 m / 4675'), an alpine spur that reaches north from Carter Dome toward Zeta Pass and the main summits of the Carter Range, supports several arctic-alpine species. Other areas of alpine vegetation are found along the Rainbow Ridge and in Carter Notch, which conventionally divides the Carter Range from the Wildcat Ridge (11-15). The entire range, including Wildcat Mountain, may be accessed via the Appalachian Trail from Pinkham Notch, or through Carter Notch via the 19 Mile Brook Trail. Carter Notch itself, at about 1000 m (3300') is below the treeline and not alpine in the classic sense, but its periglacial rockslides and steep cliffs support populations of silver-

7   Cogbill (1996)

ling (*Paronychia argyrocoma*), diapensia, and a highly visible population of American marten (*Martes americana*) (11-16).

Mahoosuc Range, Maine

The Mahoosuc Range straddles the New Hampshire–Maine border south of Lake Umbagog and north of the Carter Range. The range culminates at Old Speck (1270 m / 4167'), a forested mountain with a cleared summit and a lookout platform. Most of the alpine tundra is found on the "west peak" of Goose Eye Mountain (1152 m / 3780'), as well as on its subsidiary East and North Peaks (11-17). Another large occurrence of alpine vegetation may be found on Baldpate Mountain (1116 m / 3662'). Goose Eye Mountain may be reached via the Success Trail and the Appalachian Trail, which leave Success Pond Road out of Berlin, New Hampshire. Baldpate Mountain may be accessed by traveling north on the Appalachian Trail from Grafton Notch, which leaves the east side of ME SR-26. Old Speck is easily reached by traveling south on the Appalachian Trail from Grafton Notch. A good example of a subalpine bog may be found on the East Peak of Baldpate.

11-17 The Mahoosuc Range extends north and east from the Carter Range. North Peak of Goose Eye Mountain from the East Peak with Old Speck in the distance, Mahoosuc Range, New Hampshire. (DW)

11-18 Most of the interesting alpine territory in the Mahoosuc Range may be found on Goose Eye Mountain and in the Mahoosuc bogs. Main Summit of Goose Eye Mountain, Mahoosuc Range, New Hampshire. (DW)

Bakeapple, or cloudberry (*Rubus chamaemorus*), a characteristic single-fruited raspberry of the far north, occurs in the subalpine bogs of the Mahoosucs. The "Eyebrow" on Old Speck supports a population of the subalpine, talus-dwelling silverling (11-18).

Western Mountains, Maine

Various authors, as well as the Maine legislature, have tried over the years to adequately describe the various mountains and ridgelines of interior Maine, including a half-successful attempt to name all of the major mountains of Maine the "Longfellow Mountains" after Henry Wadsworth Longfellow. In any case, the alpine mountains of Maine fall nicely into three groups: the Mahoosuc Range (which are essentially the northern extension of New Hampshire's White Mountains and are described earlier); Katahdin and surrounding peaks in Baxter State Park (Ch. 8), and a cluster of four large mountains south of Flagstaff Lake and southeast of Rangeley Lake. These include Saddleback Mountain (1256 m / 4120'), which extends above treeline as a broad ridge extending north to the Saddleback Horn; Mount Abraham (1231 m / 4040'), the most remote and least-traveled of the four mountains; Sugarloaf Mountain (1291 m / 4237'); and Bigelow Mountain (1263 m / 4145'). Saddleback and Sugarloaf have been developed as major eponymous ski areas. In the 1960s and 1970s,

11-19 Bigelow Mountain in Dead River Township, Maine, includes West Peak (top) and Avery Peak (bottom). (MJ)

11-20 Recent research indicates that the shrubs on Bigelow Mountain are advancing upslope, possibly in response to moderating climate factors. Flagstaff Lake and the Dead River Valley dominate the view north. (MJ)

Bigelow Mountain was also slated to become the site of a world-class ski resort, and even factored in to the town of Flagstaff's bid to host the winter Olympics. Instead, fortunately, the range was set aside with the passage of the Bigelow Preserve Act of 1976. Bigelow and Abraham are clearly visible from the peak of Sugarloaf. Bigelow Mountain comprises the alpine summits West Peak (the high point) and Avery Peak (11-19). From Bigelow, the view north encompasses Flagstaff Lake, the Dead River valley, and the smaller summits of the upper Kennebec watershed (11-20). Portions of Saddleback, Sugarloaf, and Bigelow are traversed by the Appalachian Trail (AT). Sugarloaf is reached from the AT by a 1 km (0.6 mi) spur trail. Mount Abraham may be accessed from the AT along the 3 km (2 mi) Mount Abraham Side Trail. The four summits may be traversed in 2 to 4 days, but all are worth many days of exploration. Saddleback is formed from granodiorite as part of the Redington pluton, which formed approximately 410 million years ago. These four peaks support a heath-dominated subalpine vegetation including sheep laurel (*Kalmia angustifolia*), pale bog laurel (*Kalmia polifolia*), mountain cranberry, and Labrador tea (*Rhododendron groenlandicum*) (11-21).

Mount Desert, Maine

One hundred and sixty kilometers (100 mi) northeast of Portland, Maine, windswept Mount Desert Island extends into the Gulf of Maine like rocky lungs breathing life into Maine's Downeast coast. The highlands of Mount Desert Island are perhaps the most southerly place that subalpine habitats come into breathtaking juxtaposition with the cold Atlantic. Although not quite high enough or far enough north to host the arctic-alpine communities discussed throughout much of this book, the mountains of Mount Desert and Acadia National Park have their own flavor of mountain wilderness worth exploring. Although the park can feel packed to the gills on summer weekends, a visit off-season or a sunrise hike can bring a certain inner peace possible only here where the mountains meet the ocean and the quiet solitude of a treeless peak are only a few quick miles from the comfort of a hot breakfast (or cold beer) in Bar Harbor.

The tallest, and most famous, among the bald mountaintops of Acadia is Cadillac (466 m / 1530′). Like Whiteface in New York, Mansfield in Vermont, and Washington in New Hampshire, the wildness of Cadillac was long ago[8] replaced by a road, cars, and throngs of people. Cadillac is the highest point on the Atlantic coast (within 25 miles of the coast) north of the Yucatán Peninsula. The roadless, treeless summits of Acadia National Park (e.g., Dorr, Champlain, Sargent, and Penobscot) can be accessed via an extensive network of trails connecting the peaks with a series of Carriage Roads, the National Park Loop Road, and ME SR 3 and SR 198. Primarily composed of granite, the *roches moutonnées* and U-shaped valleys present throughout the park give evidence of the important role that glaciers had in shaping the island. Severe fires in October 1947 also had an important role in shaping the landscape, destroying much of the spruce–fir forest on the eastern portion of the island. Most of the treeless peaks in Acadia are what the state of Maine characterizes as the "Rocky Summit Heath" community type, which, although not truly alpine, features mountain cranberry, three toothed cinquefoil (*Sibbaldia tridentata*), and black crowberry.

11-21 Saddleback Mountain in western Maine supports a small but memorable area of open alpine tundra. Many subalpine species, such as bog laurel, are abundant near the treeline. (JM)

11-22 Many of Acadia National Park's rocky summits support noteworthy subalpine species. Many of the hills, such as Cadillac Mountain, bear the scars of fires in the 19th century. (JH)

Common bearberry (*Arctostaphylos uva-ursi*), alpine bilberry, and Greenland switchwort (*Minuartia groenlandica*) can also be found on several of the mountaintops (11-22).

Cape Breton Highlands, Nova Scotia

Cape Breton Island forms the northern part of the province of Nova Scotia; the Cape Breton Highlands extend across portions of Inverness and Victoria Counties and hug the ocean at the far northeastern edge of the island. Geologically, the Cape Breton Highlands are a southern extension of the escarpment that forms the Table Mountains of southwestern

8   The Cadillac Mountain Road opened in 1931.

Newfoundland (Ch. 12). The Highlands are similar in height and overall extent to the Table Mountains, rising to a high point at indistinct White Hill (532 m / 1745'). The entire massif was protected as Cape Breton Highlands National Park in 1936. Although the scenic, ocean-front Cabot Trail drive is very popular, the obscure summit of White Hill is not. It is partly accessible via the Lakes of the Islands Trail, a 2-day, 42 km (26 mi) round-trip. The White Hill summit is subalpine, clad in black spruce (*Picea mariana*) and heaths. Most of the Highlands' plateau is covered by low, boggy, boreal scrub and heath, and nowhere do the Highlands give way to characteristic alpine plant communities of the type described throughout much of this book. However, several interesting, alpine plant species reach across the Cabot Strait from Newfoundland to occur in scattered localities in the northwestern Highlands. A handful of exposed ledges support locally rare alpine species like bearberry willow (*Salix uva-ursi*), diapensia, purple mountain heather, and Lapland rosebay. A series of steep-walled canyons, river valleys, and waterfalls support alpine species typically found in alpine seepages, wet rocks, and cirques, such as mountain sorrel (*Oxyria digyna*) and alpine timothy (*Phleum alpinum*);[9] and calcareous ledges in Corney Brook Gorge provide habitat for common butterwort (*Pinguicula vulgaris*), rare saxifrages and willows.[10] Most of the interesting alpine plants are found in the gorges along the northern face of the Highlands, including Corney Brook Gorge (where purple- and yellow-mountain saxifrage, net-veined willow [*Salix reticulata*], and Lapland rosebay occur on calcareous ledges), Lockhart Brook near the Salmon River (where diapensia and mountain heath have been observed), waterfalls in the upper Cheticamp River Gorge, and Big Southwest Brook (mountain firmoss [*Huperzia appressa*] and multi-rayed goldenrod [*Solidago multiradiata*]).[11] Rock outcrops and talus in the Highlands provide habitat for the Gaspé shrew (*Sorex gaspensis*), which is limited to high-elevation habitats in Gaspésie (Ch. 10), New Brunswick, and Cape Breton Island.

---

9   Roland and Zinck (1998)
10   *Salix vestita* occurs at its southeasternmost station, and *S. reticulata* is also present.
11   *Solidago multiradiata* (Blaney and Mazerolle 2010)

Southern Long Range Mountains
NEWFOUNDLAND
12

Little Codroy Pond, Southern Long Range Mountains, Newfoundland. (IATNL)

Map 12-1. Table Mountain near Channel Port aux Basques, Southern Long Range Mountains, Newfoundland.

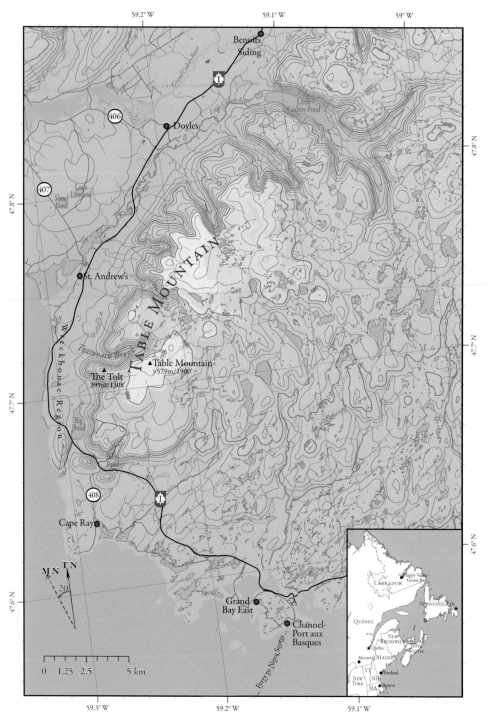

Southern Long Range Mountains,
NEWFOUNDLAND

12-1 Towering Table Mountain greets visitors arriving on Newfoundland by ferry at Port aux Basques. Pictured: the view from the coast at Wreckhouse, a stretch famous for its terrible winds. (IATNL)

It's probably not possible to explore all of alpine Newfoundland in one life, and if afforded only one trip there, proceed directly to the Bay of Islands (Ch. 13) or the Gros Morne/Northern Long Range region (Ch. 14 and 15) (arguably Newfoundland's most spectacular mountains). But if you do choose to speed north to those higher mountains, hope for thick fog as you pass through the Wreckhouse region. Otherwise, you'll be tempted to explore towering Table Mountain, the dramatic gateway to the Southern Long Range (12-1). The diverse mountains of the Southern Long Range, and adjacent Buchans Plateau, are distinctly alpine. Their western façades are decorated by lofty, hanging cirques. The high plateaus encompass critical calving areas for the La Poile caribou herd (12-2) and provide habitat for the southernmost populations of arctic hare (*Lepus arcticus*) and rock ptarmigan (*Lagopus muta*) in North America. Furthermore, the coast-facing crags support diverse assemblages of arctic-alpine and northern coastal vegetation. Still, much of this region is a bewildering mosaic of poorly drained strangmoor,[1] fens, and spruce–fir woodland, broken by innumerable rocky alpine ridgelines, knobs, and outcrops (12-3). Within this broad area of about 9000 km² (3200 mi²), numerous large massifs account for most of the relatively rare alpine elements. We focus primarily on two of these massifs: Table Mountain near Wreckhouse (north of Port aux Basques), and the Annieopsquotch Mountains on the road to Burgeo.

Table Mountain is perhaps the most scenic, accessible, and generally spectacular range in

12-2 The high plateaus surrounding the La Poile River, on the eastern slope of the Long Range, provide critical breeding habitat for Newfoundland caribou. (NR and IATNL)

---

1   *Strangmoor* is German for "string bog."

12-3 Much of the interior southern Long Range is a bewildering mosaic of bogs, fens, and tuckamore, broken by countless small summits. (NR)

12-4 A series of alpine cirques on the western face of Table Mountain make it the most visually stunning of the various ranges in the southern Long Range. Trainvain Valley, Newfoundland. (IATNL)

12-5 The Annieopsquotch Mountains, on the road to Burgeo, are remote and isolated. (MJ)

the Southern Long Range, owing to a series of great alpine cirques and hanging valleys that indent the western escarpment, as well as the mountain's close proximity to the blue-gray Gulf of St. Lawrence (12-4). Conversely, the Annieopsquotch Mountains are lumpish and wild, dull-edged and rounded, more typical of the topography of interior Newfoundland, lacking prominent cirques and other features associated with alpine glaciation. The blue ocean lies at the foot of the Table Mountains, the interior position of the Annieopsquotch Mountains lends them a remote and isolated feeling (12-5).

In 1979, the Annieopsquotch Mountains were connected to the Trans-Canada Highway by a long spur-road that wove down to Burgeo, one-time home of the writer, Farley Mowat. Before then, the Annieopsquotch were a little-known and seldom-traveled, far-flung mountain range on the edge of the Long Range Mountains. But by chance, the new road, Route 480, called the "Caribou Trail," cut within a kilometer of treeline on Annieopsquotch's western flank, providing access to local hunters, snowmobilers, hikers, and the occasional adventurer not on a beeline for Gros Morne. *Annieopsquotch* reportedly means "terrible rocks," in Mi'kmaq, which was enough to pique our interest.

Lay of the Land

At the broad scale, alpine and subarctic elements occur throughout the Southern Long Range from Cape Ray (near Port aux Basques) at least as far north as Grand Lake, which lies due east of the prominent Bay of Islands. Table Mountain, mentioned earlier, is the southernmost extension of the Long Range in Newfoundland and the most spectacular of the many peaks and ranges in the Southern Long Range Mountains. Beginning near Cape Ray, the mountain rises abruptly from the coastal plain to the peak of Cook Stone and extends at most 28 km (17 mi) along Newfoundland's extreme southwestern coast. About a half-dozen massive cirques, replete with chaotic rockslides and talus fields, dissect the western front. The most scenic of these are Trainvain Brook, which runs westward through a subalpine valley below the distinctive promontory known locally as The Tolt, which itself is an easily identifiable landmark visible from the Trans-Canada Highway as well as the plateau of Table Mountain (12-6). At the northeast end of the massif is Little Codroy Pond, which lies at the bottom of a stunning and wild glacial valley (12-7).

The Annieopsquotch Mountains are a northeast-trending hulk that is sandwiched between the Lloyds River (a major fault) and Victoria Lake, just east of Highway 480, about 100 km (60 mi) northeast of Table Mountain. The valleys, lakes, and summits of the Annie-opsquotch are generally nameless (12-8). It is not a large range, at most about 16 km (10 mi) in length from the southern end at Route 480 and one quarter that in width.

Although we profile these ranges as representative of the rest of the region, there are many other alpine summits, crags, fellfields, and subalpine bogs within this large region. Some notable areas include the summits surrounding the La Poile River, north and east of La Poile Bay. East of the La Poile Highlands are the Blue Hills of Couteau (both are comprised primarily of granite and basalt). South of the Blue Hills, uplands surrounding Connoire Bay support several interesting alpine species. To the north of the Annieopsquotch Mountains, an upland area east of Grand Lake is known as Buchans Plateau. Many small peaks rise to stony, windswept alpine barrens.

Access

The southern escarpment of Table Mountain lies only fifteen minutes from the Port aux Basques ferry terminal, cresting at a dramatic granitic ridge about 2.5 km (1.5 mi) from the Trans-Canada Highway (Route 1). The most straightforward access is via the old tower road, which leaves the Trans-Canada Highway just west of the junction with Route 408 and proceeds directly to the treeline about 2 km (1.2 mi) from the highway. From treeline, a dirt track runs west to a Canadian Coast Guard Marine Communications station, and off-road vehicle (ORV) trails lead along the ridgeline and northeast toward Trainvain

12-6 *The Tolt (and its arête) on Table Mountain is an easily recognized landmark, visible from the coastal highways near Wreckhouse, Newfoundland. (IATNL)*

12-7 *Little Codroy Pond lies at the bottom of a stunning and wild glacial valley. (IATNL)*

12-8 *Most of the peaks and ridges of the Annieopsquotch Mountains are nameless. (MJ)*

12-9 *The deep cirques of Table Mountain make walking the ridge a fairly sinuous endeavor. (IATNL)*

Brook and Little Codroy Pond. The glacial valley of Little Codroy Pond is accessible via a 3 km (2 mi) trail that leaves from the east side of the Trans-Canada Highway approximately 5 km (3 miles) north of the community of Doyles. Halfway between Cape Ray and Little Codroy Pond the mountaintops can also be accessed via the International Appalachian Trail/Starlight Trail near the community of St. Andrews. There is also a gravel road that leads to the top of the mountains near South Branch. Once above treeline, walking is generally straightforward, although the deep indentations of the alpine cirques makes walking the ridgeline a sinuous venture (12-9).

The Annieopsquotch Mountains are best accessed from Route 480 just south of the Lloyd's River crossing. There are game trails and lightly tread footpaths that lead southeast into the mountains. Even if you go astray, or are unable to find a path to begin with, the worst scramble through **tuckamore** (Newfoundland's term for krummholz, or stunted subalpine conifer forest, see Ch. 5) may be straight forward or nightmarish. There are some steep cliffs facing the highway, but you can avoid these if you plan your path from the road. Once above treeline, as on Table Mountain, the terrain is mostly gentle and rolling, with some steep rocky outcrops and deeper subalpine valleys.

Much of the remaining area of the southern Long Range Mountains is difficult to access on foot, although the La Poile Highlands can be accessed from the Gulf of St. Lawrence at La Poile Bay near the outport town of La Poile, or overland from Route 480, and the Buchans Plateau may be accessed from the town of Buchans itself.

Geology

In a broad sense, the Southern Long Range Mountains are composed primarily of granitic rocks of Ordovician origin (around 500 million years ago (mya) in the early Paleozoic). Table Mountain—generally considered the southern terminus of the Long Range of Newfoundland—is composed primarily of Ordovician and Silurian granites, and is geologically similar to the Cape Breton Highlands of Nova Scotia, which are briefly discussed in Chapter 11.

12-10 *Most of the southern Long Range Mountains are composed of granite and gneiss. The gabbro and diabase of the Annieopsquotch Mountains, pictured here, is relatively uncommon. (MJ)*

The Table Mountain massif lies just west of the Cape Ray Fault, which marks the boundaries of the ancient supercontinents Laurentia (west) and Gondwana (east) (see Ch. 2). This fault also marks the boundary between Newfoundland's "Western" and "Central" geologic zones.

The Annieopsquotch Mountains also border this major boundary, and are composed primarily of mafic rocks such as gabbro and diabase (12-10).

12-11 The treeless plateaus of the Long Range Mountains support a variety of bogs, fens, marshes, and pools. (NR)

Ecology

Vegetation — The treeless plateaus of Table Mountain and the La Poile Highlands are poorly drained, and thus support a variety of subalpine bogs, fens, marshes, and pools (12-11). The bogs are dominated by common boreal heath plants such as early lowbush blueberry (*Vaccinium angustifolium*) and Labrador tea (*Rhododendron groenlandicum*). Stark yellow clumps of marsh-marigold (*Caltha palustris*), not typically found in alpine environments, are widely distributed in the bogs, with tufted clubrush (*Trichophorum cespitosum*), leatherleaf (*Chamaedaphne calyculata*), and Canada burnet or bottlebrush (*Sanguisorba canadensis*), which also grows around the ferry-yard at Port aux Basques and is a characteristic flower of Newfoundland's west coast wetlands. On well-drained sites in the plateau's interior, balsam fir (*Abies balsamea*) and green alder (*Alnus viridis*) form thickets of tuckamore, around and under which grow ubiquitous boreal herbs that are found commonly together: Canada mayflower (*Maianthemum canadense*), starflower (*Trientalis borealis*), and bunchberry (or "crackerberry" in Newfoundland, *Cornus canadensis*), which extend onto the drier tundra in many places. Mountain fly honeysuckle (*Lonicera villosa*) and shining rose (*Rosa nitida*) are frequent (12-12).

12-12 Shining rose (top) and mountain fly-honeysuckle (bottom) are common near the treeline ecotone. (IATNL and MJ)

12-13 Bearberry willow occurs on exposed summits throughout the southern Long Range, as it does here on Table Mountain. (LW)

The exposed, rocky ridgeline is carpeted with familiar and widespread alpine plants such as diapensia (*Diapensia lapponica*), alpine azalea (*Loiseleuria procumbens*), alpine bearberry (*Arctous alpina*), mountain cranberry ("partridgeberry" to Newfoundlanders, *Vaccinium vitis-idaea*), alpine bilberry (*V. uliginosum*), black crowberry ("blackberry" to Newfoundlanders, *Empetrum nigrum*), three-toothed cinquefoil (*Sibbaldia tridentata*), bearberry willow (*Salix uva-ursi*) and a variety of the three-leaved rattlesnake root (*Nabalus trifoliolatus*) (12-13).

The west-facing ravines, in particular, the headwalls of the various cirques, support the

12-14 Among the more common plants in the Annieopsquotch Mountains are (clockwise from top left) hoary rock moss and common juniper, cottongrass, pale bog laurel, pitcher plant, alpine bilberry, and leather leaf (MJ, LW, and CE)

arctic-alpine plants listed earlier as well as roseroot (*Rhodiola rosea*), northern beech fern (*Phegopteris connectilis*), mountain fir-moss (*Huperzia appressa*), common juniper (*Juniperus communis*), Chinese hemlock parsley (*Conioselinum chinense*), tall meadowrue (*Thalictrum polygamum*), as well as rarities such as the mountain sorrel (*Oxyria digyna*). Other alpine cirque- or snowbed-specialists may occur here, such as alpine cudweed

12-15 Black bears are important predators of caribou calves in the southern Long Range. (NR)

(*Omalotheca supina*) and sibbaldia (*Sibbaldia procumbens*), although snow doesn't persist as long in the southern Long Range Mountains as in Gros Morne National Park (Ch. 14) or the Highlands of St. John (Ch. 16).

Elegant sunburst lichen (*Xanthoria elegans*) occurs on the exposed west-facing cliffs and cirque headwalls. In protected hollows between exposed boulders, hoary rock moss (*Racomitrium lanuginosum*) is abundant.

The Annieopsquotch Mountains appear to be floristically different from Table Mountain, but have not been thoroughly surveyed or inventoried. Our cursory surveys there in 2010 revealed summit and ridgeline vegetation dominated by common juniper, leatherleaf, alpine azalea, alpine bearberry, common bearberry (*Arctostaphylos uva-ursi*), bog rosemary (*Andromeda polifolia*), alpine bilberry, mountain cranberry, sheep laurel (*Kalmia angustifolia*), Labrador tea, black crowberry, and three-toothed cinquefoil. Diapensia occurs on the most exposed sites. In protected thickets, green alder and balsam fir are the most common species. In poorly drained sites, tufted clubrush, Canada burnet, and pitcher plant

12-16 The southern Long Range Mountains provides the southernmost habitats for arctic hare anywhere in the world. Here, an adult is pictured in the Annieopsquotch Mountains. (CE)

12-17 Marshy areas and snowbeds throughout the southern Long Range Mountains show evidence of browsing by Newfoundland meadow voles. (CE)

(*Sarracenia purpurea*) are abundant, with scattered occurrences of royal fern (*Osmunda regalis*), round-leaved sundew (*Drosera rotundifolia*) and butterwort (*Pinguicula vulgaris*) (12-14). The Annieopsquotch Mountains, like most of the Southern Long Range Mountains, apparently lack the deep, pronounced alpine cirques that are common on Table Mountain; correspondingly, cliff and talus habitats and their associated plants appear to be restricted.

Fauna — Of the 29 species of terrestrial mammals on the island of Newfoundland, 15 are native, and of these, 9 are unique as endemic subspecies. The Table Mountain massif provides important upland habitat for the La Poile woodland caribou (*Rangifer tarandus*)

12-18 Arctic fritillary (left) and black-banded orange (right) are frequent in the southern Long Range in summer. (NC and CE)

12-19 Long-lipped tiger beetle is one of many species of beetle likely to be encountered at treeline in the southern Long Range. (MJ)

herd, which calves on the barren ground and muskeg that stretches east of the Table Mountain massif. Black bears (*Ursus americanus*) range above treeline on the massif, as do coyotes (*Canis latrans*) and red fox (*Vulpes vulpes*). Bear and coyotes are known to be important predators of woodland caribou in southwestern Newfoundland (12-15).[2] Arctic hare occur on the Table Mountain plateau and historically occupied the Long Range plateau east at least as far as La Poile, the Blue Hills of Couteau, and Connoire Bay (collectively, its southernmost natural occurrence anywhere in the world) (12-16).[3] Marshy areas and snowbeds throughout the range show evidence of winter activity by meadow vole (*Microtus pennsylvanicus*) (12-17), the only native small rodent on Newfoundland and an important prey item for Newfoundland marten (*Martes americana atrata*), which have been reduced to population centers around Grand and Little Grand Lakes and Red Indian Lake.

Table Mountain is the southernmost station of rock ptarmigan in North America. Introduced American toads (*Anaxyrus americanus*) and green frogs (*Lithobates clamitans*) occur to treeline, and the former probably breed in pools and marshes above treeline. The Annieopsquotch Mountains similarly provide habitat for woodland caribou, moose, and arctic hare, as well as rock ptarmigan.

Arctic-alpine fingernail clams (*Sphaerium nitidum*) occur in the alpine ponds of the Annieopsquotch, as also observed on other Newfoundland ranges such as Blow Me Down Mountain (Ch. 13). Numerous butterflies, including arctic fritillary (*Boloria chariclea*) and black-banded orange (*Epelis truncataria*), are common above treeline during the summer (12-18). Long-lipped tiger beetle (*Cicindella longilabris*), common in many Newfoundland alpine areas, is also abundant in the Annieopsquotch (12-19).

---

2   Rayl (2012)
3   Bergerud (1967)

Map 12-2. Annieopsquotch Mountains, Southern Long Range Mountains, New-foundland.

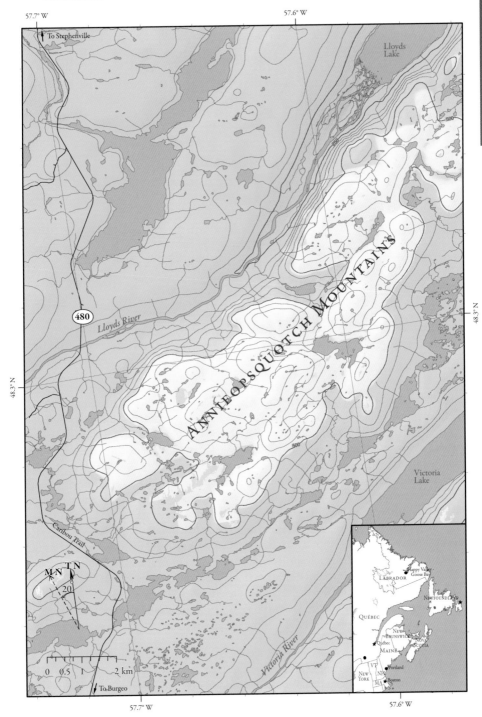

Bay of Islands
NEWFOUNDLAND 13

Tablelands from Overfalls Trail, Southeast of Trout River Pond, Newfoundland. (IATNL)

Map 13-1. Bay of Islands, Newfoundland.

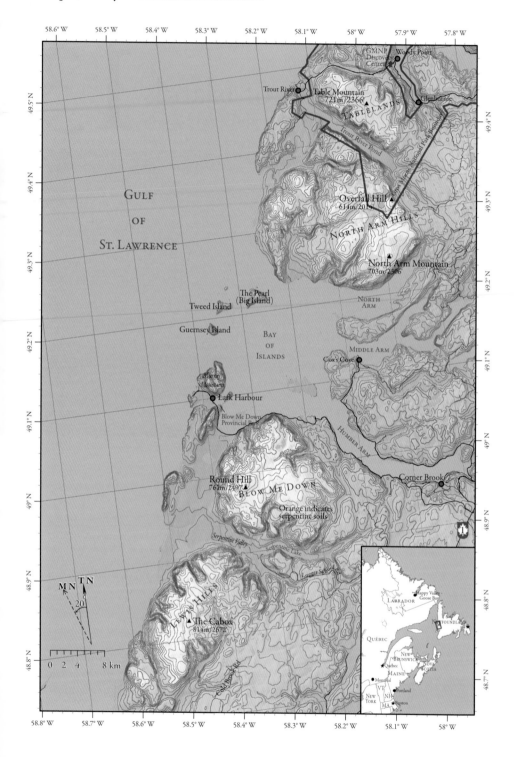

Bay of Islands, NEWFOUNDLAND

13-1 Logging roads reach into the mountains surrounding the Bay of Islands only two hours from the ferry at Port aux Basques. (MJ)

It's only about two hours from the fogbound ferry terminal at Port aux Basques up the Gulf of St. Lawrence coast, to the logging roads that lead into the mountains surrounding the Bay of Islands (13-1). There are four mountain ranges in all: broad, ocean-front mesas with enormous alpine tablelands, encompassing little-known but spectacular ranges known as the Lewis Hills, Blow Me Down Mountain, North Arm Hills, and Tablelands.

The island of Newfoundland rises to its highest point near the center of the Lewis Hills, at a windswept and lonely dome known as The Cabox. The Cabox's westerly vantage takes in the Gulf of St. Lawrence from the Port au Port Peninsula to the North Arm Hills. Several vantage points along the western rim of the Lewis Hills tableland peer down vertically to the ocean, 600 m (2000') below. These vistas are vivid testament that in Newfoundland, spectacular mountain scenery can exist at relatively low elevations (13-2).

The four mountain ranges can be divided into two pairs[1] at the Bay of Islands, itself one of the most identifiable features on the western coastline of Newfoundland.[2] The Lewis Hills and Blow Me Down Mountain form the southern block (roughly, the area southwest of the city of Corner Brook.) The northern block (northwest of Corner Brook and south of Bonne

---

1 The two southern mountains, Lewis Hills and Blow Me Down, were probably derived through the faulting of a single ophiolite feature. The North Arm Hills and the Tablelands were also likely connected before faulting and glaciation.

2 The Bay of Islands is named for the tall islands guarding its mouth, as well as several islands inland. When M.L. Fernald arrived at Lark Harbour, his host welcomed him to the "Bay of Hilands," an equally suitable name, as the fjord is bounded on the south by the great Blow Me Down massif and to the north by the North Arm Hills.

13-2 *The vistas of the Lewis Hills provide spectacular mountain scenery at elevations lower than 1000 m. (IATNL)*

13-3 *Each of the major mountains of the Bay of Islands supports thousands of hectares of boreal and alpine wilderness. (IATNL)*

13-4 *Extensive exposures of serpentine give the Bay of Islands mountains a Martian appearance. (CE)*

Bay near Gros Morne) includes the remote North Arm Hills and world-famous Tablelands, which are protected within Gros Morne National Park. Each mountain supports tens of thousands of hectares of boreal, subalpine, alpine, and serpentine wilderness (13-3). It's possible, for instance, to walk across the Lewis Hills and Blow Me Down ranges—a distance of 44 km (27 mi) from Cold Brook Road, north to the Bay of Islands—without crossing a road (you will, however, have to cross the Serpentine River).

The Bay of Islands mountains are unique because they are derived in large part from mantle—the hot, metal-rich sludge that forms the majority of the Earth's interior mass. In the Lewis Hills, for example, rocks may be observed that represent the boundary between oceanic crust and mantle, which typically occurs 10 km (6 mi) below the surface. In fact, the Bay of Islands mountains contain one of the best-preserved **ophiolites** exposed anywhere on the earth's surface.[3] The Bay of Islands region is known among geologists worldwide for the incredible, sprawling serpentine barrens. Extensive exposures of serpentine give the mountains a distinctive, Martian appearance that is similar only to Mont Albert, Québec (Ch. 9) and the White Hills of northern Newfoundland (see overview in Ch. 17), but is otherwise very distinct from every other eastern mountain (13-4). Serpentine outcrops rarely occur in the form of massive alpine mountain ranges the way they do here.

Lay of the Land

Lewis Hills — The Lewis Hills form an irregular, flat-topped, steep-flanked mesa 24 km (14 mi) long and 10 km (6 mi) wide, which is deeply indented by giant cirques along its western and southern flanks. There are few other places in eastern North America where

---

3   Ophiolites are rock sequences that span the contact between mantle and crust. As such, they contain rocks derived from mantle, such as peridotite or serpentine, and rocks derived from crust, such as basalt. Occasionally, sedimentary rocks of the ocean floor are also found. See Chapter 2, and Coleman and Jove (1992).

such magnificent alpine scenery is juxtaposed so closely with the ocean.[4] The Lewis Hills plateau is deeply incised at its western edge by the Lewis Brook glacial valley, which provides access to the southern and western portion. The eastern edge of the plateau is sharply bounded by the Fox Island River, which provides the southern access point for the Lewis Hills section of the International Appalachian Trail. The northern slopes are bounded by Serpentine Lake and the Serpentine River, and the western escarpment drops quickly into the Gulf of St. Lawrence. The steep slopes of the Lewis Hills plateau around 500 m (1640'),

13-5 Rope Cove is among the numerous ravines have formed along the ancient contact between mantle and crust, known as the Mohorovicic Discontinuity. (MJ)

and then slope gently toward the high point at the Cabox (814 m / 2672'). There is a secondary summit southwest of the Cabox called Big Level (754 m / 2474'). Serpentine summits include Springer Hill (>570 m / >1870') in the south and the Aiguille (>478 m / 1569') in the north, as well as several (mostly unnamed) peaks and spurs above Wheeler Brook and Red Rocky Brook. Mount Barren (705 m / 2312') is a minor peak at the northwest margin of the plateau. The Cache Valley Road from Port au Mal provides access to prominent Rabbit Hill (529 m / 1736').

The captivating beauty of Lewis Hills, as on all of the Bay of Islands mountains, is most evident in the cirques and ravines that adorn the north and south faces. The most eye-catching, and unusual, and remote of these are the ocean-draining sister-cirques of Molly Ann and Rope Cove Canyons, set facing northwest on either side of Mount Barren. Molly Ann Canyon is a downright Rivendell-esque landscape of waterfalls, rivulets, snowfields, and basaltic cliffs, overlooking the ocean. Rope Cove Canyon, barely a 3 km (1.8 mi) walk across dry tundra to the northeast, could not be more starkly different from Molly Ann Canyon—or any other place on earth. Its headwall is serpentine, its eastern wall is gabbroic, and its western wall is apparently a combination of gabbros and basalts. The contact zone between mantle and crust is in plain view from the headwall of Rope Cove: serpentine, gabbro, basalt, a perfect ophiolite sequence (13-5).[5] The result is a rainbow of colored cliffs. A similar effect is found in Travertine Canyon, a tripartite cirque in the Fox Island River watershed on the massif's south face. Travertine Canyon marks the southern edge of a serpentine tongue that extends north beyond the Cabox, nearly to Molly Ann Canyon. Like Rope Cove, its headwall is mostly serpentine, although the western headwall is set in gabbroic bedrock (13-6). Several more gentle valleys round out the east side of the range before coming to the great red canyons of Wheeler Brook and Red Rocky Brook, which drain into the Serpentine River.

---

4   It only takes one firsthand encounter to become completely enamored with oceanfront mountains. Other prominent examples are found primarily in Newfoundland, including the Table Mountains (Ch. 12), Gros Morne (Ch. 14), and the Highlands of St. John (Ch. 16). Smaller but noteworthy sites include the limestone barrens of the Northern Peninsula and Strait of Belle Isle (see sections on Burnt Cape, Cloud Mountain, and the Forteau Tablelands in Ch. 17 and 21). Along the Labrador coast, the Benedict, Kiglapait, Kaumajet, and Torngat Mountains rise directly from the sea (Ch. 21). South of the Gulf of St. Lawrence, only the higher mountains of Acadia National Park come close (Ch. 11).

5   The boundary between the earth's mantle and ocean crust is called the Mohorovicic Discontinuity. Where the "Moho" is exposed at the surface, there are often exposed serpentines, derived from mantle, as well as gabbros and basalts, derived from the crust.

13-6 Travertine Canyon, like several others, is formed in serpentine and gabbroic bedrock. (LW)

13-7 Blow Me Down Mountain in the southern Bay of Islands is composed of mafic and igneous rocks. Here, Blow Me Down is viewed across the serpentine barrens of the Lewis Hills. (MJ)

Blow Me Down Mountain — Blow Me Down Mountain is separated from the Lewis Hills by the Serpentine Valley, which is about 5 km (3 mi) wide at this spot. Similar to the geologic arrangement in the Lewis Hills, Blow Me Down's eastern slopes are primarily ultramafic rock and the central mass is composed of mafic rocks like gabbro. The massif reaches its highest elevation at gabbroic Round Hill (761 m / 2497') (13-7). On the south face, Simms Brook Canyon and Red Gulch drain into the Serpentine Valley. On the north face, Blow Me Down Brook plunges over a 45 m (150') waterfall and through a massive, deeply incised, cirque-like canyon before draining into the Bay of Islands. Simms Brook and Blow Me Down Brook Canyon, like Rope Cove and Travertine Canyon, straddle the discontinuity between serpentine and gabbro in their upper reaches, and so harbor an unusual juxtaposition of plants within a small region (13-8).

North Arm Hills — North of Blow Me Down Mountain, the North Arm Hills abruptly rise from the "North Arm" of the Bay of Islands, culminating at the summit of North Arm Mountain (703 m / 2306'). The adjacent subalpine tableland is known as the St. Gregory Highlands or the Gregory Mountains. North of the height-of-land, Overfall Mountain (614 m / 2014') provides access to the Overfall, a 90 m (300') waterfall along Overfall Brook (13-9). The southern face of North Arm Hills—a vast expanse of serpentine at the edge of the Bay of Islands—is visible to boat traffic coming in to the port of Corner Brook and is clearly visible from most points on Blow Me Down Mountain. The very northern portion of the North Arm massif, including the Overfall and the southern shore of Trout River Pond, is within Gros Morne National Park.

Tablelands — The Tablelands of Gros Morne National Park are also the northernmost mountains in the Bay of Islands complex. The Tablelands are smaller than the three massifs to the south and consist primarily of peridotite,[6] with a small area of gabbroic rock exposed on the plateau. They reach their highest elevation (721 m / 2366') near the center of the massif, which is sometimes referred to as Table Mountain.[7] Just east of the Tablelands near the town of Glenburnie is the Pic à Tenerife (Peak of Tenerife; >538 m / >1766'), which has

---

6   In most cases in Newfoundland, *serpentine* is the word commonly used to describe weathered peridotite or dunite.

7   We use the name *Tablelands*, which is commonly used within Gros Morne National Park, to minimize confusion with the Table Mountains near Port aux Basques and Table Mountain near Point au Mal.

a small, scrubby, open alpine area and a spectacular view.[8] Just north of the Tablelands, the Lookout Hills rise to a plain of soggy, open tundra. The Tablelands are within Gros Morne National Park, where the great Winter House cirque on the Tablelands' north face is probably the best-known landscape feature after Gros Morne Mountain itself (13-10).

Access

Summer access to the Lewis Hills can usually be made from the south, either via the rough Cache Valley Road, which begins near the right-angle left turn on the road to Point au Mal, or via Cold Brook Road, which leaves Stephenville west of Noel's Pond.[9] The Lewis Hills can be accessed from the north via Logger School Road, which also provides access to Serpentine Lake and the southern portion of Blow Me Down, including Simms Gulch. Logger School Road leaves the Trans-Canada Highway (Route 1) about 8 km (5 mi) south of Corner Brook. The International Appalachian Trail Newfoundland Labrador (IAT-NL) has cut access trails from Cold Brook Road and Logger School Road. All three primary dirt roads in the region (Cache Valley, Cold Brook, and Logger School) are rough and require a 4-wheel drive vehicle for safety (13-11). Simms Gulch and the southern tier of Blow Me Down are accessible via the International Appalachian Trail's Blow Me Down Mountain Trail off Logger School Road. Blow Me Down can be accessed from the north, from Route 450 on the southern shore of the Bay of Islands, by way of the Blow Me Down Brook Trail 5 km (3 mi) west of Frenchman's Cove. A trail also leads from Route 450 near Coppermine Brook to Blow Me Down Head.

13-8 Simms Brook and Blow Me Down Brook reach far into the southern and northern flanks of Blow Me Down Mountain. (MJ and LW)

13-9 The Overfall is a 90 m waterfall in the North Arm Hills. (IATNL)

---

8   *Pic à Tenerife* was named by Captain James Cook in 1767 because the sharp-toothed peak reminded him of Tenerife, Canary Islands. Glenburnie, the nearby town, is at the base of a steep section of road known as "The Struggle."

9   Cache Valley Road begins near Point au Mal at 48.6953°N, 58.6563°W and ends at the Fox Island River at 48.6931°N, 58.6193°W. Cold Brook Road begins at 48.5772°N, 58.5465°W.

13-10 Winter House Canyon, a cirque on the north face of the Tablelands, is among the most well-known mountain features in Newfoundland. (MJ)

13-11 Most of the dirt logging roads reaching into the Bay of Islands mountains require high-clearance vehicles. (LW and MJ)

Parts of the North Arm Hills are accessible by high-clearance vehicle from the Goose Arm network of logging roads that begin at a number of locations, including Hughes Brook near Corner Brook, Goose Arm Road in Deer Lake, White Wash Road at the northern end of Big Bonne Bay Pond, and from Glenburnie in the western arm of Bonne Bay. They can also be reached by boat either on Trout River Pond or from the Bay of Islands. The IATNL completed construction of the Overfalls Trail from the eastern end of Trout River Pond to the Overfalls waterfall in the fall of 2010. This opened a new recreational opportunity that could result in greater access to the southern North Arm Hills and Gregory Mountains, an area currently protected only under agreement with the Corner Brook Pulp and Paper Company. The trailhead is accessible via a one-hour boat ride from the community of Trout River. The Tablelands are directly accessible from the northern, or Bonne Bay side, from the road that leads from Woody Point to Trout River. Go straight forward from here to access the massif via the spurs on either side of Winter House Canyon. The Tablelands are also accessible via a trail from Trout River Pond. A backcountry ski cabin near the Tablelands is open from January to April.

Geology

All four of the mountains profiled in this chapter are composed, to varying extents, of mantle-derived rocks (ultramafics) and associated mafic and igneous rocks. Thus, they often represent the major components of a full ophiolite suite: ultramafics, mafics, igneous

13-12 Rock formed from both crust and mantle are juxtaposed in alpine and serpentine barrens throughout the Bay of Islands mountains. (MJ)

basalts, and sedimentary ocean floor (13-12). For example, the central one third of the Lewis Hills, including the Cabox and the Big Level, is composed of mafic (magnesium and iron rich) rock, which stands out in contrast to the eastern third of the range, which is composed of serpentine. This serpentine area is most accessible in the vicinity of Springer Hill, Wheeler Brook, and Red Rocky Brook (13-13). The highlands west of Lewis Brook are mostly basaltic, as are the flanks of Mount Barren. Sedimentary layers extend from the Fox Island River onto the Lewis Hills massif at Rabbit Hill. By comparison, the western portion of Blow Me Down Mountain is largely composed of diorite, an igneous rock, though its western reaches are composed of metal-rich serpentines. The North Arm Hills are similarly composed, but the Tablelands consist mostly of serpentine.

Calcium-rich soils at multiple sites scattered throughout the region that are derived either from sedimentary rock or from the serpentinization of peridotite. The latter can result in

13-13 Red Rocky Brook drains a great serpentine canyon in the northern Lewis Hills. (MJ)

13-14 Travertine deposits, formed from the serpentinization of peridotite, are widely distributed throughout the Bay of Islands mountains. White Spot, Travertine Canyon, Lewis Hills. (MJ)

13-15 *Alpine, subalpine, and serpentine vegetation are variably distributed throughout the higher elevations in response to dramatic differences in bedrock composition, aspect, slope, and snow persistence. (IATNL)*

13-16 *The north-facing slopes of Blow Me Down Mountain support lush forests and subalpine vegetation. (MJ)*

13-17 *Some of the most extensive high-elevation serpentine barrens are found in the southern and eastern Lewis Hills. (MJ)*

travertine formation at the surface, and Travertine Canyon possesses a striking example of this phenomenon. Near the mouth of the main canyon, a bold white stripe (the "White Spot") marks an active travertine seepage, which originates from the ongoing serpentinization of ultramafic bedrock.[10] Similar travertine seepages occur in the middle reach of Blow Me Down Brook Canyon and on the north face of the Tablelands massif (13-14).

Ecology

Vegetation — Throughout the uplands of the Bay of Islands, the alpine and subalpine vegetation varies tremendously in response to the dramatic differences in bedrock composition, aspect, slope, annual snow cover, and edaphic factors governing soil development (13-15). There is often a clear line between the vegetation growing in serpentine soils and that growing in the non-serpentine soils derived from other bedrock such as gabbro or basalt. The line can be very striking, especially at high elevation and on talus and scree slopes where organic soils haven't developed. Furthermore, a distinct difference exists between the vegetation found at low-elevation, protected sites and the vegetation at high-elevation or exposed sites on both serpentine and non-serpentine soils. For example, the north-facing slopes of Blow Me Down, wet with snowmelt most of the year, have a more lush and diverse flora on every soil type than more exposed sites do (13-16). Exposed serpentine vegetation can be found on each of the four Bay of Islands summits but it is most pronounced in the southern North Arm Hills, the southern and eastern Lewis Hills, and the Tablelands (13-17).

Serpentine soils and talus at low elevations are often thickly covered by well-developed scrub forest dominated by tamarack (*Larix laricina*), known locally as "tree juniper." Although small, these may be very old. A number of temperate and boreal species are also found in these areas but generally do not occur very high onto the exposed ridges and summits. These species include the more southerly royal fern (*Osmunda regalis*), white pine (*Pinus*

---

10   Travertine is a precipitate form of calcium carbonate, $CaCO_3$. The calcareous-loving plants associated with these deposits are discussed later in this chapter.

13-18 Among the common species found on sheltered serpentine sites are (clockwise from top left) white pine, royal fern, Canada burnet, and pitcher plant. (MJ, CE, and LW)

strobus),[11] red maple (*Acer rubrum*), and rhodora (*Rhododendron canadense*), as well as species more tolerant of the higher elevations: Canada burnet (*Sanguisorba canadensis*), tufted club rush (*Trichophorum cespitosum*), and pitcher plant (*Sarracenia purpurea*) (13-18). Odd though it may seem to find bog-loving pitcher plants on barren rock, remember that these

---

11   Referred to as *Pinus strobus* forma *prostrata* by Fernald (1933).

13-19 High-elevation serpentine barrens support (clockwise from left) Aleutian maidenhair fern, serpentine sandwort, and moss campion. (CE, LW, and MJ)

13-20 Lapland rosebay may occur in stunning profusion, growing over the ochre serpentine. (CE)

plants supplement their nutrient intake by digesting arthropods in their vase-shaped leaves.

Boreal and temperate shrubs are not all that grow on the lower slopes. Plants typical of Newfoundland's alpine-serpentine tablelands also thrive at very low elevations, as evidenced by a short walk from the Tablelands parking lot in Gros Morne National Park. Between scrubby patches of tamarack and juniper (*Juniperus communis*), serpentine and reddish stitchworts (*Minuartia marcescens* and *M. rubella*), and moss campion (*Silene acaulis*) are abundant on felsenmeer, scree slopes, and talus, often found with alpine catchfly (*Silene suecica*), sea thrift (*Armeria maritima*), field sagewort (*Artemisia campestris*), pitcher plant (*Sarracenia purpurea*), and Mistassini primrose (*Primula mistassinica*). In protected areas along seepages and in boulder fields, it is easy to find abundant Aleutian maidenhair (*Adiantum aleuticum*)—a ubiquitous plant on serpentine throughout the Bay of Islands mountains.

At high elevations on exposed serpentine slopes, such as the northeastern Lewis Hills, it is common to find only the serpentine specialists: serpentine and low sandworts (*Arenaria humifusa*), moss campion, and the endemic Newfoundland chickweed (*Cerastium terraenovae*) (13-19). Frequently, Lapland rosebay (*Rhododendron lapponicum*) occurs in stunning profusion, its brilliant purple flowers contrasting with ochre serpentine, as the headwall of Rope Cove (13-20). Two willows, gray-leaved and arctic (*Salix glauca* and *S. arctica*), also occur on serpentine in these areas.[12] Tamarack may be locally abundant at high elevations, but it is generally a trailing shrub found in close association with common and creeping junipers (*J. horizontalis*).

Overall, the serpentine vegetation of the North Arm Hills is similar to that found at most elevations on the serpentine Tablelands. Again, the Aleutian maidenhair is abundant, as it is everywhere on Newfoundland serpentine. Timber oatgrass (*Danthonia intermedia*), as well as the two willows of North Arm, Blow Me Down, and Lewis Hills (gray and arctic), and the two switchworts (serpentine and reddish), are extremely common. At least two species of *Cerastium* or chickweed, corresponding to *C. terrae-novae* and *C. arvense* (a serpentine-adapted, or serpentinocolous form)[13] occur in this area. There is also alpine catchfly, moss campion, Chinese hemlock parsley (*Conioselinum chinense*), sea thrift, and Lapland rosebay. Butterwort (*Pinguicula vulgaris*) and marsh arrowgrass (*Triglochin palustris*) grow abundantly in wet ditches, poolshores, and seepages. The serpentine barrens of the North Arm Hills are rather desert-like and unvegetated in contrast to the north slope of Blow Me Down and protected areas elsewhere on the North Arm massif.

Plant communities typical of acidic summits throughout Newfoundland, Labrador, Québec, and New England[14] grow on non-serpentine soils and bedrock such as gabbro and basalt at high elevation. Examples of these plant and soil configurations can be found on the actual summit of the Cabox and Mount Barren on the Lewis Hills, Round Hill on Blow Me Down and North Arm Mountain. Diapensia (*Diapensia lapponica*), alpine azalea (*Loiseleuria procumbens*), and alpine bearberry (*Arctous alpina*) grow commonly with crowberries (*Empetrum nigrum*), alpine bilberry (*Vaccinium uliginosum*), and mountain cranberry (*V. vitis-idaea*) (13-21). These species intermingle above treeline with unusual southern species such as the tawny cottongrass (*Eriophorum virginicum*) on the portions of Blow Me Down's tableland that are composed of igneous rock.

A handful of plants unable to persist on the serpentine soils are often found in profusion on calcium-rich soils near travertine seepages. These include yellow lady's slipper (*Cypripedium parviflorum*),[15] entire-leaved mountain avens (*Dryas integrifolia*), and several mountain saxifrages (especially purple and yellow, *Saxifraga oppositifolia* and *S. aizoides*). Several travertine exposures similar to those in Blow Me Down Brook Canyon and Travertine Canyon occur on the north face of the Tablelands west of Winterhouse Canyon (13-22).

---

12   These are *Salix glauca* ssp. *callicarpaea* and *Salix arctica*, respectively, referred to as *Salix cordifolia* var. *Macounii*, and *Salix anglorum* var. *kophophylla* by Fernald (1933)

13   See Rune (1954) for a description of similar phenomena on Mont Albert.

14   The alpine vegetation of acidic soils includes many of the most well-known alpine plants, such as those profiled by Slack and Bell (2006). This group is also discussed in Chapter 5. Fernald (1933, p. 5) describes a suite of "taboo" arctic-alpine species on Lark Mountain, north of Blow Me Down and just north of the village of Lark Harbour, that were not to be collected because they "are so nearly ubiquitous at all altitudes in Newfoundland." In this instance, these included (taxonomy updated): *Huperzia selago*, *Spinulum canadense*, *Juniperus communis*, *Agrostis mertensii*, *Anthoxanthum monticola*, *Trichophorum caespitosum*, *Carex rariflora*, *C. scirpoidea*, *Juncus trifidus*, *Salix uva-ursi*, *Rubus chamaemorus*, *Viola labradorica*, *Loiseleuria procumbens*, *Arctous alpina*, *Vaccinium uliginosum*, and *Diapensia lapponica*. Later, Fernald states (pp. 51–52), "We can't do it all. Most fascinating exploration awaits the right party; but, to do effective work they must be able to climb and they should not waste time and strength on the 'taboo-list' of ubiquitous and unsignificant plants!"

15   Yellow lady's slipper, *Cypripedium parviflorum*, grows abundantly along the road shoulder near the Tablelands parking lot, one of the easiest places to see this locally common but otherwise unusual orchid.

13-21 *On exposed non-serpentine ridges, typical eastern alpine plants flourish, such as (from left) diapensia, alpine azalea, and alpine bearberry. (MJ and CE)*

In addition to the travertine sites outlined earlier, calcium-loving plant species occur sporadically and infrequently on the plateaus. On Rabbit Hill, above Cache Valley in the Lewis Hills, a typical potassic flora mixes and gives way to a flora with calcareous elements including butterwort, moss campion, and small tofieldia (*Tofieldia pusilla*). Along the eastern rim of Rope Cove in the Lewis Hills, entire-leaved mountain avens and purple saxifrage mix with common acidophiles like diapensia.

Despite the peculiarities of the serpentine landscape, the most interesting sites are the areas of late-lying snow (snowbeds) on non-serpentine soils (see Ch. 5). It is possible to find a number of unusual alpine rarities in these areas, as well as along waterfalls, rills, and springs of the great coves and canyons. Most of these species range north to the Arctic, and are extremely rare farther south in Newfoundland, although some occur in the Table Mountains near Port aux Basques (Ch. 12). These alpine rarities include alpine bittercress (*Cardamine bellidifolia*), mountain sorrel (*Oxyria digyna*), Wormskjold's alpine speedwell (*Veronica wormskjoldii*), sibbaldia (*Sibbaldia procumbens*), and alpine cudweed (*Omalotheca supina*) (13-23). Other unusual, subarctic plants undoubtedly occur in these areas. Because of late-lying snow, these plants become visible later in the summer season than do the plants found in more exposed locations.

13-22 *Calcium-loving plants found near travertine seepages include yellow lady's slipper (left), entire-leaved mountain avens (center), and purple saxifrage (right). (CE, NC, and MJ)*

13-23 Alpine cirques and snowfields in the Lewis Hills support (clockwise from top left) southern populations of alpine bittercress, mountain sorrel, sibbaldia, and Wormskjold's alpine speedwell. (MJ)

Fauna — Both the Lewis Hills and Blow Me Down provide forest and tundra habitat for woodland caribou (*Rangifer tarandus caribou*), which occur throughout the region (13-24). The Bay of Islands mountains also support moose (*Alces americanus*), which were introduced to the island on two occasions (in 1878 and 1904) and are now frequently seen at all elevations, including atypical habitat such as the alpine-serpentine barrens (13-25). Arctic hare (*Lepus arcticus*) are encountered in all areas above treeline, where they camouflage with either snow or rock depending on the season (13-26). Meadow voles (*Microtus pennsylvanicus*), infrequently seen, leave trails in the vegetation under snow, which become plainly obvious after snowmelt (13-27). Masked shrews (*Sorex cinereus*), introduced from the Maritimes, now occur throughout even the alpine portions of the Bay of Islands.

Because of their proximity to the ocean, the birds of the Bay of Islands reflect a mixture of arctic-alpine, boreal, and shore birds. Rock and willow ptarmigan (*Lagopus muta* and *L.*

13-24 Woodland caribou seasonally occur at all elevations in the Lewis Hills and Blow Me Down Mountain. A young animal is pictured near Round Hill on Blow Me Down. (MJ)

13-25 Moose range at all elevations in the Bay of Islands, and their sign may be seen well above treeline. (MJ)

13-26 Arctic hare camouflage with either snow or rock, depending on the season. (MJ)

13-27 Meadow voles leave distinctive trails in vegetation throughout the alpine areas of the Bay of Islands. Here, a vole trail has been flooded by snowmelt in the Lewis Hills. (MJ)

lagopus) occur throughout the two massifs, and their droppings are often seen among boulders in felsenmeer. Rock ptarmigan prefer the open tundra whereas willow ptarmigan are encountered closer to treeline. Rock ptarmigan lay 5 to 7 eggs and willow ptarmigan lay 6 to 9 eggs; both species lay nests concealed by shrubs and vegetation, which we have only noticed when we scare up the nesting adult. Males of the two species are easily told apart when in breeding plumage, by the willow ptarmigan's maroon-brown nape and head (13-28).

In addition to ptarmigan, horned lark (*Eremophila alpestris*), savannah sparrow (*Passerculus sandwichensis*), and white-throated sparrow (*Zonotrichia albicollis*) are common on the grassy expanses above treeline. Greater yellowlegs (*Tringa melanoleuca*) nest in the extensive solifluction pools and bogs above treeline, especially throughout the giant bog-complex north of Round Hill on Blow Me Down and similar areas on the Lewis Hills and the Tablelands. Bald eagle (*Haliaeetus leucocephalus*) are sometimes seen soaring above the canyon walls and may account for high levels of ptarmigan mortality.

At least two of Newfoundland's five species of introduced frogs occur abundantly on the Lewis Hills and Blow Me Down. American toads (*Anaxyrus americanus*), native to the mainland and introduced to western Newfoundland, can be found breeding in May and June in ponds, marshes, and string bogs at all elevations on all four mountains. Mink frogs (*Lithobates septentrionalis*), also introduced, occur in large ponds along the edge of the serpentine soils at low elevations throughout the Lewis Hills and evidently breed in small solifluction ponds above treeline on Blow Me Down (13-29). At this time, these seem to be the only frogs widespread in the Bay of Islands alpine areas.

13-28 Both willow (left) and rock ptarmigan (right) occur in the krummholz and tundra of the Bay of Islands. Remains of ptarmigan killed by coyotes or bald eagles are frequently seen. (MJ)

13-29 American toads (bottom-left) and mink frogs (bottom-right), both introduced to the island of New-foundland, occur at all elevations. The Moor on the Cabox is pictured at top. (WK and CE)

A diverse array of invertebrates occupies the alpine ponds and pools of both massifs, including several species of dragonfly, leeches, and fingernail clams. Moths, butterflies, bees, ants, stinkbugs, tiger beetles, and many spiders (wolf spiders and orbweavers) can be found abundantly on the high tundra as well as the scree slopes and summits (13-30).

13-30 Invertebrate diversity in the alpine areas of the Bay of islands is poorly understood. Long-lipped tiger beetles (top left) and their larval burrows (top right) are abundant. Wolf spiders (bottom left) occur in a variety of alpine habitats. Leeches (bottom right) are present in pools above treeline. (LW, MJ, CE)

Map 13-2. Lewis Hills, Bay of Islands, Newfoundland.

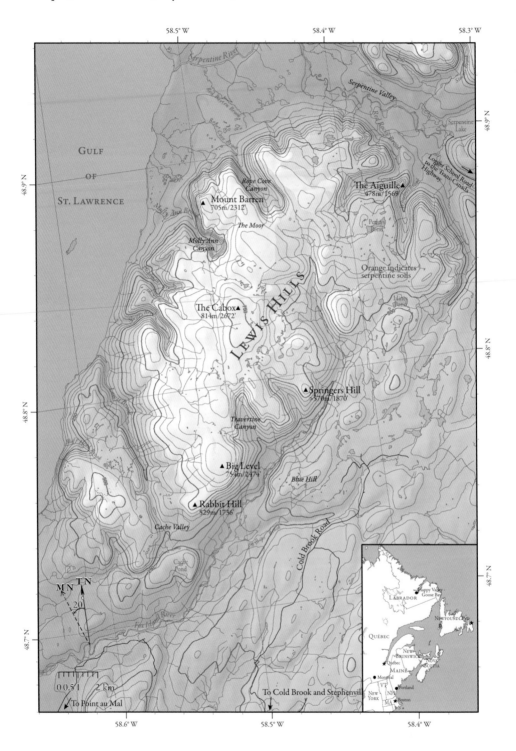

Gros Morne 14
NEWFOUNDLAND

Sunset on Gros Morne, Newfoundland. (MJ)

Map 14-1. Gros Morne Mountain and vicinity, Newfoundand.

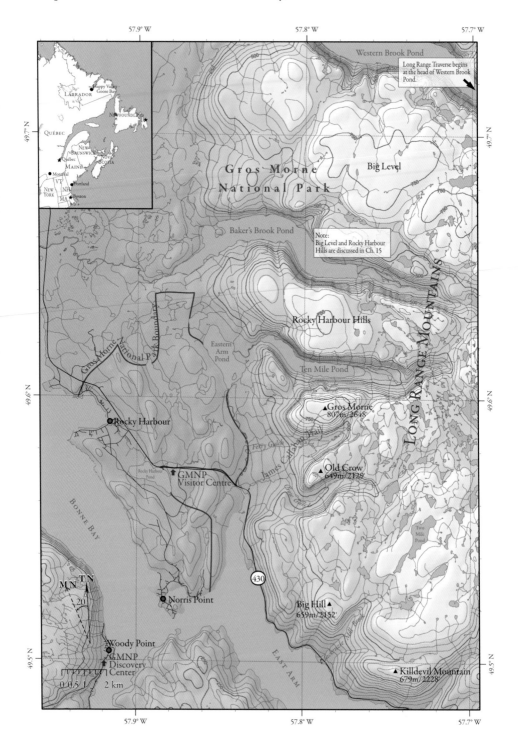

Long Range Traverse begins at the head of Western Brook Pond.

Western Brook Pond

Gros Morne National Park

Big Level

Baker's Brook Pond

Note:
Big Level and Rocky Harbour Hills are discussed in Ch. 15

Rocky Harbour Hills

LONG RANGE MOUNTAINS

Eastern Arm Pond

Ten Mile Pond

Gros Morne National Park Boundary

▲Gros Morne
807m/2648'

Rocky Harbour

Ferry Gulch

Old Crow
649m/2129'

GMNP Visitor Centre

Rocky Harbour Pond

James Callaghan Trail

BONNE BAY

430

Big Hill▲
659m/2152'

MN TN

20

Norris Point

EAST ARM

Woody Point
GMNP Discovery Center

▲Killdevil Mountain
679m/2228'

0 0.5 1   2 km

Gros Morne, NEWFOUNDLAND

14-1 Gros Morne's distinctive shape and color easily distinguish it from surrounding mountains such as Old Crow (at right) and other peaks on the Newfoundland coast. (MJ)

Gros Morne is *the* mountain of Newfoundland, as Katahdin is to Maine. Gros Morne may refer either to this dome-shaped, quartzite-capped mountain on Newfoundland's west coast, or to the surrounding Canadian National Park and UNESCO World Heritage Site that bears its name (in this chapter, we refer specifically to the mountain). Gros Morne, the mountain, is uniquely shaped and prominently displayed against the backdrop of the Long Range Mountains, and it stands apart from the surrounding sea of mountains (14-1). In this way, Gros Morne Mountain is the clear centerpiece of Gros Morne National Park's world-famous tundra-fjord landscape. Although Gros Morne is probably the most recognizable (and recognized) mountain in Newfoundland, it is not the highest. In fact, as noted in the preceding chapter, Gros Morne's quartzite summit apparently tops out about three meters (16') below the gabbroic dome of the Cabox, in the Lewis Hills about 100 km (60 mi) to the south (Ch. 13) (14-2).

Gros Morne's winsome and eerie name is sometimes taken to mean "Big Hill," and in Québec (as in the Caribbean), *morne* is taken to mean an isolated mountain or monadnock. But *morne* is also French for "sad," or "bleak," and *gros* may be translated to "great," "heavy," or "colossal." Some of these combinations result in too much of an overstatement to be poetic (think "Colossal Sadness"). Other translations nicely capture the romance and mystique of the most famous mountain on Newfoundland, which is known for its whimsical naming conventions in the face of sustained tragedy. For example, "Big Bleak" has more of a Newfoundland ring to it than "Great Somber," another translation sometimes tossed about. In any case, *morne*'s double-entendre is welcome and appropriate, but it should be noted that Gros Morne is no bleaker than any other eastern alpine mountain. Though Gros Morne is

14-2 The stony, foggy summit of Gros Morne crests about 3 m below that of the Cabox in the Lewis Hills. (MJ)

14-3 On clear days, Gros Morne's ocean views are unparalleled. (MJ)

14-4 Gros Morne's silvery-pink quartzite felsenmeer stands out in contrast to the gneiss of the older Long Range Mountains, which runs west of the coastal summits profiled in this chapter. (MJ)

14-5 Rock ptarmigan (top), arctic hare (middle), and woodland caribou (bottom) occur throughout the higher coastal summits of Gros Morne National Park, but are nowhere more easily seen than on Gros Morne Mountain itself. (MJ)

frequently windswept and cloudcapped, on a clear day, its ocean views of Bonne Bay and the Gulf of St. Lawrence are unparalleled in the East (14-3).

Gros Morne's summit is composed largely of quartzite and, from some angles, takes on a pinkish-silvery appearance, distinguishing it from the gneissic and granitic Long Range Mountains (Ch. 12 and 15), which runs the length of Newfoundland's west coast to the north and south (14-4). Gros Morne's alpine vegetation shares more in common with Mount Washington and Katahdin (Ch. 7 and 8) than the serpentine portions of the nearby Bay of Islands Mountains (Ch. 13). The frost-shattered, fairly level, alpine tableland harbors populations of rock ptarmigan (*Lagopus muta*), arctic hare (*Lepus arcticus*), and woodland caribou (*Rangifer tarandus*) (14-5). Because of a well-developed loop trail, Gros Morne is perhaps the most accessible place in North America to see these three species.

Lay of the Land

Gros Morne Mountain (807 m / 2648') is only one of about a dozen prominent summits in Gros Morne National Park. The peaks within the National Park span the geological spectrum, from the metal-rich, serpentine Tablelands (actually a part of the Bay of Islands complex profiled in Ch. 13) to gneissic portions of the Long Range and granitic peaks like Old Crow (649 m / 2129') (covered in Ch. 15). Gros Morne itself is similar in composition to the nearby summits of Big Hill (659 m / 2162'), Killdevil Mountain (679 m / 2228'), and Fang Hill (>650 m / >2200'), and collectively these are the focus of this chapter. These four mountains are sandwiched between Route 430, the coast highway known as the "Viking Trail" and the gneissic portions of the Long Range Mountains, near the southeastern edge of Gros Morne National Park.

14-6 *About 460 hectares of tundra, felsenmeer, and alpine snowbeds occur across the diverse topography of Gros Morne's broad summit. (LW)*

The summit area of Gros Morne is sprawling and immense and entirely above treeline, encompassing about 460 hectares (1135 acres) of alpine tundra, felsenmeer, and snowbeds (14-6). Killdevil Mountain has a comparatively more linear summit and a dramatic south-facing talus slope, which together support an alpine area of about 200 hectares (404 acres). Killdevil's summit basically runs east-west, with a massive south-facing and unstable scree slope, and provides long views south across Bonne Bay to the Tablelands and north to Big Hill, Gros Morne, and the interior Long Range Mountains (14-7). Big Hill is similar to Killdevil and Gros Morne but encompasses only 64 hectares (158 acres) of exposed alpine habitats, and Fang Hill even less.

14-7 *Killdevil Mountain's long ridgeline provides views of the Tablelands (left) and Big Hill (right). (MJ)*

Gros Morne Mountain is buttressed on the south by the granite monolith Old Crow, and is bounded to the north by a landlocked fjord, Ten Mile Pond (14-8). North of Gros Morne Mountain, gneissic Long Range topography asserts itself in the form of a low-relief plateau deeply incised by fjords, most of them spectacular. The deep fjord, Baker's Brook Pond,

14-8 *Ten Mile Pond, a landlocked, freshwater fjord, separates Gros Morne mountain from the Rocky Harbour Hills to the north. (MJ)*

isolates the Rocky Harbour Hills (756 m / 2480') from Big Level (795 m / 2608'), both of which support noteworthy alpine tablelands but are more properly part of the Long Range Mountains (Ch. 15).

Access

Gros Morne's summit can be easily accessed on foot via the 16 km (10 mi) James Callahan loop trail, located 7 km (4.2 mi) east of the port-town of Rocky Harbour on Route 430 (the Viking Trail). The Callahan Trail is a very popular hike in summer, but is seasonally closed to protect caribou in the spring and fall. The alpine backcountry of Gros Morne and its surroundings may be explored via the Long Range Traverse, an eastern trekking classic. The Traverse winds over trailless tundra from the east end of Western Brook Pond (a landlocked freshwater fjord deeply incised into the Long Range Mountains, discussed in Ch. 15), across portions of the Big Level and Rocky Harbour Hills to the summit of Gros Morne, where

it meets the Callahan Trail and descends to Route 430 (14-9). The multiday trek may be undertaken in four or five days with a permit from Gros Morne National Park (Parks Canada). The National Park rules and guidelines suggest that the stony summits of Killdevil, Big Hill, and Fang Hill are closed to public access without a research permit.

14-9 The Long Range traverse is a multiday trek that begins at Western Brook Pond, the famous fjord near the northern edge of Gros Morne National Park, and ends at Gros Morne Mountain. (MJ)

Geology

Gros Morne, Big Hill, Killdevil Mountain, and Fang Hill are essentially geological layer-cakes composed largely of partially metamorphosed sedimentary rock. Gros Morne itself is capped by quartzite, a metamorphosed sandstone, overlying the Precambrian gneiss and granite that is exposed on the Long Range Inlier (Ch. 12 and 15). Exposures of limestone occur on the north face. These peaks are surrounded to the south and east by granites of the Grenville portion of the Canadian Shield, and to the north by gneisses of similar age as the Grenvillian granites. Because they are derived from metamorphosed sedimentary rocks, Gros Morne and adjacent peaks are strongly different in composition from the four major mountains of the Bay of Islands (Ch. 13), which largely comprise mafic and ultramafic rock and differ in age from both the Bay of Islands and the Precambrian gneiss of the Long Range Mountains to the east and north. In both composition and age, Gros Morne, Killdevil, and Big Hill are more similar to the Highlands of St. John (Ch. 16), nearly 160 km (100 mi) to the north, than to any of the nearer mountains surrounding them.

14-10 Common alpine plants on Gros Morne and the adjacent summits include (clockwise from left) mountain firmoss, alpine bearberry, and diapensia. (MJ)

Ecology

Vegetation — The gently domed summit of Gros Morne is sparsely covered by a variation on the same arctic-alpine assemblage found on the tableland of other non-serpentine, non-limestone peaks of Newfoundland. Mountain firmoss (*Huperzia appressa*), black crowberry (*Empetrum nigrum*), alpine bearberry (*Arctous alpina*), mountain cranberry (*Vaccinium vitis-idaea*), three-toothed cinquefoil (*Sibbaldia tridentata*), and highland rush (*Juncus trifidus*) are abundant, as are extensive areas of open felsenmeer supporting a variety of mosses and lichens[1] as well as bearberry willow (*Salix uva-ursi*), alpine bilberry (*Vaccinium uliginosum*), and common juniper (*Juniperus communis*). As on most alpine summits in Newfoundland, diapensia (*Diapensia lapponica*) and alpine azalea (*Loiseleuria procumbens*) are common on the most exposed sites (14-10). Snowbeds on the northern slopes of Gros Morne's summit, directly above Ten Mile Pond, support such rare alpine species as alpine clubmoss (*Diphasiastrum alpinum*), sibbaldia (*Sibbaldia procumbens*), moss heather (*Harrimanella hypnoides*), and alpine cudweed (*Omalotheca supina*), as well as the more widespread and common

14-11 Rare snowbed species occur sparingly on Gros Morne, including (clockwise from top left) moss heather, alpine clubmoss, alpine cudweed, and sibbaldia. (MJ)

---

1   These data are from data collected by the authors between 2001 and 2010, and Anions (1994).

14-12 Regrettably, invasive species such as coltsfoot have colonized disturbed sites along the Callahan Trail on Gros Morne. (MJ)

14-13 Black and pink crowberry co-occur on Killdevil Mountain, as they do on Gros Morne and Big Hill. (MJ)

purple mountain heather (*Phyllodoce caerulea*) (14-11). Exotic plant species such as coltsfoot (*Tussilago farfara*) and common plantain (*Plantago major*) have colonized disturbed sections along the Callahan Trail (14-12).

Killdevil and Big Hill support similar alpine vegetation as found on Gros Morne Mountain, although smaller in area. Two native varieties of crowberry (black and pink, *Empetrum nigrum* and *E. eamesii*) co-occur abundantly on Killdevil as on Gros Morne Mountain (14-13). Numerous scattered tufts of purple mountain heather and long-styled sedge (*Carex stylosa*) occur along Killdevil's eastern summit ridge (see Ch. 17). The eastern ridge then winds northward along a narrow spine used heavily by caribou. At its turning point, an east-facing slope contains many rare plants. However, westward along the ridge, an area of huge boulders and caves separates the main summit from a small cliff and a smaller knoll. Some limestone is present, which provides for some interesting (and rare) plants as well as arctic-alpine species (*Euphrasia oakesii, Potentilla nivea, Saxifraga paniculata, Hedysarum alpinum, Silene acaulis,* and *Arnica griscomii*). Small tofieldia (*Tofieldia pusilla*) grows with a number of saxifrages (*Saxifraga* spp.). Two species of *Woodsia* fern—*ilvensis* and *alpina*—occur in the talus with several species of pussytoes (*Antennaria* spp.).[2]

14-14 The tuckamore or krummholz of Gros Morne and Killdevil Mountains mostly comprises balsam fir and black spruce (left). The tuckamore understory consists of bunchberry, alpine bilberry, and black crowberry (right). (MJ)

---

2   For a more detailed description of the summit of Killdevil Mountain, see Fernald (1933, pp. 89–90).

In protected areas of black spruce (*Picea mariana*) and balsam fir (*Abies balsamea*) tuckamore, both species of crowberry are abundant and occur with sheep laurel (*Kalmia angustifolia*), bunchberry (*Cornus canadensis*), glandular birch (*Betula glandulosa*), alpine bilberry, Canada burnet (*Sanguisorba canadensis*) and northern comandra (*Geocaulon lividum*) (14-14).

14-15 *A group of five rock ptarmigan huddles on the quartzite felsenmeer of Killdevil Mountain. (MJ)*

Fauna — As already noted, the summit of Gros Morne provides the best opportunity, as well as the most scenic context, in eastern North America to see woodland caribou, arctic hare, and rock ptarmigan in their native habitats (14-15). Moose (*Alces americanus*) range to treeline, as on other Newfoundland mountains, but are commonly encountered in the marshes below Gros Morne's south face (14-16). Alpine meadows on Gros Morne also support populations of Newfoundland's endemic meadow vole (*Microtus pennsylvanicus terraenovae*), the only small mammal native to the island. In addition, alpine lakes in the region support arctic char (*Salvelinus alpinus*) (14-17).[3] Smaller pools and marshes above treeline, as well as larger ponds provide breeding habitat for introduced populations of American toad (*Anaxyrus americanus*) and will likely one day support mink frogs (*Lithobates septentrionalis*), which occur throughout the nearby Bay of Islands Mountains (14-18).

14-16 *Introduced moose are abundant near treeline throughout Gros Morne National Park. (MJ)*

14-17 *Arctic char occur in ponds high on the mountains of Gros Morne National Park. (MJ)*

14-18 *American toads, introduced from Nova Scotia, now breed in treeline pools on Gros Morne. (MJ)*

---

3 Rombough et al. (1978)

Northern Long Range Mountains
NEWFOUNDLAND

15

Long Range Mountains east of Daniel's Harbour, Newfoundland. (IATNL)

Map 15-1. Northern Long Range Mountains, Newfoundland.

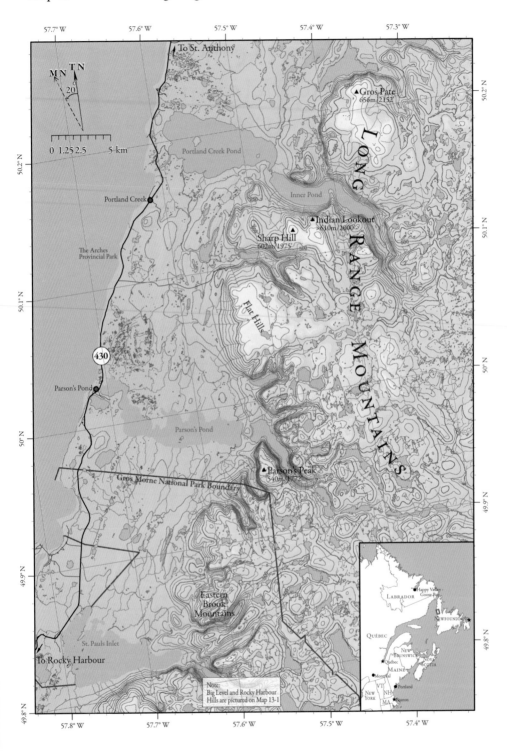

Northern Long Range Mountains, NEWFOUNDLAND

15-1 The Northern Long Range of Newfoundland encompasses an isolated exposure of the ancient gneiss and granite of the Canadian Shield. North Rim Parsons Inner Pond, Newfoundland. (IATNL)

The Long Range Mountains of Newfoundland are an extension of the mainland Appalachians, which run continuously from Alabama and Georgia before plunging dramatically into the sea at the Cape Breton Highlands in Nova Scotia (Ch. 11). The entire chain of mountains running down the western spine of Newfoundland, from the northern capes of Norse legend to Port aux Basques in the south, are often collectively referred to as the "Long Range Mountains." This broad and expansive definition does little to capture the diversity of the mountains of western Newfoundland, which vary in composition, structure, origin, and age. The broad definition includes many mountains that are geologically distinct from the core of Long Range Mountains, including those of the largely serpentine Bay of Islands (Ch. 13) and the sedimentary Highlands of St. John (Ch. 16). In this chapter, we treat the northern core section of the Long Range, a distinct block of ancient gneiss and granite. This block essentially consists of a large segment of the Grenville Province of the Canadian Shield and is known as the "Long Range Inlier," a reference to its isolation from the bulk of the Shield in Québec and Labrador (15-1). Because of the great scope of the Long Range mountain system, the Southern Long Range is (somewhat arbitrarily) treated separately in Chapter 12.

The broad plateau of the Long Range Inlier has been extensively shaped by glacial and periglacial activity (15-2). Much of its 160 km (100 mi) expanse is thickly forested with spruce and fir and impenetrable krummholz (which, as mentioned in earlier chapters, Newfoundlanders refer to as *tuckamore*). At several intervals along the western rim of the range, overlooking the Gulf of St. Lawrence as it becomes the Strait of Belle Isle, the ancient gneiss

15-2 The alpine plateaus of the Long Range Inlier have been extensively shaped by mountain glaciers, such as this valley above Portland Creek. (IATNL)

15-3 Billion-year-old gneiss projects above the treeline at intervals all the way from Gros Morne Mountain to Canada Bay. Portland Creek, Inner Pond, Newfoundland. (IATNL)

15-4 The rims of Western Brook Pond support alpine tundra. (MA)

projects high enough to exclude the subalpine forests (15-3). Here its ridges and crags support alpine tundra and other distinct alpine habitats. Several of these alpine areas are small, and most are relatively inaccessible. However, the southwestern edge of the Inlier encompasses two large alpine areas that support a variety of diverse alpine communities.

Rocky Harbour Hills, Big Level, and the North Rim — The more southern of these alpine tablelands begins within the boundaries of Gros Morne National Park adjacent to, and surrounding, the quartzite cap of Gros Morne itself. Here, the Long Range Inlier is cut into clean sections by breathtaking, landlocked fjords, known (with characteristic Newfoundland understatement) as *ponds*. These glacial incisions result in three seemingly distinct alpine highlands: the Rocky Harbour Hills, Big Level, and the North and South Rims of Western Brook Pond (15-4). These highlands are composed of the old, grey gneiss of the Canadian Shield, which stand in contrast to Gros Morne's younger, pink, quartzite felsenmeer.

Flat Hills, Indian Lookout, and Gros Pate — Thirty-two kilometers (20 mi) to the north of the Western Brook Pond region, the Flat Hills, Indian Lookout, and Gros Pate make up the second major alpine area of the Inlier (15-5). These are also deeply incised by glacial action and are generally regarded to be distinct alpine areas.

The large alpine areas on the Long Range Inlier offer quintessential Newfoundland scenery and an unusual perspective on this geologically rich coastline, with their steep cliffs plunging into the deep waters of isolated glacial fjords (15-6). On clear days in the early summer the late-lying snow fields on Gros Morne (to the south, Ch. 14) and the Highlands of St. John (to the north, Ch. 16) reflect afternoon sun. It's easy to romanticize this landscape, but more often than not, strong winds blow off the water and low clouds cloak the mountains, making navigation by sight difficult (15-7). Trekking across saturated tundra and through the neck-deep tuckamore can be arduous. Tuckamore can often be thick enough to allow walking on top of it—though crashing through into hidden crevices is a real hazard. Still, chances are good—as good as on Gros Morne—that on these lonely, rocky plateaus you just might see the silhouette of a caribou (*Rangifer tarandus*) or ptarmigan (*Lagopus* spp.) moving through the mountain fog (15-8).

Lay of the Land

The best introduction to the Long Range Inlier may be from the top of Gros Morne Mountain, which is geologically distinct and treated separately in this book (Ch. 14). Looking north across Ten Mile Pond from Gros Morne's summit fellfields, the deeply incised, rounded topography of the Long Range Inlier is clearly visible in most directions. Just north of Gros Morne, the Rocky Harbour Hills (706 m / 2317') rise above the treeline. As is often the case in Newfoundland, partly cloudy skies will cast speckled light over these hills, making the already varied greens, blues, browns, and grays of the alpine meadows, bogs, lakes, and rock all the more lively.

The view from Gros Morne takes in only the edges of the Long Range Inlier, which goes on to stretch 160 km (100 mi) north, as far as Canada Bay. Not all of the Inlier is alpine, although it may appear so from the summit of Gros Morne Mountain. Only this southwestern corner—essentially one really large alpine plateau divided by a series of landlocked fjords and glacial valleys—supports extensive alpine ecosystems. The Inlier actually reaches around east of Gros Morne to the granitic Old Crow (558 m / 1830'), the southernmost alpine summit of the Long Range Inlier. North, beyond the Rocky Harbour Hills, rises the Big Level (795 m / 2608'), the high elevation alpine plateau wedged between Baker's Brook Pond and Western Brook Pond, two magnificent landlocked fjords. The plateau continues on the northern side of Western Brook Pond where it is known as the "North Rim," though much of this section of the plateau lies below the treeline. The plateau remains below treeline until it breaks out briefly at the Eastern Brook Mountains, about 12 km (7 mi) east of Cow Head, and then again at the Flat Hills, east of Arches Provincial Park (15-9). Sixteen kilometers (10 mi) farther north, the rocky peak of Gros Pate (656 m / 2152') stands as the last major summit above the trees as the plateau falls below the treeline for most of the remainder of its journey northward. North of Gros Pate, but west of the Inlier itself, are geologically distinct sedimentary highlands that are discussed elsewhere (the Highlands of St. John are profiled in Ch. 16 and Blue Mountain is discussed in Ch. 17). The Inlier remains below the

15-5 The mountains east of Parsons Pond are the first major outcropping of alpine tundra north of the Big Level in Gros Morne National Park. West Brook Gulch, Newfoundland. (IATNL)

15-6 The large alpine areas on the Long Range Inlier offer quintessential Newfoundland scenery with their steep gneissic cliffs plunging into the deep waters of isolated glacial fjords. Hiker at Western Brook Pond. (MA)

15-7 More often than not, strong winds blow off the ocean from the west and thick, low clouds cloak the tablelands, making navigation by sight difficult. Indian Lookout, Newfoundland. (MJ)

15-8 *The woodland caribou of the Northern Long Range Mountains frequent all of the major alpine tundra areas. Parsons Pond Inner Pond, New-foundland. (IATNL)*

15-9 *The alpine tundra surrounding Indian Lookout and the Flat Hills is relatively poorly known. (IATNL)*

15-10 *Western Brook Pond is 164 m deep, with cliff walls extending 600 m above the water's surface. The winds of Western Brook are almost as famous as those of Wreckhouse (Ch. 12). (MD)*

trees for most of the rest of its journey up the coast with only small isolated patches of alpine above the forest. Rounding out the Inlier at its northernmost edge, however, is Cloud Mountain,[1] which, although geologically related, is more than 100 km (60 mi) from Gros Pate, separated by a sea of forest and taiga. We discuss Cloud Mountain separately (Ch. 17).

Access

It goes without saying that most of the alpine portion of the Long Range Inlier is difficult to access, given the limited roads along the Northern Peninsula and the thick forest and tuckamore in the lowlands. There are, however, two straightforward ways to explore the various alpine wildlands of the Long Range Inlier: from within Gros Morne National Park, and by using the International Appalachian Trail's trail system between Parsons Pond and Portland Creek.

The prominent Long Range alpine areas within Gros Morne National Park are accessible by way of two multiday, unmarked, backcountry wilderness expeditions known as the Long Range Traverse and the North Rim Traverse. These routes provide access to Big Level and the Rocky Harbour Hills and the North Rim of Western Brook Pond, respectively. Both routes are wet, steep, long, and difficult and require extensive backcountry experience and navigational skills.

The alpine area on the North Rim can be reached by a very long dayhike from the Western Brook Pond parking area or in a two-day hike by camping at Snug Harbour on the shore of Western Brook Pond, just at the foot of the Inlier. In the early summer, before Parks Canada puts in the bridge over Western Brook Pond, this requires a nasty, cold river crossing.

Alternatively, hikers can begin either traverse from the eastern shore of Western Brook Pond, after a 16 km (10 mi) boat tour of the fjord and its towering cliffs. The boat ride alone is worth the trip. The pond is now fresh water and is 164 m (538') deep at its deepest point,

---

1   The region surrounding the Highlands of St. John, the northern Long Range, and Cloud Mountain is discussed in Chapter 16.

with cliff walls extending to 600 m (1000') above the surface of the water (15-10). The fjords all along the west coast north of Bonne Bay were at one time "true" saltwater fjords connected to the sea. Isostatic rebound of the land from the weight of the Pleistocene glaciers has isolated them. Reservations and a permit are required to be dropped off at the dock on the east side of the lake to begin either traverse. Following a very steep 4 km (2.4 mi) trek out of the gorge on an old, unmarked trail, the Long Range Plateau is reached, affording astounding views down into Western Brook Pond and across the vast alpine tableland.

15-11 The North Rim Traverse and Long Range Traverse diverge at the top of Western Brook Pond. (NR)

From here, the trail disappears, and the two traverse routes split (15-11). Long Range Traverse hikers will usually head south across the eastern edge of the Big Level, lower in elevation than the expansive alpine tundra and snowbeds typically associated with the Big Level. Hikers then pass the eastern edge of the Rocky Harbour Hills and Ten Mile Pond to complete their journey on Gros Morne's Callahan Trail. Several primitive campsites are situated along the route. The North Rim traverse breaks northwest at the head of the Western Brook Pond and traverses an unmarked 27 km (16 mi) around the north side of the fjord. Although shorter, much of this traverse is below the treeline and involves a difficult bushwhack through thick tuckamore. These two wilderness traverses can be combined, foregoing the boat ride, and hikers usually travel from north

15-12 The mountains east of Parsons Pond are now accessible by boat and the IATNL's Devil's Bite Trail. (IATNL)

15-13 Looking north at Portland Creek Inner Pond, Blow Me Down Mountain, Gros Pate, and Blue Mountain, from unnamed peak. (IATNL)

to south, beginning at the Western Brook Pond parking area and ending at Gros Morne. However, the boat ride provides an unusual angle on the fjord and is an experience its own right.

To attempt either traverse, or to hike anywhere in Gros Morne National Park's extensive backcountry, visitors are required to make reservations, obtain a permit and a VHF radio-transmitter (so they can be located in the event of an accident), and complete a pre-trip orientation the day before embarking. The orientation involves an informational video and the requirement that you demonstrate proficiency at map and compass and GPS skills. Wilderness hiking within the National Park is currently open from July through October. For additional information, visit the National Park website or contact park officials.

15-14 East Brook Pond is one of many landlocked fjords on the western face of the Northern Long Range. (IATNL)

15-15 Tufted clubrush (top) and bakeapple (bottom) are common in bogs on the alpine tundra of the Northern Long Range Mountains. (MJ and IATNL)

North of Gros Morne National Park on coastal Route 430, the next best place to access the alpine highlands of the Long Range Inlier is near Parsons Pond (15-12) and at the Flat Hills, where the International Appalachian Trail Newfoundland and Labrador (IATNL) has developed trails into the region. The IATNL's 40 km (24 mi) Indian Lookout Trail loop extends from the base of the Flat Hills, around Portland Creek's Southwest Feeder Gulch, providing access to the extensive Flat Hills tableland and summit (640 m / 2100') as well as Sharp Hill (602 m / 1975') and Indian Lookout (540 m / 1772'), which afford views over Portland Creek Inner Pond, another landlocked fjord (15-13). To access the IATNL trails from the Viking Trail (Route 430), turn east onto Five Mile Road, which is located halfway between the towns of Parsons Pond and Portland Creek, about 1.6 km (1 mile) south of Arches Provincial Park. This gravel road wanders a bumpy 8 km (5 mi) inland toward the base of the Flat Hills. The first few miles are relatively easy-going, but completing the distance to the trailhead might require some walking, or a high-clearance vehicle. The trail system can also be accessed by boat via Parsons Pond and Portland Creek Pond.

Geology

The Long Range Inlier is the easternmost outcropping of the Grenville Province of the Canadian Shield, and is the oldest rock in Newfoundland (about 1.25 billion years). This 160 km (100 mi) long swath of metamorphic rock is primarily gneiss with some more recent igneous intrusions, and it is roughly bounded by the quartzite-capped Gros Morne and Killdevil Mountains to the south (Ch. 14) and Canada Bay to the north.

As the inlier runs northward, just west of the large homogenous block of ancient rock, the metamorphic rocks are sometimes overlain by younger, sedimentary layers. Where these outcrop large enough and high enough to support their own alpine tundra, we have profiled them elsewhere (e.g., the Highlands of St. John in Ch. 16 and Blue Mountain in Ch. 17).

15-16 Beachhead iris (left) and shining rose (right) are easily noticed, if not particularly abundant. Flat Hills, Newfoundland. (IATNL)

Numerous fjords indent the Inlier on its western face (15-14). Following the deglaciation of Newfoundland 10,000 years ago, the land rebounded (a phenomenon known as "isostatic rebound") as it was freed from the weight of its massive glacial burden. The rebound isolated several fjords from the ocean, creating deep freshwater lakes. Western Brook Pond, one of the most prominent landmarks mentioned in the preceding text, is thus stranded about 5 km (3 mi) from the coast.

Ecology

Vegetation — Because the rocks of the Long Range Inlier are primarily gneiss and granite, the alpine plateaus of this region support a flora similar to that found on nearby Gros Morne Mountain (which is silica-rich) as well as potassic summits throughout the eastern alpine region. The flora of the Inlier alpine area within Gros Morne National Park, particularly the Big Level, has been well-studied,[2] with 92 species catalogued in the alpine zone of the plateau (73 of these in snowbeds), including 21 species considered rare in Newfoundland.[3]

The windswept alpine ridges on the Big Level are covered with a typical alpine ridgetop heath and lichen community characterized by pink and black crowberry (*Empetrum eamesii* and *E. nigrum*) and diapensia (*Diapensia lapponica*) along with other typical species including Bigelow's sedge (*Carex bigelowii*), highland rush (*Juncus trifidus*), alpine azalea (*Loiseleuria procumbens*), and alpine bilberry (*Vaccinium uliginosum*). In wetter areas, two alpine bog communities occur: alpine bogs dominated by tufted clubrush ("deergrass" in Newfoundand, *Trichophorum cespitosum*), bakeapple (*Rubus chamaemorus*) and several species of *Carex* (15-15), and an unusual hummocky heath community dominated by crowberry and Pickering's reedgrass (*Calamagrostis pickeringii*).[4] These hummock communities, which are common throughout the plateau, particularly on its eastern side, form unusual patterned ground that is thought to result from a combination of a gentle slope, high moisture content, and thick snow cover.[5] Most notable on the Big Level are the extensive snowbeds, which harbor many species rare in Newfoundland including Lachenal's sedge (*Carex lachenalii*), Labrador willow (*Salix argyrocarpa*), Wormskjold's alpine speedwell (*Veronica wormskjoldii*), alpine cudweed (*Omalotheca supina*), and moss heather (*Harrimanella hypnoides*). Beachhead iris (*Iris*

---

2   e.g., Anions (1994); Bouchard et al. (1994); Brouillet et al. (1998)
3   Brouillet et al. (1998)
4   Brouillet et al. (1998)
5   Brouillet et al. (1998)

15-17 Common boreal species include black crowberry, bunchberry, and Canada mayflower, growing here around a caribou skull. Indian Lookout, Newfoundland. (IATNL)

15-18 Large herds of woodland caribou congregate on the tundra of the Flat Hills to calve. (IATNL)

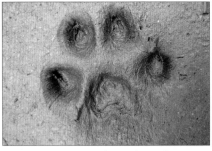

15-19 Lynx range to treeline in the Northern Long Range Mountains. (NR)

hookeri) and shining rose (*Rosa nitida*) are occasional in wet seepages (15-16). Farther inland, the very rare mountain fern (*Thelypteris quelpaertensis*) occurs disjunct from the nearest sites in Britsh Columbia, Alaska, and Washington.

By comparison, the Flat Hills have received little attention from botanists, though surveys of the area suggest the flora is quite similar to the Big Level areas farther south. The exposed headlands of the Flat Hills themselves, and near Indian Lookout, are thickly carpeted by crowberry, diapensia, alpine azalea, and highland rush. Oakes' eyebright (*Euphrasia oakesii*) and other easily overlooked rarities are present. In the Flat Hills, as on Big Level and the Rocky Harbour Hills, common boreal species are abundant under tuckamore and in the protected lee of large rocks. Here, crowberry, bunchberry, and Canada mayflower (*Maianthemum canadense*) dominate (15-17).

Fauna — All of the alpine areas of the Long Range Inlier support characteristic alpine fauna found throughout Newfoundland. Herds of caribou use the high elevation areas for calving and are often present throughout the rest of the year (15-18). The arctic hare (*Lepus arcticus*) can be found in alpine areas throughout the Long Range, along with both rock ptarmigan (*Lagopus muta*) and willow ptarmigan (*L. lagopus*). With increasing frequency, Canada geese (*Branta canadensis*) are seen foraging on the tundra. Several lowland terrestrial species also occasionally make their way up to the alpine areas including black bear (*Ursus americanus*), lynx (*Lynx canadensis*), coyote (*Canis latrans*, an island newcomer), and moose (*Alces americanus*) (15-19). Amphibians seem to be less abundant and diverse in this part of Newfoundland than in the Southern Long Range, the Bay of Islands, and Gros Morne, although this hasn't been demonstrated. Arctic char (*Salvelinus alpinus*) may be found in landlocked lakes throughout the mountains, and Atlantic salmon (*Salmo salar*), arctic char, and brook trout (*Salvelinus fontinalis*) occur in the deep landlocked fjords.[6]

---

6   Parks Canada (1986)

Highlands of St. John 16
NEWFOUNDLAND

North Summit, Highlands of St. John, Newfoundland. (MJ)

Map 16-1. Highlands of St. John, Newfoundand.

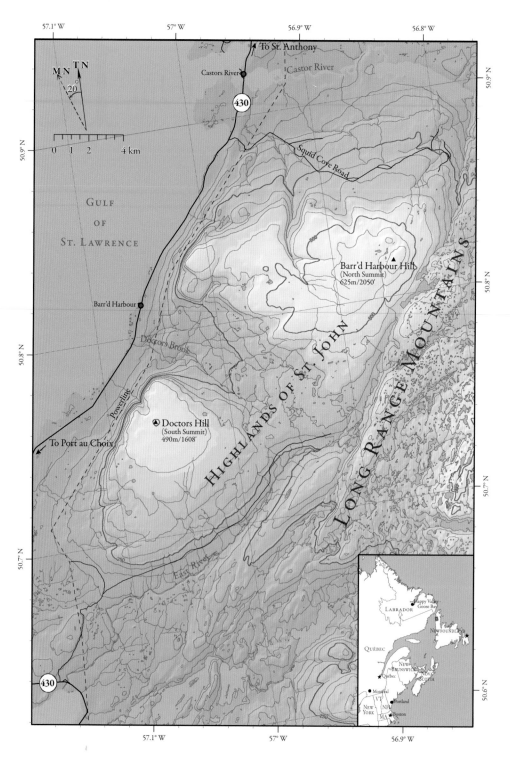

Highlands of St. John, NEWFOUNDLAND

16-1 The Highlands of St. John appear to rise directly from the sea about 130 km north of Gros Morne and the same distance south of L'Anse aux Meadows. (MJ)

On the Great Northern Peninsula of Newfoundland, life is stripped down to its basic boreal elements. The rolling coastal forests are finely dissected by sheltered ports and harbors, and punctuated by high and stony alpine tundra and coastal limestone barrens. The peninsula terminates in a series of rounded bluffs, capes, and peninsulas facing Labrador's Forteau Tablelands and the old gneiss of the Canadian Shield. There, the remains of North America's only confirmed Norse settlement were scoured by the wintry Labrador Current for 960 years before its remains were discovered in 1960 by Drs. Helge and Anne Stine Ingstad.

The Highlands of St. John appear to rise directly from the sea about halfway up the Great Northern Peninsula—132 km (80 mi) north of Gros Morne and the same distance south of the Ingstads' Viking site at L'Anse aux Meadows (16-1). This pair of dignified (if not majestic) mountains is situated only 2 to 3 km (1.2–1.8 mi.) from the coast-highway.[1] As a result, the Highlands are relatively accessible, if seldom explored on foot by hikers.

The Highlands' formal name doesn't come from the capital city of St. John's on the southeastern tip of Newfoundland (almost 450 km [300 mi] to the southeast), but instead from their exposed vantage overlooking St. John Island in the Gulf of St. Lawrence. Harvard's Dr.

---

1   The Viking Trail, Route 430, was constructed in 1967 along the coast just west of the Highlands, providing access from the Deer Lake region to the Great Northern Peninsula.

16-2 *The expansive summits are capped by quartz-rich felsenmeer. (MJ)*

16-3 *Numerous scattered gulches, canyons, and subalpine river valleys dissect the plateaus. (MJ)*

M.L. Fernald visited the Highlands of St. John during two expeditions in 1925 and 1929,[2] after having seen them from the ocean. Amid his carefully detailed plant lists are clever and humorous insights into the logistics of alpine adventure on the edge of the modern world almost a century ago.

Eight decades later, the Highlands feel only slightly closer to civilization, despite the close proximity of the 45-year-old two-lane coast highway. Numerous logging roads now cut through spindly forests of black spruce (*Picea mariana*) to reach high onto their northern and southern flanks, but it still takes some trial and error (as well as maps, aerial photos, and GPS or compass) to avoid a tedious bushwhack through timberline tuckamore. Summiting these mountains on foot, without knowledge of the caribou trails and ATV tracks that cut through the scrub, must be fairly similar to what it was like in the early 1900s, although the coast highway (Route 430) cuts off a bit of distance.

Since Fernald's visits, several botanists have explored various parts of the mountains, identifying approximately 400 species of vascular plant species, including more than 50 rare and 30 uncommon species in Newfoundland, making this one of the most interesting and diverse alpine plant sites in the East.[3]

From the coast highway, the Highlands present themselves to the northbound traveler as a jarring series of abruptly rising, banded sandstone and limestone cliffs, broadly topped with expansive summits of sparkling, whitish-gray, quartz-rich felsenmeer (16-2). These twin summits are perhaps the most striking and distinctive alpine mountains of northern Newfoundland (north of Gros Morne). Among eastern mountains, the Highlands are visually unique. Their steep western cliffs provide sweeping views of the ocean to the west—an open vista as great as that from Gros Morne or Killdevil to the south. Eastward, the silver-pink, windswept fellfields afford the occasional glimpse into the numerous scattered "gulches," "canyons," and lush subalpine river valleys that cut into the plateau and drain west to the sea (16-3). The Highlands are among the best places in the East to observe woodland caribou (*Rangifer tarandus caribou*): deep, entrenched caribou trails and wallows mar the peat

---

2   Fernald (1926; 1933). Fernald wrote several dozen accounts of alpine explorations in the early 20th century, many published in the esteemed Boston journal *Rhodora*—and they are worth the effort to find and read. He went on two multiday expeditions to the North Summit of the Highlands of St. John, both in 1925, and also took day trips to a limestone band on the South Summit or Doctors Hill, in 1925 and 1929. There he found the extremely rare *Draba pycnosperma*, which Claudia Hanel has unsuccessfully looked for on several occasions.

3   The Highlands were visited by R. Tuomikoski, A.W.H. Damman, G.W. Argus, G. McPherson, S. Hay, A. Bouchard, N. Djan-Chékar, and C. Hanel during a number of separate botanical surveys between 1949 and 2003.

and turf above treeline. Large numbers of antlers strewn across the rocky plateau suggest that in some seasons, the Highlands abound with these majestic creatures (16-4). In fact, Fernald himself noted that local guides called the eastern dome of Barr'd Harbour Hill just east of Deer Pond Gulch the "Plains of Waterloo" due to "great slaughters of caribou" that occurred in that area in the past—a shaky connection to the Belgian flats that witnessed Napoleon's defeat in 1815.

In the past decade, the Highlands have been variously proposed for either silica mining or some form of protected status,[4] giving a taste of the precarious situation several eastern mountains now find themselves in. The Highlands are also along the route of the proposed Labrador-Island Transmission Link (a high-voltage direct-current transmission line), which would run along their east side.[5] It is likely that the coming decade will see either dramatic change associated with further development or natural resource extraction, or (preferably) some resolution as to their permanent conservation status.

16-4 Large caribou herds on the Highlands are evidenced by abundant trails and strewn antlers. Here, a shed antler is overgrown by Labrador tea. (MJ)

Lay of the Land

The Highlands of St. John comprise two mountains, North and South. These mountains resemble one another from the west, and are separated by Doctors Brook, a great and

16-5 The North and South Summits of the Highlands of St. John are separated by Doctors Brook Valley. (WLD)

diverse subalpine watershed. The northernmost is known as North Summit or Barr'd Harbour Hill (625 m / 2050'), and is taller and larger than the South Summit, also known as Doctors Hill (490 m / 1608') (16-5). The summits appear similar from the highway because both peaks rise abruptly from the coast to an elevation of about 490 m before either gradually rising eastward to a distant summit (in the case of the North Summit), or sloping downward into thick spruce forest (in the case of the South Summit). The two summits eventually join together at a low plateau several miles east of the highway, enclosing the headwaters of Doctors Brook. The Highlands are bounded by the Castor's River to the north, and the East River to the east. Both mountains are topped by whitish-gray, expansive tablelands of broken quartz arenite (16-6). From a distance, these appear relatively featureless. Closer inspection, of course, reveals many broad krummholz valleys, long ridges, and undulating hills surrounding small ponds, few of which have been given names (16-7).

---

4   Dickson (2002) and WERAC (2007), respectively.
5   See a report by Nalcor Energy (2009)

16-6 *Both of the Highlands of St. John are topped by whitish-gray, expansive tablelands of broken quartz arenite. (MJ)*

16-7 *Many of the ponds, ridges, and valleys on the Highlands are unnamed. (MJ)*

16-8 *Squid Cove Road is among the logging roads that wrap around the Highlands of St. John, providing increased access to the alpine tundra. (MJ)*

Access

Like many of Newfoundland's off-the-beaten-path peaks, accessing the Highlands' alpine area is difficult and intense but relatively fast. Regardless of the route you choose, you'll be picking your way through thick spruce forest and krummholz. A growing network of logging roads north of the Highlands reaches the coast highway (Route 430) via Squid Cove Road and creeps up the hillside toward treeline on Barr'd Harbour Hill (16-8). In a sturdy, high-clearance vehicle, you can wind your way to within a kilometer or so of the rocky plateau, leaving only a couple hours' struggle through the tuckamore to reach the open barrens. In at least one case, logging roads come within 500 m of the North Summit tundra.

Similarly, a rough access road from Hawke's Bay weaves around the southeastern end of the range, mostly within about 3 km (1.8 mi) of Doctors Hill.[6] Alternatively, several overgrown, impassable access roads cut east from coastal Route 430 toward a powerline right-of-way that runs between the road and the mountains, putting the hiker a very steep 1 km (0.6 mi) bushwhack below the westernmost point of the northern plateau.[7]

After fighting your way to the rocky top of either summit, you'll find the gently sloping plateau seems to go on forever. The northern peak extends a full 10 km (6 mi) east of the highway, and the southern peak, 6 km (3.6 mi), allowing many days of undisturbed alpine exploration. For the time being, that is. If the Highlands and their surrounding old growth forests are not given protected status, it is unclear how the aggressive logging will influence the wild character of the mountains (16-9).

---

6   Dickson (2002)

7   There are some trails beyond the powelines that skirt the western edge of the mountains, but at least the southern one (accessing Doctors Hill) is extremely wet, and one still has to cut though the forest up the side of the mountain. The trick is to go up the valley a bit more and ascend the northern flank, rather than attacking the northwestern edge just past the cliffs. The northern flank has some moose trails going right through the tuckamore.

Geology

The Highlands are composed of Cambrian sedimentary bedrock, which is much younger and entirely distinct from the adjacent gneiss of the Long Range Mountains that trail away from the Highlands toward Cloud Mountain on the eastern shore of the Great Northern Peninsula. The Highlands' characteristic banded western cliffs are made up of several geological formations. The thick band of lower cliffs belongs to the Bradore Formation (a mixture of sandstone, conglomerate, and shale). Above this rests the Forteau Formation, thinner layers of shale and limestone. The limestones of the Forteau Formation support unusual limestone alpine vegetation on the slopes and cliffs, described in greater depth later. The Highlands are topped by Hawke Bay Formation quartz arenite, a metamorphosed sandstone.[8] This rock type creates the great felsenmeer of the plateaus and the striking frost polygons throughout the alpine tundra. The calcareous bedrock of the northwestern lower slopes (facing the highway) belongs to the Watts Bight Formation.

Several researchers have concluded that the Highlands were unglaciated during the most recent glacial events of the late Pleistocene, although more recent work has identified clear evidence of glaciation on the plateaus, including gneissic erratics from Labrador and glacial striae (16-10).[9]

Ecology

Vegetation — The windswept, white, felsenmeer of quartz arenite supports what eventually comes to seem like a very familiar and common alpine flora: three-toothed cinquefoil (*Sibbaldia tridentata*), black crowberry (*Empetrum nigrum*), alpine bearberry (*Arctous alpina*), mountain cranberry or partridgeberry (*Vaccinium vitis-idaea*), alpine bil-

16-9 The Highlands of St. John and the surrounding old-growth spruce forest are suitable and appropriate conservation targets. (MJ)

16-10 Several researchers have concluded that the Highlands of St. John escaped glaciation during the most recent glacial advance. (MJ)

16-11 Frost boils (top) on the felsenmeer are eventually colonized by common juniper, black spruce, and black crowberry (bottom). (MJ)

8   Dickson (2002)
9   Dickson (2002)

16-12 Alpine sweetgrass (top) and alpine bearberry (bottom) were among the "taboo" species listed by M.L. Fernald as hardly worth the time to collect. It grows abundantly on both summits. (MJ)

16-13 Red baneberry (left) and purple avens (right) are among the many plants found along subalpine streams on the Highlands. (MJ)

berry (*Vaccinium uliginosum*), and Labrador tea (*Rhododendron groenlandicum*) dominate the open rockfields, with scattered patches of alpine azalea (*Loiseleuria procumbens*), diapensia (*Diapensia lapponica*), mosses, and lichens. In places on the tableland, the felsenmeer is pocked with polygons and circles of vegetation, some filled with crowberry and others by black spruce or common juniper (*Juniperus communis*). Among these are occasional frostboils of clay and gravel (16-11).

On many of Newfoundland's rounded summits, and true here too, the tableland does not support the bulk of the alpine plant diversity. Rather, most of the unusual alpine species in this area occur in subalpine valleys and along seepages on limestone cliffs. In both of his descriptions of the Highlands of St. John, Fernald describes how unimpressed he was with the floristic diversity of the tablelands themselves. In 1926[10] he noted, "We were on a vast tableland of hopelessly sterile quartzite and tried to be enthusiastic over *Juncus trifidus, Luzula spicata, Anthoxanthum monticola*, and the other ubiquitous plants of the most silicious mountain-crests." Later, describing the plateau,[11] he complains upon reaching it, "The open tableland itself was a disappointment, too dry for any but the most extreme xerophytes of any bleak and arid silicious summit. In a few places, where a damp sphagnous carpet occurred, the plants were the most ordinary of species of any wet peaty lowland in New England or even the Southeastern States" (16-12).

By contrast, vegetation completely distinct from the alpine tableland persists on the slopes along the many subalpine brooks. Here, there is a tall *Angelica* and several species of small willowherbs (*Epilobium*), including *E. hornemannii* and *E. lactiflorum*. Clasping-leaved twisted-stalk (*Streptopus amplexifolius*), alpine bistort (*Polygonum viviparum*), purple avens

---

10   Fernald (1926)
11   Fernald (1933)

(*Geum rivale*), red baneberry (*Actaea rubra*) and Chinese hemlock parsley (*Conioselinum chinense*) are common (16-13), with some of the nearly ubiquitous boreal species such as starflower (*Trientalis borealis*), Canada mayflower (*Maianthemum canadense*), and others. Several plants of restricted range in Newfoundland can be found growing locally at the steep heads of brooks, and in snowbeds in the deeper ravines. These include alpine clubmoss (*Diphasiastrum alpinum*), alpine ladyfern (*Athyrium alpestre*), snowbed willow (*Salix herbacea*), mountain chickweed (*Cerastium cerastoides*), purple mountain heather (*Phyllodoce caerulea*), sibbaldia (*Sibbaldia procumbens*), Lachenal's sedge (*Carex lachenalii*), alpine mountain sorrel (*Oxyria digyna*), and the long-styled sedge (*Carex stylosa*)—all plants of restricted range in Newfoundland (16-14).

16-14 Alpine snowbeds on the Highlands support numerous rare alpine species, such as snowbed willow. (MJ)

The cliffs also host several interesting and rare species including Chinese firmoss (*Huperzia miyoshiana*), smooth woodsia (*Woodsia glabella*), arctic bluegrass (*Poa arctica*), tufted alpine saxifrage (*Saxifraga cespitosa*), alpine rock cress (*Arabis alpina*), alpine groundsel (*Packera pauciflora*), and Rocky Mountain willowherb (*E. saximontanum*) along with more widespread species like roseroot (*Rhodiola rosea*) (16-15).

16-15 Alpine cliffs in the Highlands of St. John support rare species such as Chinese firmoss (top) and ubiquitous cliff species such as roseroot (bottom). (MJ)

The Highlands were thought to support the only known Newfoundland population of Fernald's milkvetch (*Astragalus robbinsii* var. *fernaldii*), the variety of Robbins' milkvetch endemic to the Strait of Belle Isle.[12] The taxonomic status of this variety (and the Highlands' station) was up for debate, however, and was examined by Paul Sokoloff in his M.Sc. thesis.[13] After genetic and morphological analysis of many milkvetch populations in Newfoundland, southeastern Labrador, and extreme northeastern Québec, he determined that most material called Fernald's milkvetch is a minor variant of elegant milkvetch (*As-*

12   COSEWIC (1997)
13   Sokoloff (2010)

16-16 Woodland caribou may be frequently observed on the plateaus of the Highlands of St. John. (MJ)

16-17 Moose are frequent at all elevations, including treeline meadows and tuckamore. (IATNL)

16-18 White-lipped snails are locally abundant in wet meadows and cliffs near treeline. (MJ)

tragalus eucosmus). Sokoloff also concluded that that this particular population was misidentified and is actually Robbins' milkvetch (*Astragalus robbinsii* var. *minor*), a much rarer species than elegant milkvetch.[14]

In protected areas under black spruce krummholz, starflower, bunchberry (*Cornus canadensis*), northern comandra (*Geocaulon lividum*), bakeapple (*Rubus chamaemorus*) and Canada mayflower are common. Especially on the eastern portion of North Summit above its central canyon, small bogs fill depressions on the tableland where tufted clubrush (*Trichophorum cespitosum*) and bakeapple (cloudberry) grow thick.

The Highlands host an unusually vast array of regionally rare plants between the major habitat types—the alpine tablelands, the high wet slopes, and steep cliffs, making them an ideal target for long-term conservation.

Fauna — The fauna of the Highlands of St. John has not been systematically studied, with the exception of ongoing caribou predator-prey studies[15] on the Great Northern Peninsula and province-wide assessments of arctic hare distribution.[16] Caribou are clearly in evidence throughout the plateaus, and long trails cut through the wet sod, sometimes running parallel in many lines (16-16). In addition to a highly visible population of woodland caribou, which range along the Northern Peninsula as far as St. Barbe,[17] the plateaus harbor an isolated population of arctic hare (*Lepus arcticus*).[18] Moose (*Alces americanus*) are common to treeline and above (16-17).

---

14 Maunder (2011)
15 e.g., Rayl (2012)
16 Bergerud (1967)
17 Rayl (2012)
18 Bergerud (1967)

As on the Newfoundland summits already profiled, such as Gros Morne to the south, rock ptarmigan (*Lagopus muta*) are also commonly observed on the Highlands.

The white-lipped snail (*Cepaea [Helix] hortensis*) is locally abundant on the higher slopes near treeline (16-18). In fact, Fernald's exploratory party in the 1920s named a portion of the slope leading to the North Summit the Helix Slide. It is also common in some years to find evidence of the vapourer or rusty tussock moth (*Orgyia antiqua*), a European species introduced to Newfoundland (16-19).

16-19 Rusty tussock moths were introduced to Newfoundland from Europe but are now a frequent sight even at high elevations. This caterpillar is feeding on bunchberry, or "crackerberry" as it's called in Newfoundland. (MJ)

Other Alpine Sites *in* Newfoundland 17

Burnt Cape, Newfoundland. (MJ)

Map 17-1. Other alpine sites in Newfoundland.

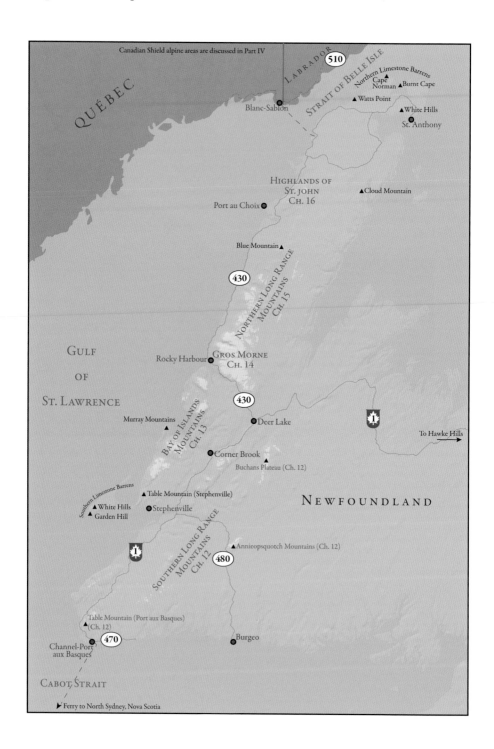

Other Alpine Sites in Newfoundland

17-1 Hawke Hill Ecological Reserve protects a portion of the Avalon Peninsula's alpine ecosystems. (MK and DD)

Hawke Hills

When traversing the center of the Avalon Peninsula in southeastern Newfoundland along the Trans-Canada Highway (Route 1), travelers pass very close to the Hawke Hills, a rounded alpine ridge with an assortment of communications towers. A short ascent along a gravel access road provides access to the towers. Visitors will be greeted more often than not by fog or a stiff blast of wind, recalling some of the important reasons this area is alpine despite its low elevation of approximately 300 m (984') (see Ch. 4). The open hillcrest provides a good vantage point to observe "fingers" of fog creeping inland along the valleys from the Atlantic Ocean, a phenomenon locally referred to as "the white hand," an eerie name, apt for this lonely, windblown, oceanfront hill.

An off-road vehicle (ORV) trail leads from the towers to the Hawke Hill Ecological Reserve, which protects a portion of this unique alpine ecosystem, purportedly the easternmost alpine site in mainland North America (17-1). The treeless mosaic of rocky ground and wetlands stretches for approximately 80 km (50 mi) to the southwest to the southernmost tip of the Avalon Peninsula, but the Hawke Hills have the highest elevation and are the most alpine in character.[1]

---

1 Visitors to this area should note, however, that the Avalon Wilderness Area was established in the central portion of these barrens to protect the main population of caribou on the Avalon Peninsula.

17-2 Pink crowberry, three-toothed cinquefoil, diapensia, and tufted clubrush are common above treeline in the Hawke Hills region. (MK and DD)

17-3 Table Mountain near Stephenville, formerly the site of a U.S. Air Force radar station, provides distant views of the Lewis Hills (see Ch. 13). (MA)

All of the characteristic alpine phenomena are visible here: rocky frost-patterned ground interspersed with low mats of black and pink (Eames') crowberry (*Empetrum nigrum* and *E. eamesii*), diapensia (*Diapensia lapponica*), three-toothed cinquefoil (*Sibbaldia tridentata*), and common alpine graminoids such as tufted clubrush (*Trichophorum cespitosum*) (17-2). Other alpine plants found here include alpine azalea (*Loiseleuria procumbens*) and highland rush (*Juncus trifidus*). Saxicolous (crustose) lichens are abundant on the many glacial erratics strewn about the tundra. These include target lichen (*Arctoparmelia centrifuga*), which grows in concentric rings; the alpine bloodspot (*Ophioparma ventosa*), a pale grey-green lichen with bright red fruiting bodies; black or dark gray rock tripe (*Umbilicaria* and *Lasallia* spp.) and the familiar (and ubiquitous) map lichen (*Rhizocarpon geographicum*) (see Ch. 5). On the subalpine flanks of the Hawke Hills, lowbush blueberries (*Vaccinium angustifolium* and *V. boreale*) abound in the early fall. In late summer and fall, wild-picked blueberries and other local berries are sold along the Trans Canada Highway at pullout spots.

Southern Limestone Barrens

Table Mountain (>381 m / >1250') is a limestone ridge located east of the road from Stephenville to Point au Mal, close to Newfoundland's southwestern corner and just south of the Lewis Hills portion of the Bay of Islands complex (Ch. 13).[2] Table Mountain is accessible by a gated communications road that leads to the more southerly of its indistinct summits. In the vicinity of Stephenville, Table Mountain is locally known as "Pinetree Ridge," a reference to the U.S. Air Force surveillance radar station (part of the Cold War–era Pinetree Line) that occupied its summit from the early 1950s until 1971. Remains of the radar site are still visible around the southern end of the ridge.

A rocky monolith near the sea, Table Mountain supports a fascinating example of alpine tundra between the Port au Port Peninsula in the south and the Bay of Islands in the north,

---

2   At least three large alpine mountains in Newfoundland are named *Table Mountain* or some variation thereof. The southernmost, near Port aux Basques (part of the Southern Long Range Mountains) is mostly granitic and is profiled in Chapter 12. The more northerly example (known commonly as the *Tablelands*, but also known as *Table Mountain*) is composed largely of serpentine and forms the northern end of the Bay of Islands complex (Ch. 13).

and is strikingly different from other large mountains nearby, such as the Lewis Hills a few kilometers to the north (17-3). About 40 km (25 mi) west of Table Mountain, the Garden Hills and White Hills (>354 m / >1160') comprise rocky limestone barrens on either side of Route 460. The latter of these share their name with the serpentine White Hills near St. Anthony, at the far-northern tip of the island of Newfoundland (discussed later in this chapter).

The limestone has weathered to fine gravel on Table Mountain and in the Garden Hills (the portion of the White Hills to the west of Route 460, the road encircling the Port au Port Peninsula). Strips of gravel barrens alternate with strips of calcareous heath and shallow valleys filled with spruce-dominated tuckamore. The flora is very similar to that of the limestone barrens on the Great Northern Peninsula or the limestone cliffs near the Bay of Islands, with numerous rare calciphile plant species such as entire-leaved mountain avens (*Dryas integrifolia*) and purple mountain saxifrage (*Saxifraga oppositifolia*). Other noteworthy species found on Table Mountain include smooth woodsia (*Woodsia glabella*), bright-green spleenwort (*Asplenium viride*), and narrowleaf arnica (*Arnica angustifolia*), all three of which range west to the Rocky Mountains (17-4). Table Mountain supports the southernmost outliers of two arctic species listed as "Endangered" in the Province of Newfoundland and Labrador: Low northern rockcress (*Braya humilis*) (17-5) and Mackenzie's sweetvetch (*Hedysarum boreale* ssp. *mackenziei*).

Murray Mountains

Northwest of Blow Me Down Mountain (Ch. 13), immediately adjacent to of the village of Lark Harbour, are Mount Maud, Murray Peak, and Lark Mountain—the Murray Mountains, which rise to about 320 m (1500'). Lark Mountain and Murray Peak support an interesting juxtaposition of alpine plants with more southerly plants typical of the Atlantic coastal plain. Just north of the Murray Mountains, the Bay of Islands itself—the major land-

17-4 Table Mountain's limestone slopes support a variety of unusual alpine plant species, including smooth woodsia (top), brightgreen spleenwort (middle), and narrow-leaved arnica (bottom). (MA)

17-5 Table Mountain also supports the southernmost outlier of low northern rockcress. (MA)

17-6 The Murray Mountains, viewed from Blow Me Down Peak. From left to right are: Virgin Mountain, Murray Peak and Lark Mountain, and Mount Maud (top). Guernsey and Tweed, at the mouth of the Bay of Islands, are visible in the winter image, taken from the North Arm Hills. (IATNL)

mark of the region—harbors interesting alpine communities on the islands of Guernsey (321 m / 1053') and Tweed (220 m / 720'). Other nearby ridges, such as Virgin Mountain and Little Port Head, are also crowned with alpine summits. The South Head Lighthouse trail and the Little Port Head trails (constructed by the Outer Bay of Islands Enhancement Committee) provide access to the alpine zone of the Murray Mountains' Little Port Head, and have some stunning ocean views (17-6).

Blue Mountain

Blue Mountain (647 m / 2124') is a standalone summit set apart from the Long Range Mountains. It lies about 25 km (15 mi) inland from the coast highway and too far from Gros Morne National Park to receive much foot traffic. Blue Mountain may be accessed via logging and camp roads that run east from the Viking Trail (Route 430) in the vicinity of Hawke's Bay. The logging roads nearly encircle the mountain, and on the north side of the hill, it is possible to drive within easy walking distance of the summit. The topography and geology of this mountain are so similar to the Highlands of St. John, that it could be Highlands' lost little brother (it lies only 30 km [20 mi] to the south). On the northern and western slopes of the hill, below the felsenmeer summit of the Hawke Bay Formation are a number of small brook ravines and cliffs composed of limestone outcrops of the Forteau Formation. These areas are also late snowbeds and support both calciphiles and snowbed flora. The area has not been exhaustively surveyed and a comprehensive flora of the mountain would be an exciting contribution to Newfoundland botany.

Cloud Mountain

The gneissic Long Range Mountains trail away from the Highlands of St. John (Ch. 16) toward Cloud Mountain (337 m / 1105'), a low summit due west of the small town of Roddickton. Cloud Mountain is not road-accessible, but may be accessed via boat from Rod-

17-7 The White Hills form two distinct serpentine plateaus near the port town of St. Anthony on the Great Northern Peninsula. (MJ)

17-8 The serpentine plateaus of the White Hills are heavily vegetated by hoary rock moss (detail at right). (MJ)

17-9 Dominant shrubs on the serpentine felsen-meer include common juniper (top, middle) and Lapland rosebay (bottom). (LW and MJ)

dickton. Its higher reaches support several rare plants including alpine bittercress (*Cardamine bellidifolia*). Like Blue Mountain to the southwest and the White Hills to the north, Cloud Mountain has been only cursorily studied, and a thorough floristic study would complement those already undertaken in Gros Morne National Park, on the Highlands of St. John (Ch. 16), and in the Northern Limestone Barrens (discussed later in this chapter).

White Hills

Newfoundland's Great Northern Peninsula makes its most important contribution to the regional alpine flora either on the cliffs of the Highlands of St. John (Ch. 16) or the limestone barrens of Cape Norman and Burnt Cape. However, if not for the White Hills near St. Anthony, the peninsula would be almost completely devoid of the serpentine flora that is so well represented in the Bay of Islands region to the south (Ch. 13). The White Hills form two distinct plateaus immediately west of the northern port town of St. Anthony, the western of which is essentially bisected into two small plateaus by a shallow drainage that trends southeast (17-7). Combined, these plateaus contain about 60 km² (23 mi²) of exposed serpentine soils. The highest of these plateaus (the easternmost) rises only to about 320 m (>1050') but nevertheless provides conditions similar to the much higher serpentine-alpine ecosystems found in the Bay of Islands. The plateaus of the White Hills can be accessed through a mosaic of dense spruce forest and shrubby clearcuts (with only one or two river crossings) from Route 430 (the Viking Trail), leaving from the junction with the road

17-10 The serpentine felsenmeer of the White Hills (top-left) supports numerous interesting spiders, including this furrow spider (bottom) and thinlegged spider (right). (LW and MJ)

to St. Lunaire-Griquet. The serpentine plateaus exhibit well-preserved frost features such as sorted soils and stone nets, which are heavily vegetated by hoary rock moss (*Racomitrium lanuginosum*) (17-8). As on the more southern serpentine mountains in the Bay of Islands region, common juniper (*Juniperus communis*) is a dominant shrub, occurring in twisted krummholz form, and Lapland rosebay (*Rhododendron lapponicum*) is also abundant (17-9). Siberian sea thrift (*Armeria maritima*) and Chinese hemlock parsley (*Conioselinum chinense*) are common. Many small pools dot the treeless barrens, and numerous interesting spiders such as furrow spiders (*Larinioides* sp.) and thinlegged wolf spiders (*Pardosa* sp.) are frequent among the serpentine felsenmeer, which is strewn with erratic boulders thickly vegetated in lichens (17-10). As is currently the case with Cloud Mountain and Blue Mountain (this chapter), the White Hills await a thorough examination of their flora and fauna.

17-11 At a dozen sites on the northwest coast of Newfoundland's Northern Peninsula, the coastal scrub gives way to limestone barrens. (MJ)

Northern Limestone Barrens

In about a dozen places between Bellburns and Cook's Harbour on Newfoundland's Great Northern Peninsula, the coastal forest and scrub give way to open, frost-patterned barrens on limestone bedrock (17-11). In many ways these areas are similar to the Southern Limestone Barrens (discussed briefly earlier). Although few of the northern barrens rise to impressive heights, they are subjected to similar ecological phenomena as the alpine areas including strong winds, reduced growing season, and severe frost-action in the soil. Correspondingly, they tend to support disjunct arctic-alpine plants in the same way as the true alpine mountains.

Unlike most of the alpine mountains in the East, the vegetation of the limestone barrens is composed largely of calciphiles, including many of the same species found on Mont Logan, Québec (Ch. 9), Table Mountain near Port au Port (this chapter, earlier), Killdevil Mountain (Ch. 14), the Highlands of St. John (Ch. 16), and portions of the Forteau Tablelands (Ch. 21). Because of their relative accessibility, the limestone barrens provide ample opportunity to study and learn the limestone alpine plants of the eastern region.

Many of the coastal barrens are accessible from the Viking Trail (Route 430), which also provides access to Gros Morne, the Long Range Mountains, and the Highlands of St. John. Visiting any of these great mountains means driving near, or across, wonderful examples of limestone habitat. The far-northern barrens, for example, Watt's Point, Cape Norman, and Burnt Cape, are road-accessible from Route 430 between Eddie's Cove East and St. Anthony. Watt's Point is accessible from both the north and south, but only with difficulty

17-12 The dull grey limestone contrasts sharply with the dark Gulf of St. Lawrence and the elegant sunburst lichen that often colonizes exposed areas. Burnt Cape, Newfoundland. (MJ)

and a high clearance vehicle. ORVs are permitted on the old roadbed, which was once the main highway on the northern end of the Great Northern Peninsula. Of the southern sites, Bellburns and Port au Choix are the most extensive, but noteworthy exposed limestone barrens can also be found from Anchor Point to Eddies Cove East. In these areas, construction of the Viking Trail in the 1960s severely degraded much of the original habitat. At the Port au Choix National Historic Site, the Burnt Cape Ecological Reserve, and the White Rocks Walking Trail in Flowers Cove, the features and flora of the limestone barrens are interpreted for the public.[3]

In our opinion, the most stunning sites are Cape Norman and Burnt Cape, two of the northernmost sites, located about 13 km (8 mi) apart at the absolute tip of the Northern

---

3   An excellent, recent guide to the Newfoundland limestone barrens is Burzynski and Marceau (undated). The Limestone Barrens Habitat Stewardship Program hosts a fantastic website (www.limestonebarrens. ca), created by M. Brazil and maintained by J. E. Maunder that is rich in information and includes many links to additional resources about Newfoundland's limestone barrens.

17-13 Among the four saxifrages common on exposed limestone are yellow mountain (top) and tufted (bottom). (MK, DD, and MJ)

17-14 Arctic bladderpod ranges into the high arctic and is known from several coastal limestone barrens in northern Newfoundland. (MJ)

Peninsula.[4] Here, calcium-loving alpine species abound. Dull-grey limestone contrasts sharply with the dark sea and bright orange elegant sunburst lichen (*Xanthoria elegans*) (17-12). The cliffs of Labrador's Forteau Tablelands (on the Canadian Shield) are visible across the Strait of Belle Isle on clear days. Perhaps the more unique and historically significant is Burnt Cape, located just west of the town of Raleigh at Newfoundland's far northern tip, but Cape Norman is nearly as interesting. At Cape Norman the northerly feel is inescapable: not a single tree, not even a stunted patch of tuckamore, as far as the eye can see. Burnt Cape, and the more westerly Watts Point, are protected as ecological reserves. All are worth a few days' visit in June or July, at a bare minimum.

In addition to widespread alpine species like alpine timothy (*Phleum alpinum*), three-toothed cinquefoil, alpine azalea, Lapland rosebay, and alpine bearberry (*Arctous alpina*), which occur on exposed ledges above the ocean, these barrens support at least four saxifrages including yellow (*Saxifraga aizoides*), white (*S. paniculata*), and purple mountain (*S. oppositifolia*), and the distinctive tufted saxifrage (*S. cespitosa*) (17-13). Roseroot (*Rhodiola rosea*), Siberian sea thrift, alpine meadowrue (*Thalictrum alpinum*), net-veined willow (*Salix reticulata*), soapberry (*Shepherdia canadensis*), and moss campion (*Silene acaulis*) are abundant. Arctic bladderpod (*Physaria arctica*) is widespread, though formerly thought to be restricted to a few sites (17-14). In damp areas of poor drainage, Canada burnet (*Sanguisorba canadensis*), northern spikemoss (*Selaginella selaginoides*), and common butterwort (*Pinguicula vulgaris*) are frequent. In addition to these, Burnt Cape also hosts a handful of arctic or endemic rarities, including velvetbells (*Bartsia alpina*), a parasite on

---

4   Burnt Cape was among the numerous limestone sites visited by Merritt Lyndon Fernald and colleagues in 1925. Fernald (1926) is absolutely an essential reference for anyone exploring or studying the limestone vegetation of northern Newfoundland. Cape Norman, to the west, was visited by members of Fernald's party during the same summer.

17-15 Burnt Cape, Newfoundland, supports the southernmost populations of the arctic species velvetbells (left) and dwarf hawksbeard (right). (MJ)

willowheaths and other species; dwarf hawks-beard (*Askellia nana*), known from the higher mountains of northern Labrador (Ch. 21) (17-15); and the Burnt Cape or pretty cinque-foil (*Potentilla pulchella*), formerly considered (by M.L. Fernald and others) to be an endemic species (as *P. usticapensis*) but highly disjunct from the main arctic range even as *P. pulchella*. Three vascular plant species are endemic to Newfoundland: Long's Braya (*Braya longii*), Fernald's Braya (*Braya fernaldii*), and barrens willow (*Salix jejuna*). All three are restricted to parts of the northern limestone barrens be-tween Port au Choix and Burnt Cape and are listed as threatened or endangered under both provincial and federal legislation (17-16).

In many areas, signs of moose (*Alces ameri-canus*) and woodland caribou (*Rangifer taran-dus*) are abundant. Horned lark (*Eremophila alpestris*) and short-eared owl (*Asio flammeus*) nest on the exposed barrens.

17-16 Fernald's braya (top) occurs only on the northern limestone barrens and is protected under provincial and federal legislation. The Burnt Cape cinquefoil (bottom), once thought to only occur at Burnt Cape, is now thought to be the pretty cique-foil of the high arctic. (MJ)

Other Alpine Sites in Newfoundland — 241

Monts Groulx (Uapishka)
QUÉBEC
18

Mont Jauffret, Monts Groulx, Québec. (MJ)

Map 18-1. Western Monts Groulx, Québec.

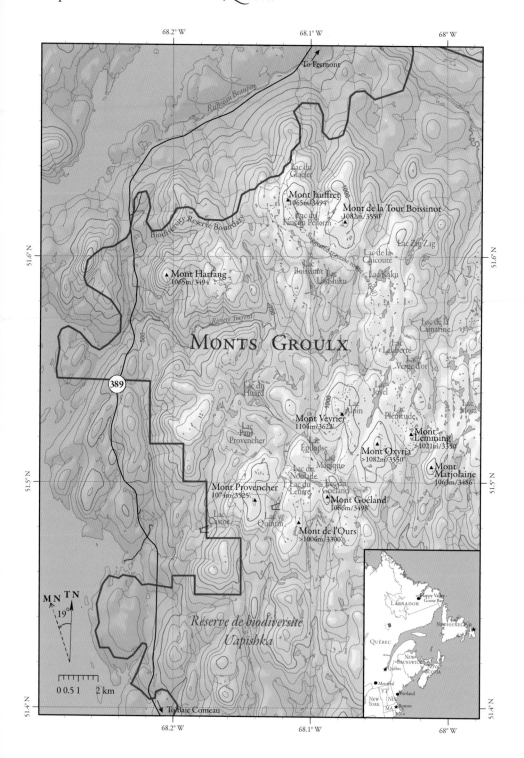

To Fermont

Ruisseau Beaupin

Lac du Glacier

Mont Jauffret
1065m/3494'

Mont de la Tour Boissinot
1082m/3550'

Biodiversity Reserve Boundary

Lac du Rancon Pèlerin

Lac Zig Zag

Lac de la Chicouté

Lac Kaku

Mont Harfang
1065m/3494'

Lac Boissinot

Lac Ussishiku

Rivière Torrent

Lac de la Camarine

MONTS GROULX

Lac Laliberté

Lac Verge d'or

Lac Joyel

Mont Veyrier
1104m/3622'

Lac Alpin

Lac Plénitude

Lac Mora

Mont Lemming
1021m/3350'

Lac du Huard

389

Lac Paul-Provencher

Lac Épilobe

Lac Magique

Mont Oxyria
>1082m/3550'

Mont Marjolaine
1063m/3486'

Lac du Nontain

Lac du Leurre

Lac du Goéland

Mont Goéland
1066m/3498'

Mont Provencher
1074m/3525'

Lac Castor

Lac Quintin

Mont de l'Ours
>1006m/3300'

MN T N

19°

Réserve de biodiversité
Uapishka

0 0.5 1   2 km

To Baie Comeau

LABRADOR

Happy Valley
Goose Bay

NEWFOUND'D

QUÉBEC

NEW BRUNSWICK

Québec

NOVA SCOTIA

Montreal

Portland

NEW YORK

VT

NH

MA

Boston

Monts Groulx (Uapishka), Québec

18-1 Massif (Mont) Provencher as viewed from the southwest, surrounded by old-growth black spruce and aspen. (MJ)

In the spruce-studded, gray-green heart of boreal Québec, 720 km (450 mi) north of Québec City, the Monts Groulx huddle above the spruce forest like some basking, brooding creature that just heaved itself up out of the adjacent Manicouagan meteor crater (18-1). "Monts Groulx" is the formally recognized name for the massif, but the name may not persist much longer. The Monts Groulx were named in honor of Lionel-Adolphe Groulx (1878–1967), a Québecois priest and nationalist, in November 1967, commemorating Groulx's death in May.[1] The original Innu name *Uapishka*, which means something between "rocky summits covered in snow"[2] and "white mountains," is gaining popularity (18-2). The Monts Groulx encompass the largest area of low-latitude alpine tundra east of the Rocky Mountains; its rounded summits rise above the surrounding spruce and fir for some 90 km (54 mi.) from east to west.

18-2 The Innu name for the Monts Groulx is Ua-pishka, *which may mean "white mountains" or "rocky summits covered in snow." (MJ)*

The Monts Groulx became road-accessible only in the 1980s following the construction of the Québec–Labrador Highway (Route

1   Rémillard (1978)
2   Drapeau (1994)

18-3 Route 389, the Québec-Labrador Highway, traverses interior Québec from the port town of Baie-Comeau to the iron towns of Fermont and Labrador City, skirting the Monts Groulx about 350 km from the coast. (MJ)

18-4 The Daniel Johnson dam, constructed in the 1960s and named for Québec's 20th premier, is 214 m high and floods the valley of the Manicouagan impact crater. (MJ)

389), built to connect the hydro-electric infrastructure of the Québec North Shore to the iron mines near the Labrador border (18-3). Thirty years later, the Monts Groulx are now hemmed in on all sides by burgeoning frontier industry, including sprawling deforestation to the south, iron mines to the north, the Labrador Highway to the west, and two major railroads to the east—but despite the industrial latticework, this is still wild country—a real frontier. Here you can count on finding only whatever luxuries and comforts you bring with you.

Ninety kilometers (54 mi) from west to east and a kilometer high, the mountain range is about the size and extent of Rhode Island and about four times as high above the sea. These mountains are old. The Monts Groulx lie within the Grenville Province of the Canadian Shield,[3] which formed during orogenic events more than one billion years ago (Ch. 2). Grenville Province includes a broad band of extremely old metamorphic bedrock extending from the Adirondacks in New York to interior and coastal Labrador.

Immediately southwest of the massif lies the Manicouagan impact crater, which was formed near the height of the Triassic period 214 million years ago when the mid-latitudes were forested by a great diversity of newly evolved conifers, cycads, and lycopsids. The enormous, ancient crater—now impounded by Hydro-Québec's Daniel Johnson dam[4]—provides some context for the ancient Monts Groulx plateau (18-4). The rocks of the Monts Groulx were already older than the Appalachians are today when a 5 km-wide meteor fragment struck earth 50 km (30 mi) to the southwest, creating an impact crater more than 100 km (60 mi) wide (the fifth largest on earth and one of the oldest known). The meteor apparently formed from the breakup of a large comet or asteroid, the larger pieces of which left a linear crater chain spanning more than 4400 km (2640 mi) across what is today Ukraine, France, Québec, North Dakota, and Manitoba. Today, the visible crater[5] is sharply demarcated by the boundaries of the Manicouagan Reservoir (18-5).

---

3   The Canadian Shield itself is far and away the largest exposure of Precambrian bedrock on earth. The Shield is North America's stable foundation; it is generally several hundred million to several billion years older than the bedrock of surrounding regions and has yielded the world's oldest known rocks. The Grenville Province is the youngest section of the Shield. A tectonically isolated region of Grenville material forms the alpine Long Range Mountains of western Newfoundland.

4   The Daniel Johnson arch and buttress dam and "Manic Cinq" generating facility are located near km 218 on Route 389. The Dam is a 214 m (702') monolith and the 8th tallest dam in North America.

5   The Manicouagan crater is the brecciated (fractured) zone surrounding the meteor impact site, subsequently eroded by Pleistocene glaciation.

Lay of the Land

The most accessible portion of the vast Monts Groulx plateau is the high ridge that borders the Québec–Labrador Highway (Route 389) near the Manicouagan Reservoir between highway kilometer marks 335 and 365. From much of this alpine region, the enormous ring-shaped reservoir is clearly visible. Here, the plateau rises to its highest elevation, forming an arc of three sentinel-summits. These are the summits known best to the handful of hikers, skiers, and *randonneurs* who explore the Monts Groulx regularly: Jauffret, Veyrier (18-6), and Provencher.[6] As they are all named in some way for Catholic bishops, their names reflect southern Québec's catholic heritage but do little to capture the stark, lonesome grandeur of the plateau (unless one finds in their stony visage a quality of prayerful discipline). Neither, one might argue, do Washington, Madison, or Jefferson (Ch. 7), or other summits named for mortal men.

18-5 *The Manicouagan impact crater, flooded by the Daniel Johnson dam, is clearly visible in this image of the aurora borealis from space. (NASA)*

18-6 *Mont Veyrier, photographed from the north, is the highest mountain in the Monts Groulx and part of a crescent of summits nearest the highway and most often explored by adventurers. (LW)*

Mont Jauffret (1065 m / 3494'), at the northwest edge of the Monts Groulx plateau, forms a discrete alpine island with the imposing Mont de la Tour Boissinot (>1082 m / >3550'), named for a fire tower that existed on the summit until the 1980s. Mont Jauffret's big, rolling summit provides spectacular views across the interior plateau toward Mont Veyrier, as well as northwest along the great north arm of the Manicouagan Reservoir as far as the now-defunct Lac Jeannine iron mine and the neighboring ghost town of Gagnon (18-7).[7] Mont de la Chicoutai lies east of Tour Boissinot across a wooded valley of white spruce, and Mont Oxyria is just east of the summit of Mont Veyrier (1104 m / 3622'), the highest mountain in the Monts Groulx. Mont de l'Ours rises opposite Massif Provencher (1074 m / 3525') on the east side of Lac Quintin. Near Mont de l'Ours, the landscape is marked with many large lakes and talus slopes. East of Mont Oxyria, the alpine tableland continues unbroken beyond Mont Lucie, where the scenery becomes more rugged, with precipitous cliffs

---

6   (1) Mont Jauffret is named for Francois-Antoine Jauffret (1833–1902), bishop of Bayonne, France; (2) Mont Veyrier, highest in the Uapishka, is named for Veyrier, Switzerland, the hometown of Louis Babel (1826–1912), a missionary in the Sanguenay and interior North Shore (for which the mountain at the center of the Manicouagan Reservoir is named); (3) Mont Provencher is named for Joseph-Norbert Provencher (1787–1853), a Québécois bishop and missionary.

7   The iron mine at Gagnon closed in 1985 and the outpost town was dismantled. Today, the townsite is overgrown with alder, but the paved and painted blacktop and divided median attests to the town's former presence. Southeast of town, the mine may be accessed by a dirt road. Alluvial deposits of fine tailings—easily visible from Monts Jauffret and Boissinot—provide alkaline habitat for *Dryas drummondii* and *Minuartia rubella*, arctic-alpine wildflowers not recorded from the Groulx massif.

18-7 Mont Jauffret provides views west across the northern arm of the Manicouagan Reservoir (top), including the (mostly) abandoned Lac Jeannine iron mine near the ghost town of Gagnon (bottom). (MJ)

and broken talus unlike the rolling terrain of the western plateau.[8] The eastern portion of the massif is bisected by the Cartier Company's iron train, which runs from Port Cartier on the Québec North Shore as far as the mines of Fermont near the Labrador border.

Access

Three major trails provide straightforward access to the western mountains, constructed over the past 30 years by *Les Amis des Monts Groulx*, the mountain advocacy alliance of *randonneurs*, woodsmen, and agency officials most closely tied to the Monts Groulx region.[9] Two of these provide access to the main portion of the massif near Mont Provencher and Mont Jauffret (18-8). These leave from small parking lots on Route 389 at km 335 and 365, respectively. A third trail near km 352, constructed in 2010, provides access to isolated Mont Harfang,[10] which rises prominently above the Québec-Labrador Highway.

18-8 Two footpaths, constructed by Les Amis des Monts Groulx (a unique and important mountain association), lead from Route 389 to the alpine tundra on Monts Provencher and Jauffret. (LW and MJ)

---

8   Mont Oxyria is apparently named for the alpine plant mountain sorrel (*Oxyria digyna*), which occurs in alpine snowfields and ravines as far south as Mount Washington.

9   *Les Amis des Monts Groulx* were instrumental in obtaining protected status for the western part of the range and continue to be very active in the Uapishka region. The group has a governing board and hosts an annual trail maintenance weekend on Labor Day—the *Grand Corvée*. Representatives from the group believe strongly that the Uapishka should be protected, but left as wilderness with minimal trails and infrastructure, rather than being designated as a national park with the associated development, "improved" access, crowds, and regulation that often accompanies such status. Their trails reach only to treeline and they advocate low impact, diffuse, and responsible use of the alpine zone. When visiting the region, respect their wishes to leave the mountains wild. See monts-groulx.ca for more information.

10   "*Harfang des neiges*" means "snowy owl" in French, although the original meaning of "harfang" may have come from the Swedish for "rabbit-eater." It is the bird of Québec.

18-9 *Mont Harfang stands alone near the Québec–Labrador Highway. The view from its summit encompasses many thousands of acres of virgin high-elevation spruce forest. (MJ)*

The Mont Provencher Trail originates near Camp Nomade[11] and ascends 3.2 km (1.9 mi) to Lac Castor before ascending another kilometer through forest to treeline. A rough trail leads from the treeline to the long summit of Mont Provencher before dropping into white spruce on the mountain's eastern flank. In total, the distance is about 6 km (3.7 mi) from highway to summit. On Provencher's east face, the light path runs through bogs to Lac Quintin, where a small shelter was constructed in the 1980s. No developed trails lead beyond Lac Quintin.

18-10 *The interior plateau of the Monts Groulx is a seldom-traveled expanse of tundra, bogs, subalpine spruce forest, and rocky outcrops. (MJ)*

The Mont Jauffret Trail leaves the highway at Camp Matsheshu[12] and ascends 5.2 km (3.2 mi) to treeline near Lac du Glacier. A rough trail through the tundra, marked by cairns and *inukshuks* (stone figures), then ascends about 1.8 km (1.1 mi) to the summit of Mont Jauffret. No developed trails lead beyond Mont Jauffret. Both the Provencher and Jauffret trails are relatively steep and rough, gaining approximately 700 m (2300′).

The Mont Harfang Trail leaves Route 389 from a parking area near km 352 and ascends 3 km (1.9 mi) to the rocky summit of Mont Harfang. It's a relatively short and easy day hike, and the summit provides views of the Monts Groulx Plateau to the east and the Manicouagan Reservoir to the west, including the prominent summit of Mont Manic (18-9).

Although these three western summits are relatively accessible from Route 389, the interior of the Monts Groulx plateau still requires backcountry navigation skills. The vast, little-known, alpine tableland east of Monts Jauffret and Provencher can be accessed by overland

---

11   Camp Nomade is the home of Jacques Duhoux, who was instrumental in developing the trail system on Mont Provencher in 1986.

12   Camp Matsheshu (Innu for "fox") is the home of Michel Denis, who has guided in the area for more than 25 years. Together with Jacques Duhoux and *Les Amis des Monts Groulx*, he constructed the earliest recreational footpaths on the mountain.

compass navigation or GPS, although only a minority of visitors venture beyond the western summits. Those who do mostly follow a vaguely arcing traverse, laid out by *Les Amis des Monts Groulx*, that is bounded to the south and north by the Provencher and Jauffret access trails and skirts the summits of Jauffret, Veyrier, and Provencher. The traverse may be completed in three days of hiking (less if skiing in optimal conditions), but extra time provides the adventurer or naturalist time to explore the adjacent summits (18-10).

The railroad of the Québec-Cartier Iron Company bisects the Monts Groulx beyond Mont Lucie, and the hills to the east of the railroad are unknown to most and seldom climbed. The peripheral eastern front of the Monts Groulx lies outside of the currently protected area and will possibly be subjected to future mining development.

Geology

The majority of the Monts Groulx bedrock is gabbronorite, a mafic igneous rock comprising high levels of magnesium and iron found in association with gabbros and ultramafic rock. Diorites and metagabbronorites are found in association with gabbronorite (18-11). To a lesser extent, the massif comprises anorthosite (mafic igneous rock) in the north and gneiss/paragneiss in the east (metamorphic rocks).[13] The Monts Groulx are centrally located within the Grenville Province of the Canadian Shield and are geologically closer to the Adirondacks of New York (Ch. 21) and the Mealy Mountains of Labrador (Ch. 20) than to the nearby Monts Otish (Ch. 19). Erratic boulders, likely originating on the Monts Groulx itself or immediately to the north, are strewn prominently across the alpine areas, relics of the last glaciation (18-12). In many places on exposed tundra, severe freeze-thaw cycles have exposed boils of mineral soil (18-13).

18-11 The majority of the western Monts Groulx is composed of ancient igneous rocks, such as diorite. (MJ)

18-12 Erratic boulders, carelessly discarded by the Laurentide Ice Sheet, are strewn throughout the tundra of the Monts Groulx. (MJ)

Ecology

The tableland of the Monts Groulx is a singular and complex alpine environment unique in eastern America. The relatively low relief of the tableland itself, combined with poor drainage in the wake of extreme glaciation, and the apparent absence of fire for several hundred years, has resulted in an aesthetically rewarding mosaic of high tundra, muskeg, and marsh, as well as park-like old-growth white spruce (*Picea glauca*) in protected subalpine valleys (18-14). The overall effect is almost western. Even without the sharp summit promontories characteristic of the Rockies, the juxtaposition of open spruce forest and thick tundra appears similar to the foreground of classic photos of the Maroon Bells or Trail Ridge in Colorado.

Vegetation — The Monts Groulx are positioned about 300 km (180 mi.) north of an important ecological boundary (around 49°N in the Canadian subarctic) between black spruce (*Picea mariana*) dominated forests to the north and balsam fir–dominated forests to the south. The subalpine, old-growth white spruce of the Monts Groulx may in fact be relict forests resulting from earlier balsam fir (*Abies balsamea*) forest range expansions (in which white spruce was a common associated species).[14] During a warm period earlier in the Holocene, balsam fir and white spruce may have established themselves in the vicinity of the Monts Groulx, at 51°N a region currently dominated by black spruce. Subsequent cooling may have deteriorated the growing conditions for balsam fir, and simultaneously, increased fire frequency at low elevations led to a shift toward black spruce dominance, except where fire is extremely rare, such as the Monts Groulx tableland (where fire occurs only at intervals >1000 years). At the treeline, the most cold-tolerant species, white spruce, is able to persist as massive canopy trees, but balsam fir and black spruce, at the same elevations, are relegated to trailing shrubs. Today,

18-13 *Severe freeze-thaw cycles have created frost polygons and "boils" of sorted soils, which are quickly colonized by such alpine plants as Greenland stitchwort. (MJ)*

18-14 *The alpine area of the Monts Groulx is a mosaic of high tundra, bogs, fens, marshes, and subalpine white spruce forests. (MJ)*

18-15 *Cold-tolerant white spruce outsize the other conifer species at treeline. (MJ)*

14   de Lafontaine and Payette (2010)

18-16 Treeline krummholz supports many species of moss and other bryophytes including (clock-wise from top-left) Pleurozium schreberi, Hylocomium splendens *and* Ptilium crista-castrensis, *and several species of* Sphagnum. *(MJ)*

white spruce dominates the upper, subalpine slopes of the entire Monts Groulx as high as 1000 m (3300') in elevation, and balsam fir often trails near the ground at the feet of giant spruce (18-15).

Above 1000 m, the white spruce parkland gives way to wet tundra through an extensive transition zone consisting of balsam fir and black spruce krummholz with glandular birch (*Betula glandulosa*) and green alder (*Alnus viridis*). These scrubby treeline ecosystems support extremely thick moss groundcover (18-16).[15] The krummholz also shelters typical boreal assemblages of graminoids and forbs. Oak fern (*Gymnocarpium* sp.), blue-bead lily (*Clintonia borealis*), starflower (*Trientalis borealis*), bunchberry (*Cornus canadensis*), bakeapple or cloudberry (*Rubus chamaemorus*), wood sorrel (*Oxalis montana*), and goldthread

18-17 In protected areas on the tundra near treeline grow oak fern (left), bakeapple (center), and heartleaved twayblade (right), although the latter is easily overlooked. (MJ)

---

15   Common mosses include *Pleurozium schreberi*, *Sphagnum* spp., and *Dicranum* spp. (de Lafontaine and Payette 2010).

18-18 *Widespread shrubs of the birch-alder communities include squashberry (left) and Labrador willow (right, with alpine bilberry). (MJ)*

(*Coptis trifolia*) are abundant in sheltered areas around the treeline ecotone, and heartleaved twayblade (*Listera cordata*) is frequent but often overlooked (18-17). On well-drained slopes, or vegetated talus, it is common to find balsam fir, skunk currant (*Ribes glandulosum*), squashberry (*Viburnum edule*), Labrador willow (*Salix argyrocarpa*), green alder, Bartram's shadbush or serviceberry (*Amelanchier bartramiana*; the only shadbush common in eastern alpine areas), glandular birch, and heart-leaved paper birch (*Betula cordifolia*) (18-18).[16]

Occasionally, these krummholz communities merge either with canopy stands of white spruce, or vegetation more typical of alpine snowbeds. These mosaics of scrub and tundra are frequently interrupted by expansive wetlands and waterbodies. Innumerable lakes, ponds, pools, marshes, bogs, and fens add color to the landscape: aquamarine in late-winter, as runoff and meltwater pools on the surface of iced-up ponds; green and white in summer, as tall cottongrass goes to seed among beds of tufted clubrush; yellow and red in the fall, when these same sedges take on their distinctive, fleeting autumn colors (18-19).

18-19 *In alpine bogs, various sedges and grasses take on fleeting but distinctive autumn colors. (MJ)*

18-20 *Diapensia (left) is widespread and abundant on exposed ridges, but alpine azalea (right) seems to be less common—although the two species occur together on hundreds of sites throughout the eastern alpine areas. (MJ)*

Only the most exposed sites—the very summits themselves, and certain exposed ridges—support a heath-rush-lichen community similar to that found on other potassic or acidic summits in the region. Alpine sweetgrass (*Anthoxanthum monticola*), diapensia (*Diapensia lapponica*), black crowberry (*Empetrum nigrum*), alpine bilberry (*Vaccinium uliginosum*), mountain cranberry (*Vaccinium vitis-idaea*), alpine bearberry (*Arctous alpina*), Bigelow's sedge (*Carex bigelowii*), and Lapland rosebay (*Rhododendron lapponicum*) are dominant

---

16 See also de Lafontaine and Payette (2010).

18-21 Bearberry willow (yellow foliage) occurs sparingly on summit gravels, sometimes with alpine bearberry (red foliage), its namesake. In autumn, the foliage of these species is strikingly different. (MJ)

on the highest summits (as they are on most eastern alpine summits). Alpine azalea (*Loiseleuria procumbens*) is rare on Mont Jauffret and probably occurs on other high summits (18-20). Bearberry willow (*Salix uva-ursi*) occurs at scattered stations near the various summits, often with the unrelated alpine bearberry (18-21). Bakeapple and tufted clubrush (*Trichophorum cespitosum*) are abundant on damp summit tundra.

On most alpine ranges—certainly those with higher relief: the Presidential Range (Ch. 7), or Katahdin (Ch. 8), or Mont Logan in Québec (Ch. 9)—areas of late-lying snow stand out in relative contrast to the extensive felsenmeer and alpine meadows. On the Monts Groulx, there are far more intermediate areas, with the exception of the high summits. In fact, a major part of the plateau resembles a typical eastern alpine snowbed community (see Ch. 5). Above 1000 m (3300'), snowbeds and snowbed vegetation are extremely common. On high-elevation, south-facing sites near disturbed soils, it is common to find creeping sibbaldia (*Sibbaldia procumbens*) and dwarf willow (*Salix herbacea*), the classic snowbed species of the Canadian arctic (18-22). Norwegian cudweed (*Omalotheca norvegica*)—a regionally rare plant, known also from the British Isles, Greenland, western Newfoundland, and the Gaspé—is common along rivulets and creeks that drain or flow through large snowbeds, often with clustered lady's mantle (*Alchemilla glomerulans*), arctic sweet coltsfoot (*Petasites frigidus*), and northern willow (*Salix arctophila*) (18-23). Several species of willow-herb, including Hornemann's (*Epilobium hornemanni*) and alpine (*E. anagallidifolium*) and fireweed (*Chamerion angustifolium*, the largest willow-herb above treeline), occur in seepage areas in and adjacent to snowbed communities. Several local rarities occur in the snowbed areas, including alpine bittercress (*Cardamine bellidifolia*), which occurs on a cliff/snowbed system on Mont de la Tour Boissinot and probably occurs at similar sites throughout the massif. Long beech fern (*Phegop-*

18-22 Sibbaldia (left) and snowbed willow (right)—perhaps the two most diagnostic species of alpine snowbeds in North America—co-occur abundantly throughout the Monts Groulx alpine areas. (MJ)

18-23 Norwegian cudweed (left) and arctic-loving willow (right) are locally common on wet tundra and along creeks in alpine snowbeds. (MJ)

teris connectilis), mountain woodfern (*Dryopteris campyloptera*), and large-leaved goldenrod (*Solidago macrophylla*)—all of which are common in most of the alpine habitats on the Monts Groulx—are also common constituents of the extensive snowbeds, occurring with occasional northeastern paintbrush (*Castilleja sepentrionalis*) (18-24).

With the exception of occasional cliff-faces near the taller summits, there is relatively little exposed rock. In most summit areas, lichens compose nearly half of the total ground cover. The lichens in these areas are typically common species such as blackfooted lichen (*Cladonia stygia*) and witch's hair (*Alectoria* spp.).

Fauna — The alpine landscapes of the Monts Groulx more than resemble the middle elevations of the western mountains, they are functionally similar in that their great wilderness still harbors most of its original mammal species. Wolves (*Canis lupus*), lynx (*Lynx canadensis*), marten (*Martes americana*), bear (*Ursus americanus*), moose (*Alces americanus*), and woodland caribou (*Rangifer tarandus*)—all species extirpated from most of New England and southern Québec during the 18th and 19th centuries—still occur on the Monts Groulx at most elevations (18-25).

The mountain range may also provide habitat for wolverine (*Gulo gulo*), the iconic scavenger of the far-north, but this may be wishful thinking.[17] Wolverines are clearly native to the region—pelts were regularly shipped from Québec and Labrador ports until the 1920s—but their decline and apparent extirpation from Québec were both well-documented between the 1920s to the 1970s. Wolverines persist in western Canada from the Yukon as far east as Ontario, but there have been no verified records east of Hudson Bay since 1978, when a

---

17   Johnson et al. (2012)

18-24 Long beech fern (top left), mountain woodfern (top right), and largeleaf goldenrod (bottom left) are abundant on sheltered alpine sites, including snowbeds. Northeastern paintbrush (pinkish form shown at bottom right) is less common. (MJ)

18-25 Woodland caribou still occupy the high elevations of the Monts Groulx, despite widespread population collapse by the mid-20th century. Here, a adult is pictured in a spring rain near Mont Boissinot. (MJ)

bounty was paid to an Innu hunter at Schefferville, near the Labrador border. That particular wolverine had been killed somewhere to the north and was reportedly photographed.[18] In the decades since, there have been more than 30 reported sightings of wolverine north of the Gulf of St. Lawrence in Québec and Labrador, several of which were made by apparently informed and qualified field biologists and trappers. However, anecdotal information of this sort is notoriously unreliable. For a cynic, it's tempting to dismiss these accounts out of hand, but there are other indications that the Monts Groulx may be an important part of the story of wolverine recovery in eastern North America. It is now clear that if the wolverine was not extirpated from the Québec-Labrador peninsula by the late 1970s, it was at least severely depleted.

Recently, a meta-analysis of confirmed wolverine data from across North America[19] suggested that the original range of wolverines was closely tied to tundra, subalpine, and taiga habitats. Specifically, it is thought that the presence of boulderfields, avalanche debris, and late-lying snow facilitate successful denning by female wolverines. If recent observations of

18   Moisan (1996)
19   Aubry et al. (2007)

Québec-Labrador wolverines were not fishers (*Martes pennanti*), or porcupines (*Erethizon dorsatum*), or black bears, but were correctly identified and reported in good faith, it seems likely that the source of the animal(s) was one of the mountain plateaus in the region; if not the Monts Groulx, then possibly the Monts Otish to the west (Ch. 19) or the Montagnes Blanches to the southwest (Ch. 21). Given that wolverines regularly traverse 150 kilometers (90 mi) per week in Glacier National Park, Montana—climbing to the high passes of the Continental Divide in winter and ascending 1500 m (5000') cirque-walls in 90 minutes—it's possible to imagine wolverines roaming the wilderness between the Monts Otish and Monts Groulx plateaus and remaining undetected.[20]

So continues to unfold the opaque and sad story of wolverines in eastern North America. Wolverines may no longer occur in eastern North America, although they could certainly return. It seems like common sense that when that day comes, there should be ideal habitat waiting. Given the extent of industrial development on all sides of the Monts Groulx—and a new plan by Québec to construct a road near the Monts Otish—immediate planning is necessary. Under any scenario, the Monts Groulx's excellent and extensive alpine and subalpine habitats will likely factor strongly into future recovery efforts for wolverine in eastern Canada.

One hypothesis for the wolverine collapse is the regional collapse of caribou (a primary food source), which occurred during the first half of the 20th century; another is the collapse of the wolf population that provided carrion for the wolverine to scavenge.[21] Perhaps the best evidence that wolverine are extirpated from Québec is that although caribou have increased following the collapse, there has not been a concurrent increase in reported wolverine sightings, and no confirmed observations, though we might expect a lag between prey and predator recovery.

Camera-trapping by Beyond Ktaadn from 2010 to 2011 above treeline in the vicinity of Monts Jauffret, Tour Boissinot, and de la Chicoutai revealed extensive use of the summit tundra by black bear, American pine marten, porcupine, and snowshoe hare (*Lepus americanus*). Gray wolf, red fox (*Vulpes vulpes*), moose, river otter (*Lontra canadensis*), mink (*Mustela vison*), and red squirrel (*Tamiasciurus hudsonicus*) were detected above treeline in low numbers. Beavers (*Castor canadensis*) were not detected on cameras, but several relatively recent beaver dams and lodges currently exist at elevations more than 1000 m (3300'), well above treeline. Groundhog (*Marmota monax*) and lynx, common along low elevation portions of Route 389 and detected during concurrent low-elevation camera surveys, have also not been detected on the tundra. Northern bog lemming (*Synaptomys borealis*) and rock vole (*Microtus chrotorrhinus*) probably occur throughout the tundra area.

The alpine avian diversity of the Monts Groulx is noteworthy. Common loons (*Gavia immer*) are frequently seen on the largest lakes, such as Lac Boissinot and Lac Joyel,[22] and it's possible that the larger lakes provide breeding habitat for Barrow's goldeneye (*Bucephala islandica*).[23]

---

20   Present or not, a recovery plan is currently in place for the eastern population of the species. Fortin et al. (2005).

21   Fortin et al. (2005)

22   Several times we've heard loons calling from treeline lakes such as Lac Joyel and Lac Boissinot.

23   Barrow's goldeneye ranges primarily in the western Canadian Arctic and the North Shore of Québec; a single adult was photographed at Lac du Glacier in 2010.

18-26 Willow ptarmigan occur on all of the major alpine summits, one of their southernmost stations on the eastern North American mainland. (MJ)

Perhaps due to the proximity of the great Manicouagan Reservoir, several coastal species are seen—for example—common tern (*Sterna hirundo*) and several gulls (*Larus* spp.).

Because of the size and extent of typically subarctic ecosystems, a number of northern birds occur on the massif. Many of these birds are common to alpine and subalpine areas throughout our eastern region: gray jay (*Perisoreus canadensis*), yellow-rumped warbler (*Setophaga coronata*), ruby-crowned kinglet (*Regulus calendula*), white-throated sparrow (*Zonotrichia albicollis*), dark-eyed junco (*Junco hyemalis*), and robin (*Turdus migratorius*) are relatively abundant. Others are more generally known from grassland habitats and tundra habitats, such as horned lark (*Eremophila alpestris*).[24]

Several breeding species that are typically known from farther north also occur on Monts Groulx, suggesting strongly that the size and extent of the alpine landscape sufficiently mimics the northern tundra to attract arctic birds. Willow ptarmigan (*Lagopus lagopus*) (18-26), American pipit (*Anthus rubescens*), and rough-legged hawk (*Buteo lagopus*) are three of these. Gray-cheeked thrush (*Catharus minimus*) and white-crowned sparrow (*Zonotrichia leucophrys*) are two additional species found on the Monts Groulx which generally breed farther north. Lapland longspurs (*Calcarius lapponicus*) and snow buntings (*Plectrophenax nivalis*), which we have observed on the Groulx tundra in late May, probably represent migrating individuals still a few hundred miles from their breeding range. But overall, it's clear that the Monts Groulx massif provides important southern habitat for disjunct subarctic species.

The entire Monts Groulx massif lies beyond the range of most reptiles. Garter snakes are observed frequently on the Côte Nord as far as Sept Îles, but have not been documented in the Québec interior. In Monts Groulx field trips spanning ten years, we haven't observed any species of snake. Amphibians, on the other hand, are another matter. It's becoming fairly well documented that as many as nine species of eastern amphibian tolerate, occupy, and breed in alpine tundra environments in the eastern region.[25] We have found spotted salamander (*Ambystoma maculatum*) and wood frog (*Lithobates sylvaticus*) breeding commonly in roadside pools along Route 389.[26] Northern leopard frogs (*Lithobates pipiens*) and mink frogs

---

24   Horned lark were detected on Mont Jauffret during our camera-trapping in the summer of 2010.

25   Jones and Smyers (2010) documented wood frogs, American toads, and spring peepers in the Lakes of the Clouds on Mt. Washington, New Hampshire.

26   Spotted salamanders have not been documented north of the Manic Cinq hydroelectric complex (50.6°N, but see Hébert 2005), but our surveys in 2010 indicated that they are apparently common as far north as 51°N (Jones and Willey 2011).

18-27 Hudson Bay toads, a colorful form of the American toad, are common above treeline. They breed in alpine ponds in June, up to a month after low-elevation populations. (MJ)

(*Lithobates septentrionalis*), as well as wood frogs and American toads (*Anaxyrus americanus*), are common in central Labrador, and the latter three probably occupy the whole of Québec and Labrador more or less continuously.[27] Two other species of salamander have been documented either in the vicinity of the Monts Groulx, or farther north: the blue-spotted (*A. laterale*), which, like the Northern leopard frog, has been reported from the Churchill Valley of Labrador, and the two-lined salamander (*Eurycea bislineata*), reported from locations around Route 389 as well as Île Levasseur in the center of Reservoir Manicouagan. On the high tundra of the Monts Groulx, we have observed spring peepers (*Pseudacris crucifer*), wood frogs, and toads.

18-28 Spring azures are often abundant on the Monts Groulx tundra in June. (MJ)

These are Hudson Bay toads (formerly subspecies *copei*), a little-known, colorful variety that flirted with full subspecies-level recognition until the 1970s, when allozyme analysis suggested they were identical to the American toad of the Great Lakes and New England (18-27). The Hudson Bay toad is stockier than the southern races and much more colorful. In marked contrast to toads in the United States, which themselves are extremely variable, Hudson Bay toads are mottled with sharp whites, blacks, and brick-reds. The ecological preferences of Hudson Bay toads haven't been well-reported, but they appear to fill a niche similar to southern individuals, exploiting temporary waterbodies in early spring (on the Monts Groulx tundra, in June, after ice-out) and overwintering on land in sheltered pockets of white spruce. On the Monts Groulx, toads breed in large ponds such as Lac du Glacier, Lac Boissinot, and Lac Quintin as well as in smaller, marshy ponds, fens, and solifluction pools. Toads may use stagnant backwaters of perennial streams, especially in areas inaccessible to brook trout or red trout, which are common at all elevations on the massif.[28]

The Lepidoptera (butterflies and moths) of the Monts Groulx are generally boreal and subarctic species. Azures (*Celastrina* spp.) are abundant in June (18-28). At least two species of fritillary (Freija and arctic) are common in June and July. Plume moths are commonly ob-

---

27   Northern leopard frogs in the Churchill Valley may represent a relict population.

28   The Québec red trout (*Salvelinus alpinus*) is a landlocked form of arctic char related to blueback trout and Sunapee trout. Sea-run arctic char occur in the lower Manicouagan, but not in the vicinity of the Uapishka.

Monts Groulx (Uapishka), Québec — 259

18-29 Black flies may be the only insect you remember from a trip to the Monts Groulx in July or August—although cool or windy weather will keep them at bay even during the height of their season. (MJ)

served, as are sulphurs (*Colias philodice*) and whites (*Pieris* spp.).

Several species of Odonata (dragonflies and damselflies) breed in semi-permanent and permanent waterbodies. Several occur in the pools on the highest summits, including the large-bodied darners (*Anax* spp.). Limnephilids (caddisflies), and other families occur in most pools and ponds at all elevations.

In addition to the innocuous species listed earlier, it bears repeating that black flies also occur in the Monts Groulx, where they put the hordes of northern Maine to shame. The flies can be mind-numbing in July and August (18-29), but less so in the fall and spring.

Map 18-2. Monts Groulx massif, Québec.

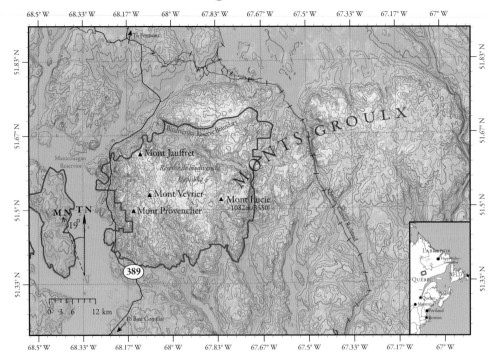

Monts Otish 19
QUÉBEC

Monts Otish, Québec. (GD)

Map 19-1. High peaks of the Monts Otish, Québec.

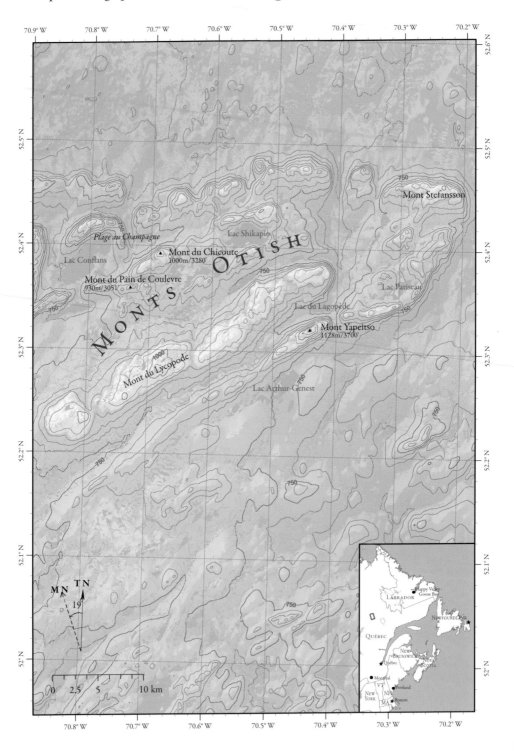

Mont Stefansson

Plage au Champagne

Lac Shikapio

Lac Conflans

Mont du Chicoute
1000m/3280'

Mont du Pain de Couleuvre
930m/3051'

Lac Pariseau

Lac du Lagopède

Mont Yapeitso
1128m/3700'

M O N T S   O T I S H

750

750

750

1000

Mont du Lycopode

Lac Arthur-Genest

750

750

750

750

750

750

MN   TN

19

0   2.5   5   10 km

LABRADOR
Happy Valley
Goose Bay

NEWFOUNDLAND

QUÉBEC

NEW
BRUNSWICK
Québec

NOVA
SCOTIA

Montréal
VT
NH
Portland

NEW
YORK
MA
Boston

Monts Otish, Québec

19-1 The alpine tundra of the Monts Otish is perhaps the least-known and most isolated of the alpine ecosystems profiled in this book. Mont Yapeitso, Monts Otish, Québec. (GD)

The remote knobs, peaks, and ridges of Québec's Monts Otish lie 200 km (120 mi) west-northwest of the Monts Groulx (Ch. 18) and about the same distance south of Hydro-Québec's Trans-Taiga Highway, which supports the hydroelectric installations of the James Bay watersheds. The alpine tundra of the Monts Otish is perhaps the least-known and most isolated of the alpine ecosystems profiled in this book, with the exception of some of the Labrador coast ranges discussed in Chapter 21. The Monts Otish complex comprises several dozen distinct, boomerang-shaped or linear summits. Diverse alpine habitats occur throughout the Monts Otish but are clustered around the highest peak of Mont Yapeitso at the far-eastern edge of the sprawling range (19-1). Throughout this mountainous region, relatively pristine alpine ecosystems are distributed over an area exceeding 4000 km^2 (1544 mi^2)—a large area. The tundra is much less contiguous than the massive alpine tableland of the Monts Groulx. The Monts Otish lie at a geophysical crossroads. They form the watershed divide between James and Hudson Bays and the Gulf of St. Lawrence and lie at the boundary of two major geologic provinces of the Canadian Shield, the Grenville and Superior provinces.[1] Furthermore, the Monts Otish are situated at the border of the closed-canopy black spruce (*Picea mariana*) forest and the lichen woodland zone, a major ecological divide in central Québec (19-2).

---

1   de Lafontaine and Payette (2010)

The high peaks region surrounding Mont Yapeitso in the Monts Otish has been proposed as part of the 11,000 km^2 (4247 mi^2) *Réserve de biodiversité projetée Albanel-Témiscamie-Otish*[2] (Albanel-Témiscamie-Otish Biodiversity Reserve), which the Québec government is working with Cree leaders to establish. However, because the Monts Otish are situated near a major geologic boundary, the rocks contain mineable iron and uranium ore as well as kimberlite, a rock type associated with diamonds elsewhere in Canada. Multiple areas about 80 to 100 km (48 to 60 mi) west of the high peaks have been proposed or slated for additional gold and uranium mining, as well as diamond extraction.[3] Most of the current emphasis is on developing the Renard diamond mine due west of the Mont Yapeitso region,[4] but interest in mining the region's uranium deposits increased during the early 2000s, corresponding to a jump in uranium prices.[5] The Otish region has been thrown into conflict as Strateco Resources, Inc. recently proposed the Matoush exploratory uranium mine project southwest of the Monts Otish. The Misstisini Cree Nation strongly opposes the plan, and the future of the project remains to be determined.[6]

19-2 The Monts Otish form the watershed divide between James Bay and the Gulf of St. Lawrence, and also lie at the boundary of two major divisions of the Canadian Shield and the Canadian boreal forest. (GD)

19-3 The actual extent and configuration of the Monts Otish did not become known to western scientists until the mid-1900s. (GD)

The Monts Otish range has a storied history among European colonists, too. Because the region was so difficult to access, the exact location and extent of the mountain range was not precisely understood for centuries. Cartographers predicted that the major rivers emerging from wild, interior Québec must originate in a hypothesized mountain range. Referred to vaguely as the *Monts Washish*, the theoretical mountain range was drawn ten times larger than its actual size on maps of the 19th century. Not until the mid-20th century did the true extent and location of the range become known (19-3).[7]

---

2   MDDEP (2008)
3   Strateco (2009); SWM (2010)
4   SWM (2010)
5   Perreault (2007)
6   Although the Monts Otish are the site of aggressive new natural resource exploration and development, resource extraction history in the vicinity of Lake Mistassini (Québec's largest natural lake, which lies to the southwest of the Monts Otish) dates back half a century. Gold and copper mines were developed north of Lake Mistassini in the 1950s, but these had closed by the 1980s and in any case were not located near the higher mountains of the Otish.
7   The first scientific exploration of the Monts Otish was undertaken in 1949 by botanists Jacques Rousseau and René Pomerleau, who accessed the mountains by floatplane. Rousseau suffered a heart attack on this expedition (Rousseau 1959).

In addition to their botanical and industrial significance, the Monts Otish are culturally significant as well. For centuries, the mountains have been an important area for the Cree Nation of Eeyou Istschee, whose members live primarily in communities 200 to 300 km (120 to 180 mi) to the south, but who hunt, trap and fish and maintain a degree of stewardship in the region.

19-4 Mont Yapeitso is the highest peak in the Monts Otish range, at 1128 m. (GD)

Lay of the Land

The highest peak of the Monts Otish is Mont Yapeitso (1128 m / >3700'), the fourth highest major mountain in Québec (19-4).[8] The largest contiguous area of alpine tundra, however, is associated with Mont du Lycopode (Mount Clubmoss), which runs along a northeast-trending axis for 27 km (16 mi) beginning only 3 km (1.8 mi) north of the summit of Mont Yapeitso. These two major mountains are clearly visible from one another across a damp valley of spruce and muskeg. Scattered occurrences of alpine tundra, many in excess of several hundred hectares, may be found along the Otish range to the north, south, and west of the Mont Yapeitso region.

Many of the great rivers of central Québec arise on the higher mountains of the Monts Otish, including the Rupert, Eastmain, La Grande, Péribonka, Temiscamie, Outardes, and Manicouagan Rivers. These rivers either flow ultimately south into the temperate Gulf of St. Lawrence or north to frigid James Bay, their mouths separated by a linear distance of some 800 km (500 mi) but a hydrologic distance of nearly 4000 km (2500 mi), leading Rousseau (1959) to call the Monts Otish the "hydrographic hub of Québec."

Access

Accessing the Monts Otish is currently an expensive and time-consuming proposition. For present-day summer explorations, most visitors will arrange floatplane transport from the northern Québec village of Témiscamie, reached from Québec City and points south via the towns of Chibougamau and Mistissini. The city of Chibougamau lies about seven hours northwest of Québec City and five hours from Chicoutimi by car, and may be accessed via Route 167 from the vicinity of Lac St-Jean, Québec. About two hours beyond Chibougamau along Route 167, the Cree village of Mistissini has a welcome center and tourist accommodations, which were built at the height of the sport fishing industry in the 1970s. From Mistissini, a dirt road leads 110 km (66 mi) to the Témiscamie River.

In March 2009, Québec announced plans to construct the *Route des Monts Otish*. The 250-km (150 mi) road (also known as the Route 167 extension) will link Chibougamau and

---

8   *Yapeitso* reportedly means "bull caribou" (Rousseau 1959). Mont Yapeitso was apparently referred to as "Mont Lagopède" by Shchepanek (1973). Mont d'Iberville (a western shoulder of Labrador's Mount Caubvick) and many other summits in the Torngat Mountains on the Québec–Labrador border (Ch. 21) are higher than the Monts Otish, several hundred kilometers north of the continental treeline. Mont Jacques-Cartier in Gaspésie (Ch. 10) is also higher.

19-5 *The Monts Otish were one of the last corners of North America to be deglaciated following the most recent Laurentide ice advance, only 7000 years ago. (GD)*

Mistissini to the Renard Diamond Project north of Lake Mistassini.[9] Road construction began in 2011 and was completed in 2014. The new road may open the extremely vast wilderness area to natural resource extraction and development, but apparently will not directly disturb the high peaks of the Monts Otish.

For the immediate future, floatplane or helicopter will probably remain the only practical way to access the Otish highlands in summer. Ultimately, the development of the Albanel-Témiscamie-Otish Park may dramatically change the access to this wilderness region. The park's master plan currently proposes the development of floatplane landings and camp accommodations at *Lac Arthur-Genest*, as well as additional camps and trails at *Lac Shikapio* to connect to existing camps at *Lac Conflans* and *Lac Lagopède*, and finally, the development of yet another facility at *Lac Naococane*, in the heart of the alpine Monts Otish.[10] In addition to increasing access to these far-flung locations, the plan proposes encouraging dispersed use in the fragile alpine environment by not building trails above treeline and proposes limiting or restricting access altogether to several peaks including Mont Wepaasiu, Mont Chicouté, and parts of Mont Yapeitso for sacred or ecological reasons.[11]

Geology

As already noted, the Monts Otish lie in a geologically complex region of the Canadian Shield, near the boundary between the Grenville geologic province and the older rocks of the Superior province. The lowest, oldest layer consists of metamorphic and volcanic rock from 2.5 to 2.85 billion years old (Archaean age). The Otish group, sedimentary layers from the Proterozoic era (2.5 to 1.6 billion years old), rest atop this basement. These layers were formed after faults caused depressions in the basement rock, forming inland seas that allowed the accumulation of sedimentary deposits. The Grenvillian orogeny uplifted and metamorphosed the region (see Ch. 2). The mountains themselves are made up partly of red sandstone and conglomerates as well as gabbro dykes formed during the Grenvillian orogeny.[12]

The Monts Otish were one of the last corners of eastern North America to be released from the heavy mass of the Laurentide Ice Sheet during the mid-Holocene. Remnants of the continental ice remained here as recently as 7000 years ago (19-5).[13]

---

9   CEEA (2010)
10   Gouvernement du Québec (2005a).
11   Gouvernement du Québec (2005b).
12   Hébert (2005)
13   Dyke (2004)

19-6 Dominant summit plants in the Monts Otish include mountain firmoss, Bigelow's sedge, Lapland rosebay, and diapensia (clockwise from top-left). (LW and MJ)

Ecology

Vegetation — Despite areas of local calcareous influence, the alpine flora of the Monts Otish is in most places similar to that found on the Monts Groulx and the Montagnes Blanches (Ch. 21). The most exposed, windblown, frost-scarred summits support colonies of mountain firmoss (*Huperzia appressa*), Bigelow's sedge (*Carex bigelowii*), diapensia (*Diapensia lapponica*), and Lapland rosebay (*Rhododendron lapponicum*) (19-6).[14] Less exposed areas near the treeline (around 822 m / 2700') also support tufted clubrush (*Trichophorum cespitosum*), glandular birch (*Betula glandulosa*) (19-7), black crowberry (*Empetrum nigrum*), Labrador tea (*Rhododendron groenlandicum*), purple mountain heather (*Phyllodoce caerulea*), bakeapple (*Rubus chamaemorus*) (19-8), and alpine bilberry (*Vaccinium uliginosum*), along with the ubiquitous krummholz species balsam fir (*Abies balsamea*) and black spruce (*Picea mariana*).

---

14   Shchepanek (1973)

19-7 Glandular birch is an abundant species in birch-alder communities near treeline; note the bright orange foliage in the foreground. (GD)

Away from the highest peaks, an underlying sedimentary rock supports unusual alpine and subalpine vegetation that in some areas differs in species composition from the Monts Groulx and other ranges in central Québec.[15] Among these are several rare or disjunct species including common butterwort (*Pinguicula vulgaris*), orange agoseris (*Agoseris aurantiana*), which occurs primarily in the Cordilleran flora of the western mountains, and Robinson's hawkweed (*Hieracium robinsonii*), endemic to Québec and New England. Norwegian cudweed (*Omalotheca norvegica*) may be found in areas of late-lying snow. Although the bedrock locally yields an environment conducive to high plant diversity, some authors have suggested that late deglaciation restricted the colonization of the highlands by some species of alpine plants, such as saxifrages (*Saxifraga* spp.). This has been attributed to the large seeds inhibiting rapid dispersal and has not been thoroughly examined.[16]

Note that there is an impressive diversity of lichen species on the high peaks of the Monts Otish of which the most common is possibly witch's hair (*Alectoria ochroleuca*), bryocaulon (*Bryocaulon divergens*), one of many species of Iceland lichen (*Cetraria* spp.),[17] cup lichen (*Cladonia* spp.),[18] and the snow lichen (*Stereocaulon paschale*) (19-9).[19]

---

15   The broader region has attracted botanists for more than 200 years, but although André Michaux visited the nearby Lake Misstassini region in 1792 and many botanists followed in his footsteps, the Monts Otish remained unexplored by European scientists until botanists Jacques Rousseau and René Pomerleau explored the mountains in 1949. They were the first scientists to visit the high summits of the Otish Mountains, where they found many interesting, subarctic, boreal, and alpine plants and fungi. In 1950, Rousseau petitioned the government to establish a scientific wilderness preserve in the Monts Otish high peaks region. Since that time a number of other biologists have explored the general region, particularly around Lake Misstassini, though few ventured up to the alpine peaks (Hébert 2005). Recently, the old-growth white spruce (*Picea glauca*) forests of the Monts Otish have been intensively studied by researchers from Université Laval (e.g., de Lafontaine and Payette 2010). In addition to the many botanists who augmented Rousseau's initial flora over the past decades, M. Blondeau and J. Gagnon conducted a series of surveys as part of the National Park planning process and increased the number of known species of the region by 30%. Within the proposed park boundary, 497 species of vascular and more than 400 species of non-vascular plants have been observed (Hébert 2005).

16   Rousseau (1959)

17   e.g., *C. cucullata*, *C. islandica*, and *C. nivalis*

18   e.g., *C. amaurocraea* and *C. stygia*

19   The area hosts a variety of fungi as well. In his first expedition with Rousseau in 1949, Pomerleau found the fungi *Cortinarius herpeticus*, *Cortinarius leucopus* and *Leccinum rotundifolia*.

19-8 Bakeapple or cloudberry, a common plant of boreal and subalpine bogs as far south as New Hampshire, is common near treeline on the higher peaks of the Monts Otish. (LW and MJ)

The boreal lowland landscape south of the Monts Otish is dominated by closed-crown black spruce–feathermoss (*Ptilium crista-castrensis, Hylocomium splendens,* and *Pleurozium schreberi*) forest interspaced with monospecific jack pine (*Pinus banksiana*) stands, typical of the northern boreal zone. Of all the North American boreal species, black spruce and jack pine are the most tolerant of fire. Indeed, this area is characterized by recurrent wildfires covering large areas, which occur every 115 to 270 years on average at a given site. On the

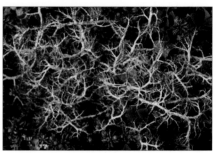

19-9 Witch's hair lichen is abundant above treeline on the Monts Otish. (MJ)

northern side of the Monts Otish, the subarctic (taiga) lowland landscape is dominated by lichen–black spruce open woodland, typical of the southern taiga zone (19-10). Wildfires ignited by natural causes occur at shorter intervals in this area (i.e., every 100 years or less). Higher fire activity combined with a colder climate was responsible for the forest opening of the taiga landscape north of the Monts Otish (19-11). Post-fire black spruce regeneration is limited by unsuitable seedbed conditions on a deep organic layer that was not disturbed by fires, which yields lower spruce density and favors dominance by lichen species in the understory. In this area, patches of closed-crown black spruce–feathermoss forest are few and far between. These patches escaped fire over the last millennia and constitute relics of the past northern expansion of the closed-crown forest.[20]

A distinct fire regime also characterizes the slopes of the Monts Otish, thanks to the orographic effect (see Ch. 4). The cold and wet climate typical of the high peaks is less conducive to wildfire, yielding a longer fire interval (500 years) with respect to the surrounding lowland landscape described earlier.[21] Consequently, high elevation forest in the Monts Otish sharply contrasts with that of the black spruce–dominated lowland (19-12). The subalpine layer (between 600 and 950 m above sea level) is dominated by a mosaic of white spruce (*Picea glauca*) or balsam fir (*Abies balsamea*) stands.[22] These subalpine stands may be of great importance in biological conservation, especially in the current context of ongoing rapid climate changes. In fact, the high peaks of the Monts Otish harbor the northernmost continen-

---

20   Pollock and Payette (2010)
21   de Lafontaine and Payette (2011)
22   de Lafontaine and Payette (2010)

19-10 *The forest landscape south of the Monts Otish is dominated by black spruce-feathermoss communities (top); but to the north a more fire-tolerant taiga landscape prevails with lichens in the understory (bottom). (MJ and LW)*

19-11 *Evidence of fire is clearly visible on the taiga landscape north of the Monts Otish. (GD)*

tal balsam fir and white spruce stands, which are more typically located in the continuous balsam fir–white birch (*Betula papyrifera*) forest zone south of latitude 49°N (350 km [210 mi] south of the Monts Otish). Of all the widespread North American boreal tree species, white spruce and balsam fir are the least adapted to fire. The persistence of these isolated stands at their northern limit is thus the consequence of a fragile equilibrium between the peculiar climate and fire regime of the subalpine layer of this area. Under a dryer climate scenario, the fire regime is expected to become more active in the subalpine zone of the Monts Otish. This could result in the collapse of the residual subalpine balsam fir–white spruce stands and their replacement by fire-tolerant black spruce stands (19-13).

Recent research suggests that fire may have been the proximate ecological factor responsible for the deforestation of the summits, and the subsequent formation of the alpine tundra on the high peaks of the Monts Otish (19-14). Summits were deforested by fire events occurring during the climatically unfavorable Little Ice Age (1550–1850 AD).[23] The harsh climatic conditions during this period prevented efficient forest regeneration. A change toward wetter and warmer climate conditions could allow the expansion of the balsam fir–white spruce forest at higher elevation, ultimately resulting in the forestation of the alpine tundra. Similar climate-wildfire-forest dynamics were also observed in the subalpine forest and alpine tundra areas of the other eastern boreal ranges discussed in this book (i.e., Monts Groulx and Montagnes Blanches).

Fauna — The Monts Otish support non-migratory herds of woodland caribou. Estimates in 2002 by Cree trappers suggest that approximately 110 caribou (*Rangifer tarandus*) inhabit the region, about two per 100 km². Moose (*Alces americanus*) are slightly more abundant and estimated at 2.5 to 5 per 100 km², and they are thought to be larger in size than those farther south in Québec.[24]

American marten (*Martes americana*), American mink (*Mustela vison*), gray wolves (*Canis lupus*), red fox (*Vulpes vulpes*), Canada lynx (*Lynx canadensis*), and black bear (*Ursus ameri-*

23   Savard (2009)
24   Rousseau (1959); Hébert (2005)

19-12 The cold and wet climate of the high peaks in the Monts Otish has allowed the development of white spruce–balsam fir communities. (GD)

canus) all occur throughout the range, though with the exception of marten, they probably venture into alpine areas infrequently. Wolverine (*Gulo gulo*), the northern phantom, was apparently present in interior Québec into the 20th century. Suitable habitat exists for the species in the form of talus slopes, avalanche trails, and late-lying spring snowfields, but there is no evidence that wolverine currently occupy the Otish Mountains (19-15).[25] Beaver (*Castor canadensis*), muskrat (*Ondatra zibethicus*), and otter (*Lontra canadensis*) occur in high-elevation watercourses and ponds, but are likely uncommon above treeline, where wetland diversity is limited.

As expected, numerous species of boreal and northern birds have been observed in the Monts Otish. Species of interest include the golden eagle (*Aquila chrysaetos*), gray-cheeked thrush (*Catharus minimus*), white-crowned sparrow (*Zonotrichia leucophrys*), redpoll (*Acanthis flammea*), and American pipit (*Anthus rubescens*).[26]

American toads (*Anaxyrus americanus*) are present throughout interior Québec, including the Monts Otish region.[27] On the neighboring Monts Groulx, toads breed in tundra pools above 1000 m (3300') following ice-out in May or June. Wood frogs (*Lithobates sylvaticus*) and spring peepers (*Pseudacris crucifer*)

19-13 Under a drier climate scenario, inreased fire frequency in the subalpine zone may result in an increase in black spruce near treeline of the Monts Otish. (GD)

19-14 Recent research suggests that fires during the Little Ice Age may have played a role in the deforestation of the summits of the Monts Otish, which led to the development of the alpine tundra. (GD)

25   Johnson et al. (2012)
26   Hébert (2005)
27   Fortin et al. (2011)

are likely present at least to the treeline and probably occur above 1000 m (3300') where suitable wetland diversity exists.

19-15 *The summit snowfields of the Monts Otish likely supported wolverine until the mid-1900s. Given the demanding space requirements of this mid-sized carnivore, the Monts Otish will likely be an important component of wolverine recovery in the future. (GD)*

Mealy Mountains 20
LABRADOR

English Mountains High Point, Mealy Mountains, Labrador. (JK)

Map 20-1. High peaks of the Mealy Mountains, Labrador.

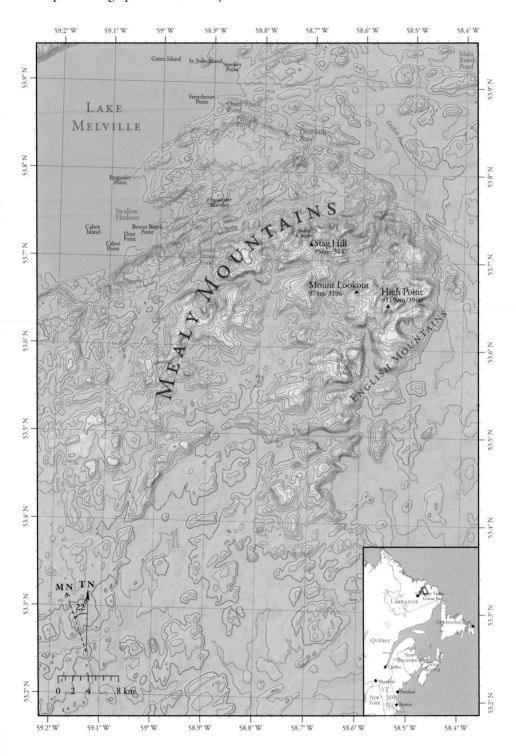

Mealy Mountains, LABRADOR

20-1 A stark, glacier-scarred landscape, Labrador's Mealy Mountains are a southern outpost of the High Sub-arctic Tundra ecoregion. (JF)

A stark, glacier-scarred, mountain landscape dominates the southern shores of Lake Melville, the great marshy inlet of the Labrador Sea that divides the interior section of Labrador into distinctly northern and southern regions. The Mealy Mountains, referred to as "*Akami-uapishku*" by the Innu, rise above 1000 m (3280') from sea-level bogs and spruce forest to tundra-clad peaks (20-1).[1] The alpine treeline in the Mealy Mountains occurs at about 600 m (1970'), about 200 m lower than the average treeline in the Monts Otish or the Monts Groulx to the west (Ch. 19 and 18).[2] Though surrounded by forest ecosystems, the alpine tundra of the Mealy Mountains forms a southern outlier of the High Subarctic Tundra ecoregion.[3] The highest alpine peaks of the Mealy Mountains, a distinct complex known as the English Mountains, are found clustered near the eastern part of the range. This subrange, historically little-traveled, forms the basis of much of the proceeding discussion.[4]

After decades of lobbying by multiple groups, in February 2010, Parks Canada announced the establishment of the 11,600 km² (about 4,500 mi² or 2.65 million acres) Mealy Mountains National Park Reserve, which will encompass much of the highlands and important supporting lowland and coastal landscapes.[5] The proposed park will be the largest in eastern

---

1	The term *mealy* was also used to describe the "mealy" look of the mountains.
2	Jacobs (2006)
3	Meades (1990)
4	Historically, during snow season, the Innu people traveled on snowshoes (*asham*), towing their supplies on toboggans (*shumin-utapanashk*) to reach traditional hunting and trading locations throughout even remote parts of the Mealy Mountains, although the higher elevations were seldom visited (Armitage 2011). Several modern parties have followed that practice and traversed long distances by snowshoe and skis.
5	See Boswell (2010). Glacial meltwater rolling off the Porcupine Hills (east of the Mealy Mountains) deposited sand, gravel, and mud along the coast, producing great sediment cliffs above Porcupine Strand, said to be the fabled "Wonderstrand" of the Viking sagas, an important historical feature. Interestingly, this 40 km (24 mi)long beach may have been deglaciated as long as 12,000 years before present, though ice persisted in marine troughs much later, until 7000 years before present (Smith et al. 2003).

20-2 *The English Mountains encompass the most dramatic scenery in the Mealy Mountains, including 600 m cliffs. (MLL)*

Canada, larger even than the Torngat Mountains National Park in northern Labrador, which was created in 2005.[6] Additionally, the provincial government of Newfoundland and Labrador has announced plans to preserve 3,100 km² within the Eagle River watershed adjacent to the Mealy Mountains, although it is yet to be determined how this region will be managed, and the Eagle River itself may not be included. However, the total protected area may rise to an astounding 15,000 km² (5,700 mi²). The new National Park allows continued traditional uses (hunting, trapping, fishing, and cutting firewood) by the Métis, Innu, and Inuit people but development and mining will no longer be permitted. The details of the proposed park boundary and the specifics of the allowed uses were hotly contested for years, but the park had broad support from a wide range of groups. The 2010 announcement was widely received as a conservation success story because of its extremely large area and multiple allowed uses.[7]

The Mealy Mountains received little attention from scientists until about 2001, when Memorial University's Labrador Highlands Research Group (LHRG) began long-term fieldwork on the massif. Prior work was limited to brief, but detailed plant surveys and caribou assessments.[8]

Lay of the Land

As mentioned, the lofty Mealy Mountains occupy the southern shores of Lake Melville in interior Labrador. The highest point in the Mealy Mountains lies within the English Mountains subgroup and reaches at least to 1190 meters (3900'), consisting of three gentle summits of about equal height. Four kilometers (2.4 mi) west of the English Mountains high point is Mount Lookout (974 m / 3196'), and adjacent Stag Hill (956 m / 3137'). Most of the remaining high peaks are unnamed. The English Mountains have the most dramatic scenery in the Mealy Mountains: several spectacular cirques and inland fjords are flanked by cliffs and steep slopes, some rising more than 600 m (2,000') above the surrounding lakes and rivers (20-2). For comparison, the western part of the Mealy Mountains reaches only to about 558 m (1832') at its highest point, and the far eastern reaches are cloaked in low-elevation spruce forest (20-3).

---

6   The towering Torngats, a section of the Arctic cordillera that reaches north to Ellesmere Island, is briefly profiled in Chapter 21. The new national park surrounding the highest Torngat peaks is approximately 9,700 km² (3,745 mi²).

7   The park has been hailed as a "new type of conservation" by the Canadian Boreal Initiative, for its ecological and cultural value, although similar multi-use provisions have been included upon the establishment of some national parks in the western Canadian Arctic, and the long-term success of these management scenarios have not been scientifically evaluated.

8   For more on the Labrador Highlands Research Group, visit http://www.mun.ca/geog/lhrg/. An important early study of the flora is presented by Gillett (1954). Important geology contributions were made by Gower and van Nostrand (1996) and Gray (1969).

20-3 The easternmost edge of the Mealy Mountains, near Cartwright, Labrador, is cloaked in black spruce forest. (MJ)

The northern foothills of the Mealy Mountains can be viewed from the 12-hour ferry trip between the coastal town of Cartwright and the regional commercial and military hub of Happy Valley–Goose Bay as it passes through Lake Melville. The final leg (Phase III) of Route 510, the Trans-Labrador Highway (TLH) (completed in 2010) was routed south and west of the Mealy Mountains in an effort to preserve both the wild character of the region and the small resident herd of woodland caribou (*Rangifer tarandus*). The mostly forested, southern foothills of the Mealy Mountains are visible at points from the new segment of the TLH, but the alpine highlands of the English Mountains remain out of sight.

Access

At any time of year, travel to the Mealy Mountains is a serious undertaking in a true wilderness setting. In the most remote parts of the range, emergency rescue could be days, and many miles, away. On the ground, distances and verticals can be greater than they appear, and are easily underestimated. In summer, or whenever there is no snow cover, the brush and string bogs in lower-lying areas could complicate and slow approach hikes or portages.

Brutal, sometimes protracted subarctic winds can pound the higher elevations of the Mealy Mountains, especially the English Mountains area. These extraordinary winds are not generally captured by the weather data for either Happy Valley–Goose Bay or Cartwright, the two nearest stations with accessible data. Snow in July is not unheard of, and in fact, the Mealy Mountains is one of the snowiest places in Labrador, receiving an annual average of more than 600 cm (236") of snow (20-4). Visitors to the Mealy Mountains between May and August must come prepared for an astounding black fly population. In summary, visitors to the Mealy Mountains should be prepared for potentially miserable and dangerous environmental conditions and must be entirely self-sufficient. It is entirely appropriate to bring satellite communications equipment, comprehensive first aid supplies, extra food and fuel, compass and detailed maps, and GPS devices.

In summer, it is easiest to access the heart of the mountains by charter helicopter or light floatplane from Happy Valley–Goose Bay. In winter and spring, several of the larger frozen lakes provide generous landing sites for ski planes. Despite the proximity of the new highway, the Mealy Mountains fortunately remain inaccessible by road. There generally are no trails, shelter, or structures of any kind in the English Mountains region, although a few private rustic cabins and shacks (know as "tilts" in local trapper terms) are scattered throughout

the lower elevations of the Mealy Mountains. There are also a handful of fly-in fishing outfitter lodge operations, located mainly on river systems in the southern and eastern areas outlying the main highlands of the Mealy Mountains.

Geology

Like the Monts Groulx (Ch. 18) to the west and the Adirondacks in New York (Ch. 21), the Mealy Mountains are composed of the Precambrian metamorphic and igneous rock that forms the Grenville Province of the Canadian Shield. The higher portions consist primarily of pinkish anorthosite, with grey to black granodiorite to the south and southwest, and granitic gneisses to the east (20-5).[9]

The glacial history of the Mealy Mountains is complex. Remnants of glacial features in the bedrock suggest two Wisconsin-aged ice sheets covered the Mealy Mountains.[10] These first moved from west to east and might have overtopped the highest point, encasing the range completely in ice. The second reached only to 700 m (2300') in the western portions of the range and 550 m (1800') in the east. Mealy Mountain peaks tend to be broad and rounded, and the limited number of cirques that do occur there face northwest or southeast, as a combined result of the rock structure and the prevailing snow deposits of the Pleistocene. These cirques are thought to have been created before the most recent glaciations, due to the limited distribution of alpine moraines, although morainal material is locally abundant.[11]

20-4 July snows present a challenge for this horned dandelion (top). On average the Mealy Mountains receive more than 600 cm of snow annually. (MA and MLL)

Ecology

Vegetation — The position of the Mealy Mountains on the eastern edge of the continent, near the coast, results in a regional montane climate with both continental and marine elements, producing abundant late-lying snow at high elevations.[12] Low elevation habitats surrounding the Mealy Mountains support a wide range of terrestrial habitats, from white spruce (*Picea glauca*)–dominated, closed-canopy old growth forests in deep river valleys to open-canopy spruce taiga with thick lichen ground cover on the middle slopes (20-6). Upper slopes and ridgelines support heath-dominated tundra communities, and the highest

---

9   Eade (1962)
10   Gray (1969)
11   Gray (1969)
12   Jacobs (2006)

20-5 The Mealy Mountains are part of the Grenville Province of the Canadian Shield and are composed of igneous and metamorphic rocks including pinkish anorthosite. (MA and JK)

peaks are ultimately topped by rounded, bare-rock summits with scattered lichens and alpine vegetation.

Periglacial features (e.g., frost boils, sorted circles, polygon cracks, and stone nets) occur at high elevations, and the LHRG has demonstrated the presence of permafrost.[13] Frost boils at high elevations generate "bare ground" that allows smaller, low-growing arctic-alpine plants such as such as diapensia (*Diapensia lapponica*), Greenland switchwort (*Minuartia groenlandica*), and moss campion (*Silene acaulis*) to colonize. As a result, these boils are hotspots of plant diversity that allow low-growing arctic-alpine species to establish in the closed tundra canopy (20-7).[14]

20-6 Lower elevations surrounding the Mealy Mountains support taiga-like, open-canopy white spruce with thick lichen ground cover. (LW)

13   Jacobs (2006)
14   Sutton et al. (2006)

20-7 *Frost boils support arctic-alpine species such as (clockwise from left): diapensia, Greenland switchwort, and moss campion. (LH and MJ)*

20-8 *Purple mountain heather, a widespread arctic-alpine plant, is abundant on talus and turfy tundra near treeline. (MA)*

The Mealy Mountains are home to many of the arctic-alpine plant species characteristic of granitic and gneissic alpine areas farther south, such as the Monts Groulx in Québec (Ch. 18) and Newfoundland's Northern Long Range Mountains (Ch. 15). Diapensia, alpine azalea (*Loiseleuria procumbens*), alpine bearberry (*Arctous alpina*), mountain cranberry (*Vaccinium vitis-idaea*), moss heather (*Harrimanella* [*Cassiope*] *hypnoides*), purple mountain heather (*Phyllodoce caerulea*) and bearberry willow (*Salix uva-ursi*) are common to abundant on rocky summits and exposed sites above treeline (20-8).[15] Common alpine snowbed species include sibbaldia (*Sibbaldia procumbens*) and arctic-alpine cudweed (*Omalotheca supina*) (20-9) and snowbed willow (*Salix herbacea*), which like all willows have both male and female plants (20-10). On turfy tundra and along alpine streams grows alpine speedwell (*Veronica wormskjoldii*), arctic-loving willow (*Salix arctophila*), as well as Labrador willow (*S. argyrocarpa*). Along the margins of alpine ponds grows elephanthead lousewort (*Pedicularis groenlandica*), a distinctive hallmark of alpine and arctic tundra from the Labrador coast to Rocky Mountains and Sierra Nevada (20-11). Redtip lousewort (*P. flammea*) and Labrador lousewort (*P. labradorica*) are also present.[16] Other rarer alpine elements include three-toothed cinquefoil (*Sibbaldia tridentata*), balsam groundsel (*Packera paupercula*), Norwegian cudweed (*Omalotheca norvegica*), and horned dandelion (*Taraxacum ceratophorum*). Seepage areas in the alpine also have rarities such as alpine brook saxifrage (*Saxifraga rivularis*), Allen's buttercup (*Ranunculus allenii*) and Koenigia (*Koenigia islandica*).

---

15   Gillett (1954)
16   e.g., Gillett (1954)

Other Alpine Sites
on the Canadian Shield

21

Torngat Mountains, Labrador. (LH)

Map 21-1. Other alpine sites on the Canadian Shield.

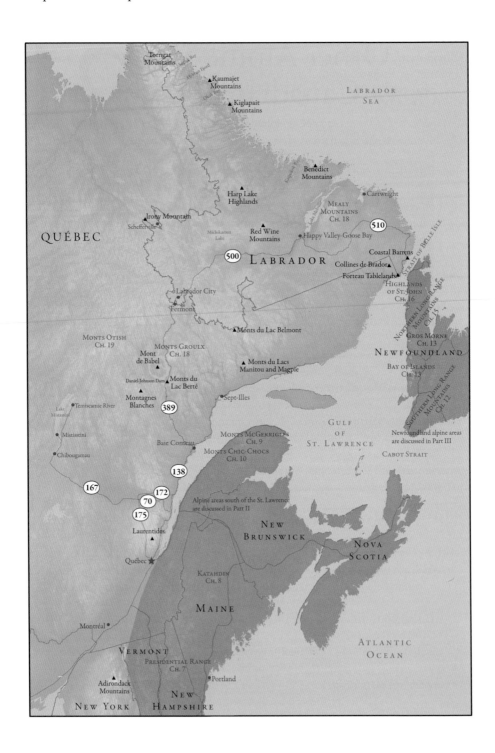

Torngat Mountains

Kaumajet Mountains

Kiglapait Mountains

LABRADOR SEA

Benedict Mountains

Harp Lake Highlands

Cartwright

MEALY MOUNTAINS CH. 18

Irony Mountain

Schefferville

QUÉBEC

Michikamau Lake

Red Wine Mountains

Happy Valley-Goose Bay

510

500

L A B R A D O R

Coastal Barrens

Collines de Brador

Forteau Tablelands

HIGHLANDS OF ST. JOHN CH. 16

STRAIT OF BELLE ISLE

Labrador City

Fermont

Monts du Lac Belmont

NORTHERN LONG RANGE MOUNTAINS CH. 15

Gros Morne CH. 13

MONTS OTISH CH. 19

MONTS GROULX CH. 18

Mont de Babel

NEWFOUNDLAND

Monts du Lacs Manitou and Magpie

BAY OF ISLANDS CH. 13

Daniel-Johnson Dam

Monts du Lac Berté

Montagnes Blanches

389

Sept-Illes

Temiscamie River

Lake Mistassini

GULF OF ST. LAWRENCE

SOUTHERN LONG RANGE MOUNTAINS CH. 12

Mistissini

Baie Comeau

MONTS MCGERRIGLES CH. 9

Newfoundland alpine areas are discussed in Part III

Chibougamau

138

MONTS CHIC-CHOCS CH. 10

CABOT STRAIT

167

172

70

175

Alpine areas south of the St. Lawrence are discussed in Part II

NEW BRUNSWICK

NOVA SCOTIA

Laurentides

Québec ★

KATAHDIN CH. 8

Montréal

MAINE

ATLANTIC OCEAN

VERMONT

PRESIDENTIAL RANGE CH. 7

Portland

Adirondack Mountains

NEW HAMPSHIRE

NEW YORK

Other Alpine Sites on the Canadian Shield

21-1 The largest alpine ecosystems in the Adirondacks occur on Mount Marcy and Algonquin Peak. Algonquin Peak from Mount Marcy, New York. (JG)

Adirondacks, New York

The Adirondacks of northeastern New York are essentially a stranded anorthosite inlier of the Canadian Shield, more closely related geologically to the Mealy Mountains of Labrador (Ch. 20) than the neighboring Green Mountains (Ch. 11), although the latter lie within sight of the Adirondacks across Lake Champlain. *Adirondack* may be a corruption of the Mohawk word *ratirontaks* ("they eat trees"), by which the Mohawks themselves may have referred to neighboring tribes.[1] The Adirondacks encompass some of the tallest mountains in eastern North America: at 1629 m (5344') Mount Marcy rises to a greater height than either Mount Lafayette in New Hampshire (Ch. 11) or Katahdin in Maine (Ch. 8). Despite their impressive height, the combined alpine area of the Adirondacks is only about 70 hectares (173 acres) distributed across more than a dozen summits, although the subalpine area is much larger.[2] Most of the alpine summits are concentrated in the "High Peaks" region surrounding Mount Marcy, southwest of Keene Valley and NY SR-73. A patch of rather isolated tundra occurs on Whiteface Mountain (1483 m / 4867'), which lies 21 km (13 mi) northeast of Lake Placid. The largest alpine areas occur on Mount Marcy and Algonquin Peak (1559 m / 5114'), with other noteworthy occurrences on Haystack (1512 m / 4960') and Mount Skylight (1501 m / 4926') (21-1). Smaller alpine areas are present on Mount Colden (1437 m / 4715'), Basin Mountain (1471 m / 4826'), Gothics (1443 m /

---

1   Donaldson (1921)

2   Carlson et al. (2011) estimated 26 hectares of alpine area using a more conservative approach than Howard (2009).

4734'), and Saddleback Mountain (1380 m / 4528').[3] The town of Lake Placid, site of the 1932 and 1980 Winter Olympics, provides a central base of operations.[4] Mount Marcy is best accessed by the Van Hoevenberg Trail, which originates at Adirondak Loj (built in the 1920s and currently operated by the Adirondack Mountain Club). Mount Skylight can be accessed via a footpath from the Four Corners north of Mount Marcy, usually done as an overnight. Algonquin Peak, the top of the MacIntyre Range, is most often accessed from the MacIntyre Range Trail, which leaves the Van Hoevenberg Trail about a mile from Adirondak Loj. Whiteface, the lone giant, is the site of a world-famous ski resort and summit auto road, both of which provide seasonal summit access. The alpine tundra of Whiteface Mountain is heavily affected by the auto road, a parking lot, and associated summit development.[5]

The Adirondack Mountains are the southernmost exposure of the Grenville Province of the Canadian Shield and are largely gneissic in composition, with igneous intrusions of granite and anorthosite. The High Peaks themselves are formed largely of Proterozoic anorthosite, an intrusive igneous rock. Although the bedrock is mostly ancient, Precambrian metamorphic and igneous rocks, the Adirondacks have been relatively recently uplifted (and subsequently dissected by erosion) within the past five million years. Thus, the tundra of the Adirondacks is underlain by potassium-rich bedrock, which supports plant species characteristic of most of the granitic and gneissic alpine summits in the northeastern United States, such as alpine bilberry (*Vaccinium uliginosum*), diapensia (*Diapensia lapponica*), and tufted clubrush (*Trichophorum cespitosum*). On Mount Marcy, a single station of single-spike sedge (*Carex scirpoidea*) is noteworthy. Mountain cranberry (*V. vitis-idaea*) is noteworthy in its absence, for it is a ubiquitous plant throughout eastern alpine and subalpine areas (even occurring on the subalpine, forested summit of Mount Greylock in Massachusetts). Purple crowberry (*Empetrum eamesii* ssp. *atropurpureum*), a rare subalpine species, has been documented from Wright Peak, Haystack Mountain, Mount Marcy, Algonquin Peak, and Mount Skylight. At least two exotic plants, Kentucky bluegrass (*Poa pratensis*) and red fescue (*Festuca rubra*), occur in the alpine on Mount Marcy.[6] Alpine plants typical of calcareous bedrock, such as saxifrages (*Saxifraga* spp.) and drabas (*Draba* spp.), which are widely distributed in the Green Mountains of Vermont, are absent from the Adirondack alpine areas, although white mountain saxifrage (*S. paniculata*) is known from the anorthosite cliffs of Wallface Mountain and purple mountain saxifrage (*S. oppositifolia*) persists along north-facing cliff walls in the Hudson River Gorge, 140 km (84 mi) from the nearest station at Smugglers Notch in the Green Mountains (Vermont). Treeline in the High Peaks region occurs at around 1300 m (4270'), ranging from 1060 m (3480') on Noonmark Mountain to over 1470 m (4820') on Mount Dix. The treeline ecosystem includes red spruce (*Picea rubens*) and balsam fir (*Abies balsamea*) in the krummholz layer and common boreal species like bunchberry (*Cornus canadensis*), goldthread (*Coptis trifolia*), and creeping snowberry (*Gaultheria hispidula*) occurring under this protective cover. Black spruce (*Picea mariana*) occurs at the alpine-treeline ecotone and on exposed summits.[7]

---

3   Carlson et al. (2011)

4   Editors' note: an added benefit of staying in Lake Placid is a trip to the Lake Placid Pub & Brewery on Mirror Lake Drive.

5   Mount Marcy, Algonquin Peak, and the other alpine summits are more extensively profiled in the ADK's Adirondack Trails High Peaks Region, in print since 1934.

6   Zika and Jenkins (1992)

7   See Slack and Bell (1993)

Laurentian Mountains (Laurentides), Québec

One hundred kilometers (60 mi) north of Québec City, the Laurentians form a dissected, subalpine tableland at the very edge of the Canadian Shield, forming a picturesque backdrop to the famous tourism region of Charlevoix. Most of the Laurentian region is forested, but several clusters of peaks rise to scrubby, subalpine summits (21-2). Generally, the tallest peaks are around 1000 m (3280'); the massif culminates at Mont Raoul Blanchard (1166 m / 3825'). The high peaks are spread throughout portions of the *Réserve Faunique des Laurentides, Parc national des Grands-Jardins*, and *Parc national des Hautes-Gorges-de-la-Rivière-Malbaie*. Although the uplands still support a herd of reintroduced woodland caribou (*Rangifer tarandus*), most of the peaks are subalpine, tapering off in thick balsam fir and green alder (*Alnus viridis*). Typical alpine plant species like diapensia, alpine azalea (*Loiseleuria procumbens*), and purple mountain heather (*Phyllodoce caerulea*) are mostly absent, although mountain cranberry, alpine bilberry, three-toothed cinquefoil (*Sibbaldia tridentata*), and black crowberry (*Empetrum nigrum*)—a suite of mostly widespread, subalpine species—are abundant (see Ch. 5).

21-2 Most of Québec's Laurentides region is forested, but several clusters of peaks rise to scrubby, subalpine summits. (LW)

21-3 Occasional jack pines, a fire-dependent species, are evidence of past fires on the mountaintops. Mont du Lac des Cygnes, Québec. (MJ)

Several of the Laurentian peaks have burned. In fact, most (if not all) of the deforested peaks are the result of post-fire forest regeneration failure. Occasional specimens of jack pine (*Pinus banksiana*), a fire-dependent species, are evidence of past fires on the mountaintops (21-3). Further complicating the ecological narrative, the Laurentian Mountains are located in the heart of the balsam fir forest zone, which is typically disturbed by eastern spruce budworm (*Choristoneura fumiferana*) outbreaks every 30 to 50 years. Whenever a wildfire occurs on a mountain summit soon after a spruce budworm outbreak (that is, before trees are reproductively mature), the peak remains deforested for a greater period of time. This process explains why one can find both forested and deforested peaks at the same elevation in this area. In addition to fires, forest clear-cutting has been another major (anthropogenic) disturbance responsible for deforestation of the subalpine summits since the late 1800s. These compounded disturbances (clear-cutting, followed by insect outbreak, followed by wildfire) are responsible for frequent openings in the boreal forest of the *Parc national des Grands-Jardins*, which support the southernmost stands of lichen woodland (similar to that of the subarctic taiga landscape).[8] The highest Laurentian peaks are generally accessible from highways that lead into the interior from Route 138, the 900 km (540 mi) route along the Québec North Shore from Québec City to Natashquan. For example, from Saint-Leon, the

---

8   e.g., Payette and Delwaide (2003)

Chemin de l'Abitibi Price leads north toward Mont Raoul Blanchard, which is heavily forested and situated in an area of heavy logging within the *Réserve Faunique des Laurentides.* Raoul Blanchard has a great western shoulder called Montagne Brulée ("Burnt Mountain," >1090 m / >3576'), which is much more exposed. Both of these mountains are difficult to access because the Abitibi-Price road is gated about 10 km (6 mi) short of Montagne Brulée, which rises above the off-limits section, and the public isn't allowed beyond the gate. However, from Baie-Saint-Paul, it's 20 km (12 mi) north on Route 381 to access Mont du Lac des Cygnes (>970 m / >3182'), a relatively well-known, subalpine peak with an extremely sturdy trail and long summit boardwalk. Close to Mont du Lac des Cygnes, the summits of Mont du Lac du Pioui (>980 m / >3215'), located to the north, and Mont du lac a Moise and Mont du Gros Ruisseau, located to the west, flirt with treeline but are essentially subalpine summits. Farther north, along the *Route du Rang B & C* from the village of Sainte-Aimé-des-Lacs, in Hautes-Gorges National Park, L'Acropole des Draveurs rises dramatically to a scrubby summit above the Rivière Malbaie.

Monts du Lacs Manitou and Magpie, Québec

About 300 km (180 mi) east of the Manicouagan Reservoir and Monts Groulx in central Québec (see Ch. 18) are a number of indistinct summits rising from a massive plateau, with topography typical of the heavily-glaciated Canadian Shield (low-relief, lacking periglacial cirques, with poorly developed "deranged" watersheds). Numerous—for the near-term, probably countless—peaks appear to reach treeline or exhibit some alpine characteristics throughout the extensive area. Some of the more prominent rise from a 6000 hectares (15,000 acre) alpine massif due west of Lac Manitou, which itself is about 50 km (30 mi) north of coastal Route 138 near *Rivière-à-la-Chaloupe.* About 50 km (30 mi) east of Lac Manitou is Lac Magpie, unmistakable from the air for its 70 km (42 mi) north-south axis. Between these great lakes lies a great subalpine plateau with many elevations more than 800 m (2500'). Farther south, adjacent to Lac Manitou are the pronounced cliffs and summit tundra of Mont Manitou. Eighty-five kilometers (53 miles) north of the Manitou–Magpie alpine areas are the highlands of the Lacs Belmont, which lie just south of Québec's border with Labrador. These high lakes lie at about 850 m (2800'), and are surrounded by several rounded alpine summits. South and west of the Lacs Belmont, straddling the Labrador border, is an indistinct subalpine tableland about 5 km (3 mi) wide. This encompasses numerous alpine summits and subalpine krummholz communities. Although these lakes experience sportfishing traffic, they remain relatively isolated and a floatplane or helicopter from Baie-Comeau is the most reasonable way to access these areas in the summer. As is the case in the Monts du Lac Berté, relatively little is known of the vegetation or ecology of these mountains, but by their proximity to the Monts Groulx (Ch. 18), they presumably bear more resemblance to these than any other mountain range profiled in this book.

Montagnes Blanches, Québec

A large, isolated alpine massif known as the Montagnes Blanches ("White Mountains") lies 320 km (200 mi) north of the Laurentian Mountains and 160 km (100 mi) south of the Monts Otish in interior Québec (Ch. 19).[9] The Montagnes Blanches are located immediate-

---

9   See de Lafontaine and Payette (2010).

ly north of timber lands owned by the Québec government and leased to the forest industry. Until now, the Montagnes Blanches have been protected from anthropogenic disturbances by their isolation even without specific conservation status (21-4). However, when flying to the area, it becomes clear that extensive clearcuts have reached the piedmont plain south of the Montagnes Blanches, and this mountain range stands out as a protected forested island in a sea of deforested landscape. The intensive forest clear-cutting to the south, and the recurrent wildfires in the lowlands to the north, contrast sharply with the forested valleys and alpine ridges of the Montagnes Blanches plateau. The Québec government's *Plan Nord*, a regional planning initiative, may "redefine" (i.e., shift farther north) the northern limit of boreal forest made available for the forest industry, and the Montagnes Blanches are at high risk of being logged in the near future if no permanent protected status is given to the range. Fortunately, there is such an ongoing project by the Québec's *Ministère du Développement Durable et des Parcs*, to create a biodiversity reserve in the area (*Réserve de biodiversité projetée des Montagnes-Blanches*).

21-4 The wild Montagnes Blanches have been protected from disturbances by their isolation from existing infrastructure, even without formal protected status. (GD)

The mountain topography of the area creates an orographic effect responsible for high precipitation (see Ch. 4), which in turn offers shelter to ancient balsam fir and black spruce forests and subalpine white spruce (*Picea glauca*) woodlands (21-5). The high-

21-5 High precipitation resulting from orographic uplift provides suitable conditions for subalpine white spruce forests, which occur throughout the Montagnes Blanches. (GD)

est peak reaches up to 1031 m (3380'), but several of the alpine summits are between 900 m (2950') and 1000 m (3280'). As is the case with the other nearby, boreal, high plateaus, Monts Groulx (Ch. 18) and Monts Otish (Ch. 19), the deforested tundra may be influenced by post-fire regeneration failure during colder climate conditions of the Little Ice Age (1550 to 1850 AD) and between 0 and 1000 AD. Above treeline, the alpine flora is similar to that found on the Monts Groulx to the northwest and typical of the interior Canadian Shield. Alpine bilberry, alpine bearberry (*Arctous alpina*), and black crowberry are abundant and are quick to colonize frost-heaved soil boils.

As noted earlier, an intermingling maze of temporary dirt roads constructed, maintained, and used by the forest companies reaches almost to the Montagnes Blanches from the south. With some luck, one could reach within 30 km (18 mi) to the south of the range in a 4x4. Using these temporary roads is not advisable; such a trip could be quite hazardous. Although largely deforested, this is a wild and remote area, and any road might close at any time due

to a lack of maintenance. There are no public gas stations, and heavily loaded logging trucks won't anticipate passenger cars without CB radios. Therefore, floatplanes (or helicopters) are the best and only reliable options to reach the Montagnes Blanches. These can be chartered from the town of Alma (a 200 km [120 mi] drive north of Québec City, where Lac Saint-Jean empties in the Saguenay River).

Monts du Lac Berté, Québec

21-6 The western edge of Québec's Monts du Lac Berté are accessible from the Québec-Labrador Highway (Route 389) through thick krumm-holz. (MJ)

The Monts du Lac Berté are a mostly sub-alpine complex just to the northeast of the enormous hydroelectric installation at Manic Cinq, itself described more thoroughly in Chapter 18 (Monts Groulx). The western edge of these mountains are partly accessible from the highway through thick krummholz (21-6). Lac Berté itself is easily found on most maps of the Manicouagan region; it lies about 18 km (11 mi) north of Manic Cinq. About 12 km (7 mi) east of the lake, a large rumpled plateau reaches 900 m (2950') in numerous places, breaking through the treeline on exposed ridges and summits. A more accessible introduction to the Lac Berté area may be found 50 km (30 mi) north of the Manic Cinq dam on Route 389. Here, an isolated massif rises to treeline only 4.3 km (2.7 mi) from Route 389, across *Ruisseau Juliette* (Juliette Creek), and is apparently capped by about 170 hectares (420 acres) of alpine tundra (about the amount found on Franconia Ridge in New Hampshire). For lack of a better name, we refer to this little-known but accessible peak as Mont Juliette (908 m / 2978'). The overland, trailless trek up to Mont Juliette would begin almost exactly where Route 389 crosses the 51st parallel. Although the uplands to the east of Lac Berté are preserved for the present within the *Réserve de biodiversité projetée du lac Berté*, Mont Juliette is excluded from the boundaries of this reserve. The vegetation of these hills is probably closely related to that found on the Monts Groulx, 64 km (40 mi) to the north.

Mont de Babel, Québec

21-7 Mont de Babel in the Manicouagan region of interior Québec is the result of shock metamor-phism from a meteor impact during the Triassic, over 200 million years ago. (MJ)

Mont de Babel (953 m / 3123') is a sub-alpine, seldom-visited mountain near the northeastern end of *Île René-Levasseur* (René-Levasseur Island), the 210,000 hectare (520,000 acre) island at the center of the Manicouagan Reservoir in central Québec (21-7). The highest point of the Babel massif is about 8 km (5 mi) from the great arm of the Reservoir known as *Baie Memory*. Mont de Babel is reportedly the result of shock metamorphism from the Triassic meteor that struck the Manicouagan region more than 200 million years

21-8 Québec's and Labrador's Forteau Tablelands terminate in cascading cliffs that terrace into the Labrador Sea. (MJ)

ago. Most of the mountain is composed of anorthosite, with an occurrence of zeolite. Mont de Babel is named for the missionary Louis Babel,[10] who surveyed the Manicouagan basin from the Québec North Shore to the Labrador border with the Innu.

Forteau Tablelands, Labrador

South and east of Québec's Collines de Brador, the Forteau Tablelands rise to four sedimentary tablelands straddling the provincial border between Blanc-Sablon, Québec and Labrador's Pinware River. Four indistinct, unnamed, "summits", from west to east, rise to >229 m (>750'), >340 m (>1117'), 319 m (1047'), and 261 m (856'), respectively. In most cases, the most visible and striking part of the Tablelands is not the summit areas, which are generally broad and featureless, but the cascading cliffs that terrace into the sea (21-8). Cal-

21-9 A number of widespread alpine and boreal species occur in the cliff communities of the Forteau Tablelands, including moss campion, strawberry, and fireweed (clockwise from top-right). (MJ)

Other Alpine Sites on the Canadian Shield — 293

21-10 Among the interesting arctic and alpine plants known from the Forteau Tablelands are redtipped louse-wort, butterwort, arctic sweet coltsfoot, and net-veined willow (clockwise from top-left). (MJ)

carous outcrops occur in the vicinity of the coast highway near Blanc-Sablon, Québec and L'Anse Amour, Labrador. These calcareous outcrops support many alpine and arctic plant species common to other eastern alpine sites in Labrador, including diapensia, moss campion (*Silene acaulis*), arctic raspberry (*Rubus arcticus*), strawberry (*Fragaria virginiana*), and fire-weed (*Chamerion angustifolium*) (21-9). Among the more interesting arctic, subarctic, and boreal species found in these areas are redtipped lousewort (*Pedicularis flammea*), common butterwort (*Pinguicula vulgaris*), alpine sweetvetch (*Hedysarum alpinum*), several cinque-foils (*Potentilla* spp.), several saxifrages (*S. paniculata*, *S. oppositifolia*, and *S. aizoides*), net-veined willow (*Salix reticulata*), and arctic sweet coltsfoot (*Petasites frigidus*) (21-10). Some invasive plant species are present in disturbed areas and along roadsides, including bird vetch (*Vicia cracca*) and dandelion (*Taraxacum officinale*) (21-11). Ravens (*Corvus corax*) nest in inaccessible nooks on cliff faces, and woodchucks (*Marmota monax*) are commonly encoun-tered on the coastal tundra (21-12).

Collines de Brador, Québec

Across the Strait of Belle Isle from the great limestone capes of northern Newfoundland (Ch. 17), on the southeastern coast of Labrador and adjacent Québec, the Collines de Brador occupy the area west of Lac Courtemanche, about 7 km (4.4 mi) north of Lac Carré on the coast highway near Fond-de-la-Baie. The Collines encompass about a half-dozen unnamed peaks in a region where any small rise culminates in alpine tundra. The Collines rise to two similar high points at unnamed peaks (>381 m; >1250'). Numerous smaller summits surround this height-of-land, most of which are unnamed and poorly explored. Underlain by gneissic rock of the Canadian Shield, these summits apparently support non-calcareous alpine vegetation, in contrast with the nearby Forteau Tablelands (discussed earlier). In these coastal borderlands, several other large massifs (mostly unnamed) rise quickly to treeline. These include, but likely aren't limited to, extensive areas west of the *Rivière Saint Paul* and west of *Ruisseau Chanion*. North of the *Rivière Bujeault*, where it flows into Indian Lake on the Québec–Labrador border, is circular system of alpine hills. Undoubtedly many other small peaks rise to treeline in this remote area.

21-11 Bird vetch (at left) is an invasive plant found along roadsides and in disturbed areas. It is pictured with alpine sweetvetch, an arctic-alpine species, near L'Anse Amour, Labrador. (MJ)

21-12 Groundhogs occur on most of the mainland alpine ranges but are particularly visible in the eastern coastal tundra of Québec and Labrador. (MJ)

Coastal Barrens, Labrador

Along the Labrador coastal highway between L'Anse Amour and Mary's Harbour, Labrador, numerous small hills are capped in alpine tundra. Examples include Tracey Hill on the west shore of Red Bay, and the unnamed rise on the old highway that leaves Route 510 west of Red Bay near Country Cat Pond (21-13). North of Red Bay, on the road to Mary's Harbour, numerous summits rise above treeline to the west, though many are

21-13 Heath-shrub-rush communities dominated by alpine bearberry (red foliage visible) occur atop the coastal hills from Red Bay to Mary's Harbour, Labrador. (MJ)

only 100 m (330') tall. These low hills are dominated by alpine species like alpine bearberry, alpine azalea, and diapensia. Woodland caribou once roamed these coastal hills, but the nearest herds are currently confined to northern Newfoundland (see Ch. 17) and the Mealy Mountains to the north (Ch. 20).

Red Wine Mountains, Labrador

21-14 The Red Wine Mountains have rounded summits with low relief and poor drainage, rising to about 880 m in interior Labrador. (CL)

The Red Wine Mountains of interior Labrador occupy an area of about 2000 km² (772 mi²), beginning about 35 km (22 mi) north of Route 500, the Trans-Labrador Highway and extending north.[11] The Red Wine massif rises to about 880 m (2900') and has rounded summits with low relief and poor drainage, typical of other large, interior mountain ranges in adjacent Québec (21-14).[12] The alpine portions are surrounded by poorly drained, black spruce forest with white spruce-balsam fir forests along river valleys. Black spruce, white spruce, balsam fir, and tamarack (*Larix laricina*) comprise the krummholz zone in interior Labrador.[13] Recent dendrological studies in the Red Wine alpine area have found submerged spruce logs in alpine ponds dating to nearly 5000 years ago, with logs in some ponds separated in age by nearly 3000 years.[14] A resident herd of several hundred woodland caribou rely on the higher elevations of the Red Wine Mountains during calving in May and June; the Red Wine herd declined by 75% during the 1980s and the 1990s. In fact, the Red Wine range gained a degree of familiarity in the scientific community because of extended, excellent caribou studies that took place there during this period.[15] During severe winters with abundant snowfall, Red Wine caribou were documented to dig through 1.2 m of snow to reach buried sedges (*Carex* spp.). The Red Wine range also encompasses the southern extent of the great George River herd, which is centered primarily on the George River and Ungava Bay to the north.[16]

Benedict Mountains, Labrador

Mount Benedict (730 m / 2400') and the surrounding Benedict Mountains lie along the Labrador coast 120 km (72 mi) north of the Mealy Mountains, about 80 km (50 mi) east-southeast of the outport town of Postville, Labrador. The northeastern face of the massif is punctuated by about a half-dozen alpine cirques. The Benedict Mountains are an intrusive granitic pluton in the Grenville province of the Canadian Shield.[17] The Benedicts—like other ranges in southern Labrador—are rounded, tundra-capped peaks. Images of the Benedict alpine area taken by Jerry Kobalenko show a tundra landscape dominated by black crowberry (*E. nigrum*), alpine bilberry, alpine bearberry, and *Cladonia* lichens, with scrubby glandular birch (*Betula glandulosa*) and krummholz tamarack, and spruce (21-15). The province of Newfoundland and Labrador has recently undertaken floristic studies of the Benedict Mountains, the results of which are hopefully forthcoming!

---

11   Schaefer et al. (1999)
12   Trindale et al. (2011)
13   Trindale et al. (2011)
14   Danek et al. (2011)
15   e.g., Brown and Theberge (1990); Schaefer et al. (1999); Schaefer et al. (2001)
16   Schaefer et al. (1999)
17   Schärer et al. (1986); Kerr et al. (1992); Rivers (1997)

Irony Mountain and the Schefferville Highlands, Québec and Labrador

Irony Mountain is the most prominent and well-known of several alpine tundra areas near the iron mining town of Schefferville, Québec.[18] Irony Mountains straddles the Labrador/Québec border about 24 km (15 mi) from Schefferville (which can be reached via passenger train from Sept-Iles, Québec), and is surrounded to the southeast by large open-pit iron mines. In this region, the treeline generally occurs between 550 m (1800') and 600 m (1970'), below which occur both open and closed lichen woodland and expansive bogs and muskeg.[19] About 7 km (4.2 mi) northeast of Irony Mountain are two low tundra ridges known as Greenbush Ridge and Highbrush Ridge in technical exploration documents, cresting just above the treeline.[20] About 18 km (11 mi) north of Irony Mountain is an alpine massif sometimes referred to as Sunny Mountain.[21] During early mine explorations and development in the 1950s and 1960s, a

21-15 The alpine tundra of the Benedict Mountains in coastal Labrador evidently supports dwarf birch (orange foliage in foreground), alpine bilberry (purple foliage in distance), tamarack (yellow conifers in distance), and crowberry (evergreen foliage in foreground). (JK)

handful of natural history observations were published on the behavior of alpine and boreal birds, including white-crowned sparrow (*Zonotrichia leucophrys*, which occurs throughout interior Labrador and adjacent Québec) and willow ptarmigan (*Lagopus lagopus*). In the years since, a few studies have been published on the flora of Irony Mountain.[22]

Kiglapait, Kaumajet, and Torngat Mountains, Labrador

The primary subject of this book has been the arctic-alpine tundra of eastern North America, focusing principally on the isolated "islands" of alpine tundra surrounded by boreal or temperate forest. In many ways, this is an arbitrary way of determining the subject matter of a book, but generally speaking, the mountains given full treatment in this book are unified by their isolation and are (in many ways) rather similar to one another. Katahdin, Gros Morne, the Monts Groulx, the Mealy Mountains, and the other ranges profiled in Chapters 7 to 20, are relict tundra ecosystems within largely forested landscapes, their upper slopes mostly forested with spruce and fir, bearing signature scars of the Pleistocene glaciers, and topped with predictable assemblages of heath-dominated alpine plant communities. The major mountains of northern Labrador—the Kiglapait, Kaumajet, and Torngat Mountains— are

18   Granberg (1994)
19   Granberg (1994); Cogbill et al. (1997)
20   Fitzgerald (1988)
21   Vellend and Waterway (1999)
22   Vellend and Waterway (1999)

21-16 *The Kaumajet Mountains of northern Labrador rise directly from the Labrador Sea to more than 1220 m. (JK)*

arguably different from the more southerly ranges.[23] These ranges are more appropriately treated in a natural history of the Arctic Cordillera, the 3000 km (1800 mi) mountain chain that runs parallel to the west coast of Greenland from northern Labrador to Ellesmere Island. They are certainly alpine—in fact, the term *alpine* in the sense of bearing ecological resemblance to the Alps, is technically more applicable to these ranges than to the southern ranges. The Torngat Mountains, for example, lie beyond the current continental treeline near Napaktok Bay,[24] and are heavily indented by active and recently active cirque glaciers.[25] Nevertheless, it would be an unacceptable oversight to exclude them from this natural history overview, and so we give them brief treatment here.

"This great peninsula of Labrador is peerless in her loneliness, and her failure hitherto to claim the attention of mankind," wrote Wilfred Grenfell in 1911 when he first identified the three major mountain groups of northern Labrador. One hundred years later, southern Labrador is marginally more connected to the western world via passenger ferry to Newfoundland and gravel highway to Québec, but northern Labrador remains one of the most remote and expensive places to visit on the North American continent.

Kiglapait Mountains — The main portion of the Kiglapait massif forms a north-facing crescent, like a buttress against the Labrador Sea, that extends about 20 km (12 mi) from west to east and is heavily indented by a dozen sharply defined glacial cirques.[26] This massif lies 60 km (38 mi) north of northern Labrador's administrative center in Nain, and comprises mostly unnamed summits that rise to about 850 m (2800'). A separate massif about 14 km (9 mi) south of the coastal crescent apparently encompasses the highest point on the

---

23   These three mountain ranges were identified as the most prominent of northern Labrador by Grenfell (1911), in whose treatise the ranges were spelled *Kiglapeits, Kaumagets*, and *Torngaks*. Grenfell went on to report that the Torngat Mountains "afford the highest and wildest scenery along the whole coast of eastern North America," a statement which remains true if only the mainland is considered.

24   Elliott and Short (1979)

25   Ives (1975)

26   According to Rogerson (2012), *Kiglapait* is Inuktitut for "saw-tooth". Odell (1933) reports *Kiglapeit* to mean "sierra" or "dog-tooth."

range, Man O' War Peak (>1035 m / 3400'). The Kiglapaits are a range of anorthosite, like the Mealy Mountains (Ch. 20) to the south.

Kaumajet Mountains — The Kaumajet Mountains lie 60 km (38 mi) south of the abandoned Moravian mission at Hebron Fjord and 80 km (50 mi) north of the Kiglapait Range.[27] The Kaumajets uniquely form two islands, Grimmington and Cod, on either side of a narrow channel known as Mugford Tickle, culminating at Brave Mountain (>1220 m / >4000'). Brave Mountain and nearby Bishop's Mitre (>1097 m / >3600') rise directly from the Labrador Sea (21-16). Noel Odell, the famed glacial geologist and participant in George Mallory and Andrew Irvine's ill-fated Everest expedition, reported that the Kaumajets "do not cover in all probability a greater area than 150 square miles, but what they lack in extent they make up for in form and the amazing abruptness with which they confront the Atlantic."[28] Like the Kiglapait Mountains, the Kaumajets are heavily ornamented with wide glacial cirques and U-shaped glacial valleys. In several locations, the cirque-floors reach to the ocean, such as on the north-facing cirques on Brave Mountain and the Bishop's Mitre. The mountain range is primarily confined to Grimmington and Cod Island but extends to the mainland in the north at Finger Hill (>1000 m / 3280').

21-17 The Torngat Mountains of northern Labrador differ in many prominent ways from the other mountains profiled in this book, but are profiled here to provide context. Ramah Bay, Torngat Mountains, Labrador. (JK)

The Kaumajets are apparently composed of gneisses and amphibolites overlain by volcanic rocks with some slates and other sedimentary deposits. Erratics of granite and amphibolite are present on the tableland, evidence of continental glaciation. The cirque between Brave Mountain and the Bishop's Mitre supports a perennial snowfield, thought by Odell (1933) to be the southernmost in all of eastern North America. Among the plants found on the higher summits is dwarf alpine hawksbeard (*Askellia nana*), an arctic composite known from the Canadian arctic archipelago and relatively few sites to the south.

Torngat Mountains — The Torngat Mountain landscape is unlike any other on the eastern North American mainland, and in many ways the Torngat Mountains are beyond the scope of this book (21-17). Certainly, the Torngats are far more extensive, tall, and complex than either the Kiglapaits or the Kaumajets (discussed earlier), a fact noted by Odell in 1933. Nonetheless, these three ranges have occasionally been grouped together since Grenfell's

---

27   According to Rogerson (2012), *Kaumajet* is Inuktitut for "shining top." Odell (1933) suggests "shining mountains" as the appropriate translation.

28   Odell (1933)

21-18 Glandular birch (left) and Labrador tea (right) are common on windswept ridges in the southern Torngats. (LH and LSC)

important 1911 account of the northern Labrador coast. "Torngat" reportedly means "the abode of spirits" or "place of spirits" in Inuktitut.[29]

Rising more than 1650 m (>5400') at Mount Caubvick, the highest point in Labrador (and the highest point between Mount Washington [Ch. 7] and Baffin Island). Other major summits are clustered along the coast north of Saglek Fjord, surrounding Nachvak Fjord, including (from south to north) Cirque Mountain (1568 m / 5144'), located only about 8 km (5 mi) northeast of Mount Caubvick; Mount Razorback (>1100 m / 3610'); Mount Tetragona (1356 m / 4449'); and the Four Peaks complex (>1340 m / >4400').[30] The Torngats differ in many ways from the ranges of southern Labrador, Newfoundland, interior Québec, and New England. For example, the Torngat Mountains lie mostly beyond the continental treeline, which reaches the Atlantic coast near Napaktok (or Black Duck) Bay (57.9°), and unlike all of the other eastern ranges, which bear the telltale signs of past glaciation but are no longer glaciated, the Torngats support numerous small, active cirque glaciers—unique among eastern mountains on the North American mainland.[31]

The Torngats are derived from a peneplain of Tertiary age, tilted to the west, so that the western flank is essentially the top of an ancient plateau. The eastern "face" is actually a dissected escarpment facing the Atlantic. Countless cirques (with and without active glaciers) and associated arêtes, endless fields of gneissic felsenmeer, and perched erratic boulders tell a complex story of Pleistocene glaciation that is still under debate.

The vegetation of the Torngat Mountains differs from the south to north, but the entire region bears affinities to the Arctic Cordillera of Baffin Island. The most exposed ridges support only scattered colonies of hoary rock moss (*Racomitrium lanuginosum*). Thick-leaf whitlowgrass (*Draba crassifolia*), rough cinquefoil (*Potentilla norvegica*), smallflower grass-of-Parnassus (*Parnassia parviflora*), and horned dandelion (*Taraxacum ceratophorum*) have been collected from the walls of Nachvak Fjord.[32] Glandular birch, northern Labrador tea (*Rhododendron tomentosum*), and black crowberry are among the most common plants on

---

29   Odell (1933); Parks Canada (2011)
30   The geography of the Torngat Mountains can be bewildering. Odell's 1933 account is fascinating and accessible, and many of the place names remain the same. Modern resources include Greg Slayden's peakbagger.com and Richard Garland's Torngats information pages (http://papabearnewyork.com/papabear/Page_Torngats.html#route_1)
31   Ives (1957)
32   Meades et al. (2000)

mountain ridges (21-18). Purple mountain heather, alpine bilberry, bog laurel (*Kalmia polifolia*), snowbed willow (*Salix herbacea*), three-toothed cinquefoil, and other widespread species of alpine habitats grow on gravel bars and alpine soils well into the interior Torngats (21-19). In the southern reaches of the Torngat region, twisted black spruce is found in sheltered river valleys, but even these isolated pockets give way to thickets of diamond-leaved willow (*Salix planifolia*), birches (*Betula* spp.), balsam poplar (*Populus balsamifera*), and green alder (*Alnus viridis*), which become the only tall vegetation in the more northern valleys (21-20).[33] Coastal elements, such as sea beach sandwort (*Honckenya peploides*), occur at the mouths of the great fjords (21-21).

The fauna of the Torngat region is primarily arctic-alpine in distribution and includes several species not known from the southern ranges. Woodland caribou range throughout the higher elevations of the Torngat Mountains during the calving period (early May to late June) and summer. Radio-telemetry has shown that the Torngat herd moves downslope into the forest-tundra of northern Québec during the autumn, and winters on the eastern shore of Ungava Bay.[34] The Torngat caribou herd overlaps spatially with the George River herd of interior northern Québec, the largest known caribou herd in the world at 600,000 to 800,000 animals.[35]

21-19 Purple mountain heather (top) and alpine bilberry (bottom), widespread alpine species known as far south as Mount Washington, occur near sea level on sheltered alpine tundra. (LH)

Diverse photographic, skeletal, and anecdotal evidence now suggests that an isolated population of grizzly bears (*Ursus arctos*) ranged the Torngats and George River region in Labrador and Québec until the early 20th century.[36] This is, quite simply, a staggering fact of real ecological significance. Recently, researchers have suggested that the extinction of the Labrador barren-ground grizzly allowed Torngat populations of black bear (*Ursus americanus*) to assume a life history reminiscent of grizzly bears: more pronounced carnivorous diet, larger body size, and larger home ranges.[37] Polar bears (*Ursus maritimus*) occur throughout

---

33   Meades (1990)
34   Schaefer and Luttich (1998)
35   Couturier et al. (1996)
36   Spiess (1976); Spiess and Cox (1976); Veitch and Harrington (1996); Loring and Spiess (2007)
37   Veitch (1992)

21-20 A stand of balsam poplar, one of few tree species to survive in the harsh Torngat climate, is visible at the center of this photograph. (LSC)

21-21 Coastal plants such as seabeach sandwort are common in the coastal fjords of the Torngat region. (LSC)

the coastal regions of the Torngat Mountains, further differentiating these mountains from those further south.

Although grizzly bears have apparently been extirpated, musk oxen (*Ovibos moschatus*) have tenuously colonized the Torngat Mountains from introduction sites in northern Québec. Most of the five recent sightings were clustered in the Komaktorvik River Valley, between Mount Tetragona and Mount Eliot (1388 m / 4553'), in the heart of the coastal Torngat ranges, but one observation occurred as far south as Voisey's Bay, closer to the Red Wine Mountains than to the Torngats.[38] Musk oxen are not native to Québec or Labrador, apparently having failed to colonize the peninsula following the retreat of the Laurentide ice sheets. At this point, it is unclear whether this arctic species, a rather interesting relict even in its native range of Greenland and the Canadian arctic archipelago, will persist in Labrador and if so, what effect it may have on vegetation structure or how it may interact with native caribou.

---

38   Chubbs and Brazil (2007)

A Conservation Vision 22

Chimney Pond, Katahdin, Maine. (MJ)

A Conservation Vision

Where from here? What path to the year 2100 will leave the eastern alpine ranges most intact?

The Northern Forest, and especially its few eastern alpine peaks, is threatened by myriad long-term and immediate changes. The negative effects of long-term trends, such as anthropogenic warming and volatile swings in precipitation, are extremely complex and potentially devastating, although the eventual results are largely unknown. Short-term or immediate threats, including wind farm development, mining, logging, heavy recreational use, invasive plant and animal species, road-building, inadequate protections for wolves and caribou, and so on, are sometimes listed in a cavalier fashion that does little to capture the actual depth, magnitude, and uncertainty of these various real and perceived dangers.

Our thesis in this book has been that the alpine mountains of eastern North America are sufficiently rare and demonstrably important to biodiversity conservation to be worthy of protection wherever they occur. The East lacks the embarrassment of alpine riches found in the West. Our scattered alpine tundra is of the heirloom variety—ancient and unique. It would (or will) require many more years of exploration to adequately understand the intricate dance of disjunct lichens, mosses, liverworts, clubmosses, firmosses, spike mosses, mustards, saxifrages, roses, heaths, beetles, ants, bees, butterflies, dragonflies, spiders, toads, frogs, tree frogs, songbirds, ducks, raptors, lemmings, and voles harbored within the more remote alpine mountains. In the meantime, and in the near term, the precautionary principle (and a reasonable sense of humility and aesthetics) dictates that the remaining unprotected mountains should be set aside with minimal or no development. This much seems clear to us.

Mountains are often logical targets for conservation anyway because there has historically been little else to do with them. Remoteness from existing infrastructure and inaccessibility caused by high-relief topography, combined with the tireless efforts of John Muir and others, resulted in great "rock and ice" parks such as Glacier, Yosemite, Banff, Jasper, Rocky Mountain, and Olympic. In fact, U.S. and Canadian national parks protect alpine habitats disproportionate to their occurrence on the landscape.[1] Mountains in general are well-represented in the park systems: many of the non-alpine National Parks such as Big Bend, Saguaro, Grand Canyon, and Great Smoky Mountains still encompass extremely high-relief, mountainous wildlands that do not lend themselves to easy development. Low-relief parks entirely free of mountains, such as Everglades, are relatively uncommon. But remoteness is no longer a protective device for the mountainous wildlands of eastern North America, and competing interests threaten their integrity, as we have seen in the preceding chapters. Adequate protection of alpine habitats, especially in the East, will require a more focused, informed, and range-specific approach. Nor, in our opinion, is remoteness the most important reason to preserve alpine mountain ranges. In the East as in the West, the high alpine mountains tend to extend across many life zones. Typically, in the East, a high mountain range will extend from the temperate forests through the Canadian (boreal) spruce–fir zone and Hudsonian krummholz to the alpine life zone. What this means is that a mountainous area in the eastern boreal forest of Québec, Labrador, or Maine is capable of sustaining the typical species of the boreal forest—black and white spruce, balsam fir, tamarack, jack pine, heart-leaved birch, mountain maple, bluebead lily, arctic sweet coltsfoot, and so on—as well as the various species characteristic of the life zones found at higher elevations (and latitudes). Even more striking, a mountain range in the more temperate regions such as New Hampshire or Vermont will support many of these boreal species, as well as many alpine obligates at higher elevations, but also support species more typical of the "northern hardwood" forests below, such as beech and yellow birch and the many species found in these forests. Hectare for hectare or acre for acre, mountains can support greater biodiversity than a comparably-sized, low-relief site nearby. The high relief of mountain systems also contain a variety of slopes and aspects that provide myriad ecological niches not as readily available in the valley below. For these reasons, 1000 hectares of low-relief boreal forest may not be as biologically rich as 1000 hectares encompassing a high mountain that breaks through the treeline. This is an important consideration when prioritizing land for protection in light of the realities of global and regional climate change because these diverse elevational and aspect gradients can operate as natural buffers against changing climate.

We hope (and anticipate) that the eastern alpine areas of New England, New York, Québec, Newfoundland, and Labrador will loom large in the coming discussions of conserva-

---

1   This isn't to say that the U.S. or Canadian national park model is the best route to preservation for the remaining unprotected alpine mountains of the East. Historically, National Park designation has been associated with its own variety of development. Although probably unfeasible in John Muir's day, other types of protections from local to international are available as alternatives today. One promising (though tenuous) model of preservation is found in the Monts Groulx, Québec (Ch. 18), where there has been limited development and there remains little need for heavy-handed bureaucracy. However, this argument can be easily dismissed because the access to the Monts Groulx has been facilitated by a major provincial highway.

tion priorities,[2] and that the coming decade will ideally see the permanent protection of the *entire* Monts Groulx, the *entire* Monts Otish (including Mont Yapeitso), the Highlands of St. John and their surrounding forests, the Lewis Hills, Blow Me Down, the North Arm Hills, Table Mountain, the White Hills, the Northern Long Range Mountains, the Benedict Mountains and other mountain ranges farther north, as well as constant improvement in the management of the alpine summits south of the Gulf of St. Lawrence in Gaspésie, Maine, New Hampshire, Vermont, and New York.[3] As we said in Chapter 1, we propose that the likelihood of this outcome will be strengthened by the existence of an informed constituency, convinced of the intrinsic and measured value of eastern alpine tundra. We hope that this book provides additional tools toward that end and, in this way, offsets the potential damage from increased use.

To make this more likely—*and this is such an important point*—we must travel to these areas responsibly. This means many things, and something slightly different in each area. In the much-loved mountains of New York, Vermont, New Hampshire, Maine, and Gaspésie, it means staying on the trails or durable rock surfaces, supporting local mountain clubs and advocating for constant improvement in the way we manage the well-loved mountains. In the remote tundra of interior Québec, coastal Labrador, and western Newfoundland, it means following the broad principles outlined by Leave No Trace: do not damage the krummholz or tundra vegetation, use campstoves instead of wood fires, avoid single-file lines on untracked tundra, pack out others' garbage as well as your own, avoid fragile habitats such as wetlands and cushion-tussock barrens, avoid the temptation to approach caribou or ptarmigan or arctic hare. The more remote eastern alpine ranges provide an interesting challenge: with no other people present to assess our backcountry ethic, how we behave is entirely a reflection of how much we value the alpine resource. This reduces the degradation to the landscape, and the more responsibly we are able to enjoy alpine resources, the less likely the need will be for heavy-handed regulations in the name of reducing additional impacts.

And there is more work to be done, we believe. It's not quite time to set the mountains aside for their own sake to be left to their own fate. We need some additional, low-impact studies and monitoring to understand the relict biodiversity of the individual mountains and how the individual species respond to the threats outlined at the head of this chapter. We need to overcome the challenging political and language barriers and improve communication among the scientists, managers, mountain groups, local residents, and recreational users across state, provincial, and international lines. With hard work and a small amount of luck, all of these ranges will be permanently protected without excessive infrastructure, and only lightly managed where necessary to strengthen or complement their resilience. It's hard to know what constitutes an appropriate vision for the eastern alpine areas a century from now. Should we simply hope that the major ranges do not degrade further? Is it reason-

---

2   Several jurisdictions are currently undergoing large-scale planning, most prominently the industry-focused *Plan Nord* in Québec, that could ultimately decide the configuration of protected areas in the boreal forest, including of many of these alpine ranges and their supporting landscapes. Fortunately, local, regional, and international conservation groups such as the Canadian Boreal Initiative, Nature Québec, Two Countries / One Forest, *Les Amis des Monts Groulx,* the Appalachian Mountain Club, Nature Conservancy, and others are working to ensure—directly or indirectly—that alpine biodiversity is considered during these planning processes.

3   The International Appalachian Trail in Newfoundland and Labrador is advancing an ambitious effort to preserve the ultramafic summits of the Bay of Islands as part of the proposed "Cabox International Appalachian Trail Park," itself part of the "Carbon Cycle and Community Stability" strategy.

able to hope for a gradual and steady improvement? The expansion of caribou south of the Gaspésie? The recolonization of Québec mountains by Ontario wolverines? The return of wolves to Newfoundland, Gaspésie, and New England? Yes, yes, and yes. May it be so.

In any case, we hope that unmarred alpine tundra persists throughout the east, supporting a full complement of subarctic and alpine wildlife and plants, for generations to come. We hope that some of you—the restless and able—will, to borrow Muir's phrase, "throw some biscuits in a sack and jump over the back fence" to see these places for yourselves. And we hope that all people who love the eastern mountains will simply delight in the fact that these wild ranges exist, offering a vision of the alpine past—and the alpine future—of eastern North America.

Literature Consulted

Adams, C.C., G.P. Burns, T.L. Hankinson, B. Moore, and N. Taylor. 1920. Plants and animals of Mount Marcy, New York, Part II. *Ecology* 1(3): 204–233.

Alcock, J. 1987. Leks and hilltopping in insects. *Journal of Natural History* 21: 319–328.

Alexander, C.P. 1940. The Presidential Range of New Hampshire as a biological environment, with particular reference to the insects. *American Midland Naturalist* 24(1): 104–132.

Allen, G.M. 1904. Fauna of New England. 3: List of the Mammalia. *Occasional Papers of the Boston Society of Natural History* 7(3): 1–35.

Allen, T.R., and Walsh, S.J. 1996. Spatial and compositional pattern of alpine treeline, Glacier National Park, Montana. *Photogrammetric Engineering & Remote Sensing* 62(11): 1261–1268.

Anichtchenko A. et al. (eds.). 2011. Carabidae of the World. http://www.carabidae.pro

Anions, M. 1994. The flora of Gros Morne National Park. Report, Contract GMR 93–021. Gros Morne National Park, Rocky Harbour, NL. 143 pp.

Antevs, E. 1932. *Alpine zone of Mount Washington Range.* Merrill and Webber; Auburn, ME. 118 pp.

Appalachian Mountain Club (AMC). 2009. Geology of the White Mountains Part 2: The Mountain Building Events. *Appalachian Mountain Club Nature Notes.*

Armitage, P. 2011. *An Assessment of Lower Churchill Project Effects on Labrador Innu Land Use and Occupancy.* Innu Nation; Sheshathiu and Natuashish, Labrador.

Aubry, K.B., K. McKelvey, and J.P. Copeland. 2007. Distribution and broadscale habitat relations of the wolverine in the contiguous United States. *Journal of Wildlife Management* 71: 2147–2158.

Atwood, J.L., C.C. Rimmer, K.P. McFarland, S.H. Tsai, and L.R. Nagy. 1996. Distribution of Bicknell's Thrush in New England and New York. *The Wilson Bulletin* 108(4): 650–661.

Austin, O.L. 1932. The Birds of Newfoundland Labrador. Publication no. 7; Memoirs of the Nuttall Ornithological Club; Cambridge, MA. 229 pp.

Baldwin, H.I. 1977. The induced timberline of Mount Monadnock, N.H. *Bulletin of the Torrey Botanical Club* 104(4): 324–333.

Barrington, D.S., and C.A. Paris. 2007. Refugia and migration in the Quaternary history of the New England flora. *Rhodora* 109: 369–386.

Bell, T., J.D. Jacobs, A. Munier, P. Leblanc, and A. Trant. 2008. Climate Change and Renewable Resources in Labrador: Looking toward 2050. Proceedings and Report of a Conference held in North West River, Labrador, 11–13 March. St.

John's: Labrador Highlands Research Group, Memorial University of Newfoundland. 96 pp.

Bergerud, A.T. 1967. The distribution and abundance of arctic hares in Newfoundland. *Canadian Field–Naturalist* 81: 242–248.

Bergerud, A.T., and W.E. Mercer. 1989. Caribou introductions in eastern North America. *Wildlife Society Bulletin* 17: 111–120.

Bertness, M.D., and R.M. Callaway. 1994. Positive interactions in communities. *Trends in Ecology and Evolution* 9: 191–193.

Blaney, S., and D. Mazerolle. 2010. Botanical fieldwork on Cape Breton Island for the Nova Scotia Species at Risk Conservation Fund. Report submitted to the Nova Scotia Species at Risk Conservation Fund; Kentville, Nova Scotia.

Bliss, L.C. 1963. Alpine plant communities of the Presidential Range, New Hampshire. *Ecology* 44(4): 678–697.

Bliss, L.C. 1963. *Alpine plant communities of the Presidential Range, New Hampshire.* Privately published; Edmonton, AB.

Böcher, J., and G. Nachman. 2010. Are environmental factors responsible for geographic variation in the sex ratio of the Greenlandic seed–bug *Nysius groenlandicus? Entomologia Experimentalis et Applicata* 134: 122–130.

Boisjoly, D. 2006. Sélection de l'habitat par le coyote, Canis latrans, dans le contexte de la conservation du caribou de la Gaspésie. Unpublished M.Sc. thesis, Université du Québec. 52pp.

Boisjoly, D., J.-P. Ouellet, and R. Courtois. 2010. Coyote habitat selection and management implications for the Gaspésie caribou. *Journal of Wildlife Management* 74(1): 3–11.

Boonstra, R., and A.R.E. Sinclair. 1984. Distribution and habitat use of caribou, *Rangifer tarandus caribou,* and moose, *Alces alces andersoni*, in the Spatsizi Plateau Wilderness Area, British Columbia. *Canadian Field–Naturalist* 98: 13–21.

Boswell, R. 2010. Huge new national park announced for Labrador. *Vancouver Sun.* February 5, 2010.

Bouchard, A., L. Brouillet, S. Hay, and P. Turcotte. 1994. The rare vascular plants of Big Level, Gros Morne National Park, Newfoundland, Canada. Report GMR 93–02. Parks Canada; Gros Morne National Park, Rocky Harbour, NL. 54 pp.

Bouchard, A., S. Hay, C. Gauvin, and Y. Bergeron. 1986. Rare vascular plants of Gros Morne National Park, Newfoundland, Canada. *Rhodora* 88(856): 481–502.

Bouchard, A., S. Hay, Y. Bergeron, and A. Leduc. 1987. Phytogeographical and life–form analysis of the vascular flora of Gros Morne National Park, Newfoundland, Canada. *Journal of Biogeography* 14: 343–358.

Bouchard, A., S. Hay, L. Brouillet, M. Jean, and I. Saucier. 1991. The Rare Vascular Plants of the Island of Newfoundland. Syllogeus No. 65. Canadian Museum of Nature; Ottawa, ON.

Boucher, D. 2000. *Évolution millénaire de la dynamique des avalanches de neige au Mont Hog's Back, Gaspésie, Québec.* Unpublished M.A. thesis, University of Laval; Sainte-Foy, QC.

Bourque, P.A. 2010. Planète Terre. Université Laval. At website: http://www2.ggl.ulaval.ca/personnel/bourque/intro.pt/planete_terre.html

Bousquet, Y. (ed.). 1991. Checklist of beetles of Canada and Alaska. Publication 1861/E. Research Branch, Agriculture Canada; Ottawa, ON.

Brisebois, D., and J. Brun. 1994. La plate–forme du Saint–Laurent et les Appalaches. In M. Hocq (ed.), *Geologie du Québec.* Les Publications du Québec; Québec, QC. 154 pp.

Brown, R.J.E. 1979. Permafrost distribution in the southern part of the discontinuous zone in Québec and Labrador. *Geographie physique et Quaternaire* 33(3):279-289.

Brown, W.K., and J.B. Theberge. 1990. The effect of extreme snowcover on feeding-site selection by woodland caribou. *Journal of Wildlife Manangement* 54(1): 161–168.

Brouillet, L., S. Hay, P. Turcotte, and A. Bouchard. 1998. La flore vasculaire alpine du plateau Big Level, au parc national du Gros–Morne, Terre–Neuve. *Géographie physique et Quaternaire* 52(2): 175–194.

Brouillet, L., and L. Lebrun. 1999. Rare vascular plants of the west coast of Newfoundland: Preliminary analysis of the distribution. Herbier Marie–Victorin, Institut de recherche en biologie végétale, Université de Montréal. Unpublished report.

Brown, R.J.E. 1979. Permafrost distribution in the southern part of the discontinuous zone in Québec and Labrador . *Geographie physique et Quaternaire* 33: 279–289.

Brown, W. K., J. Huot, P. Lamothe, S. Luttich, M. Paré, G. St. Martin, and J. B. Theberge. 1986. The distribution and movement patterns of four woodland caribou herds in Québec and Labrador. *Rangifer* Special Issue No. 1: 43–49.

BugGuide. 2011. BugGuide: Identification, images, & information for insects, spiders & their kin for the United States & Canada. http://www.bugguide.net.

Burzynski, M. 1999. *Gros Morne National Park.* Breakwater Books; St. John's, NL.

Burzynski, M., and A. Marceau. Undated. *Discovering the Limestone Barrens of Western Newfoundland.* Limestone Barrens Species at Risk Recovery Team; Memorial University Botanical Garden; St. John's, NL.

Canadian Environmental Assessment Agency (CEEA). 2010. Project to Extend Road 167 North to Otish Mountains Proposed by Transports Québec. Information Document on the Project and the Conduct of the Comprehensive Study Established Under the Canadian Environmental Assessment Act. Canadian Environmental Assessment Registry Reference Number: 10–03–54435.

Carlson, B.Z., J.S Munroe, and B. Hegman. 2011. Distribution of alpine tundra in the Adirondack Mountains of New York, U.S.A. *Arctic, Antarctic, and Alpine Research* 43(3): 331–342.

Chapco, W., and G. Litzenberger. 2002. A molecular phylogenetic study of two relict species of melanopline grasshoppers. *Genome* 45: 313–318.

Choler, P., R. Michalet, and R.M. Callaway. 2001. Facilitation and competition on gradients in alpine plant communities. *Ecology* 82(12): 3295–3308.

Chubbs, T., and J. Brazil. 2007. The occurrence of Muskoxen, *Ovibox moschatus* in Labrador. *The Canadian Field-Naturalist* 121: 81–84.

Clark, S. 1996. *Katahdin: A Guide to Baxter State Park and Katahdin.* North Country Press; Unity, ME.

Clement, R.C. 1968. White–crowed sparrow, *Zonotrichia leucophrys. Smithsonian Institution United States National Museum Bulletin* 237(Part 3): 1273–1291.

Clough, G.C., and J.J. Albright. 1987. Occurrence of the northern bog lemming, *Synaptomys borealis*, in the northeastern United States. *Canadian Field-Naturalist* 101(4): 611–613.

Cogbill, C. 1996. Alpine flora in Grafton County. *Prenanthes* 1(2): 2–4.

Cogbill, C.V., and P.S. White. 1991. The latitude–elevation relationship for spruce–fir forest and treeline along the Appalachian mountain chain. *Vegetatio* 94(2): 153–175

Cogbill, C.V., P.S. White, and S.K. Wiser. 1997. Predicting treeline elevation in the southern Appalachians. *Castanea* 62(3): 137–146.

Coleman, R.G., and C. Jove. 1992. Geological origin of serpentinites, pp. 469–494. In A.J.M. Baker, J. Proctor, and R.D. Reeves (eds.), *The Vegetation of Ultramafic (Serpentine) Soils*. Proc. 1st International Conference on Serpentine Ecology. Intercept Ltd.; Andover, Hampshire, U.K.

Collins, F.J., and M.L. Fernald. 1925. The region of Mount Logan, Gaspé Peninsula. *Geographic Review* 15: 84–91.

Commission de toponymie. 2011. *Banque de noms de lieux du Québec.* At website: http://www.toponymie.gouv.qc.ca

Cooper, S.V., P. Lesica, and D. Page-Dumroese. 1997. Plant Community Classification for Alpine Vegetation on the Beaverhead National Forest, Montana. USDA Forest Service. General Technical Report INT–GTR–362.

Copeland, J.P., K.S. McKelvey, K.B. Aubry, A. Landa, J. Persson, R.M. Inman, J. Krebs, E. Lofroth, H. Golden, J.R. Squires, A. Magoun, M.K. Schwartz, J. Wilmot, C.L. Copeland, R.E. Yates, I. Kojola, and R. May. 2010. The bioclimatic envelope of the wolverine (*Gulo gulo*): do climatic constraints limit its geographic distribution? *Canadian Journal of Zoology* 88: 233–246.

Copeland, J.P., and R.E. Yates. 2008. Wolverine population assessment in Glacier National Park: comprehensive summary update. U.S. Forest Service, Rocky Mountain Research Station; Missoula, MT. 16 pp.

COSEWIC. 1997. The Status of Fernald's Milk–vetch (Astragalus robbinsii *var.* fernaldii) *in Canada*. Committee on the Status of Endangered Wildlife in Canada; Gatineau, QC. 30 pp.

COSEWIC. 2006. COSEWIC assessment and update status report on the Gaspé shrew *Sorex gaspensis* in Canada. Committee on the Status of Endangered Wildlife in Canada; Ottawa, ON. vii + 22 pp.

Couturier, S., R. Courtois, H. Crépeau, L.-P. Rivest, and S. Luttich. 1996. Calving photocensus of the Rivière George Caribou Herd and comparison with an independent census. Proc. Sixth North American Caribou Workshop, Prince George, British Columbia, Canada, 1–4 March 1994. *Rangifer* Special Issue 9: 283–296.

Courtois, R., J.-P. Ouellet, A. Gingras, C. Dussault, L. Breton, J. Maltais. 2003. Historical changes and current distribution of caribou, *Rangifer tarandus*, in Québec. *The Canadian Field-Naturalist* 117: 399–414.

Couturier, S., D. Jean, R. Otto, and S. Rivard. 2004. Demography of the migratory tundra caribou (Rangifer tarandus) of the Nord-du-Québec region and Labrador. Ministère des Ressources naturelles, de la Faune et des Parcs; Québec, QC. 68 pp.

Damman, A.W.H. 1983. An Ecological Subdivision of the Island of Newfoundland. In G.R. South (ed.), *Biogeography and Ecology of the Island of Newfoundland*, pp. 163–206. Dr. W. Junk; The Hague, Netherlands.

Danek, M., A.B. Young, C. Laroque, and T. Bell. 2011. A preliminary dendrochronological assessment of subfossil trees from central Labrador. Poster made available by the authors to MTJ.

Danks, H.V., and J.A. Downes (eds.). 1997. Terrestrial Arthropods. In *Insects of the Yukon*. Biological Survey of Canada; Ottawa, ON.

Daubenmire, R.F. 1954. Alpine timberlines in the Americas and their interpretation. *Butler University Botanical Studies* 11: 119–136.

de Lafontaine, G., and S. Payette. 2010. The origin and dynamics of subalpine white spruce and balsam fir stands in boreal eastern North America. *Ecosystems* 2010(13): 932–947.

de Lafontaine, G., and S. Payette. 2011. Shifting zonal patterns of the southern boreal forest in eastern Canada associated with changing fire regime during the Holocene. *Quaternary Science Reviews* 30: 867–875.

Demille, J. B. 1926. Birds of Gaspé County, Québec. *The Auk* 43(4): 508–527.

Department of Environment and Conservation. 2010. *Our Wildlife: News from the Wildlife Division.* Newfoundland Department of Environment and Conservation; Corner Brook, NL.

Dickson, W.L. 2002. Silica assessment of the Hawke Bay Formation, Highlands of St. John area, Great Northern Peninsula, Newfoundland. Newfoundland Department of Mines and Energy Geological Survey, Report 01–1: 319–325.

Dodds, O.G. 1960. The economics, biology, and management of the snowshoe hare in Newfoundland. Unpublished Ph.D. dissertation, Cornell University; Ithaca, NY. 333 pp. (cited in Bergerud 1967)

Donaldson, A.L. 1921. *A History of the Adirondacks.* Century Press; New York, NY.

Downes, J. A. 1965. Adaptations of insects in the Arctic. *Annual Review of Entomology* 10: 257–274.

Drapeau, L. 1994. *Dictionnaire montagnais—français.* Presses de l'Université Laval; Sainte-Foy, QC. 762 pp.

Dullinger, S.I. Kleinbauer, H. Pauli, M. Gottfried, R. Brooker, L. Nagy, J.-P. Theurillat, J. I. Holten, O. Abdaladze, J.–L. Benito, J.-L. Borel, G. Coldea, D. Ghosn, R. Kanka, A. Merzouki, C. Klettner, P. Moiseev, U. Molau, K. Reiter, G. Rossi, A. Stanisci, M. Tomaselli, P. Unterlugauer, P. Vittoz, G. Grabherr. 2007. Weak and variable relationships between environmental severity and small-scale co-occurrence in alpine plant communities. *Journal of Ecology* 95: 1284 –1295.

Dutcher, B.H. 1903. Mammals of Mt. Katahdin, Maine. *Proceedings of the Biological Society of Washington* 16: 63–72.

Dyke, A.S. 2004. An outline of North American deglaciation with emphasis on central and northern Canada. In J. Ehlers and P.L. Gibbard (eds.), *Quaternary glaciations—Extent and Chronology*, Part II, pp. 373–424. Elsevier B.V.; Amsterdam, Netherlands.

Eade, K.E. 1962. Geology, Battle Harbour—Cartwright, Coast of Labrador, Newfoundland. Geological Survey of Canada Map 22-1962.

Eames, E.H. 1909. Notes upon the Flora of Newfoundland. *Rhodora* 11(125): 85–99.

Edwards, R.Y. 1955. The Habitat Preferences of the Boreal Phenacomys. *The Murrelet* 36(3): 35–38.

Elliott, D.L., and S.K. Short. 1979. The northern limit of trees in Labrador: A discussion. *Arctic* 32(3): 201–206.

Elliott, G.P., and Kipfmueller, K.F. 2010. Multi-scale influences of slope aspect and spatial pattern on ecotonal dynamics at upper treeline in the Southern Rocky Mountains, U.S.A. *Arctic, Antarctic, and Alpine Research* 42: 45–56.

Fernald, M.L. 1907. Soil preferences of certain alpine and sub-alpine plants. *Rhodora* 9: 149–193.

Fernald, M.L. 1911. A botanical expedition to southern Labrador. *Sarracenia* 11(4): 34–39.

Fernald, M.L. 1918. The contrast in the floras of Eastern and Western Newfoundland. *American Journal of Botany* 5(5): 237–247.

Fernald, M.L. 1926. Two summers of botanizing in Newfoundland. III. Noteworthy plants collected in Newfoundland, 1924 and 1925. *Rhodora* 28(328): 49–63, 74–87, 89–111, 115–129, 145–155, 161–178, 181–204, 210–225, 234–241.

Fernald, M.L. 1933. Recent discoveries in the Newfoundland flora. *Rhodora* 35: 1–16, 47–63, 80–107, 120–140, 161–185, 203–223, 230–247, 265–283, 298–315, 327–346, 364–386, 395–403.

Fernald, M.L. 1950. *Gray's Manual of Botany*, 8th edition. American Book; New York, NY.

Fitzgerald, J.J. 1988. Schefferville transect surface roughness data analysis. *Suffield Memorandum* No. 1232. Defense Research Establish Suffield. Ralston, AB.

Fitzgerald, N. 2002. Fernald's milk–vetch rediscovered at Barr'd Harbour highlands. *Northern Pen* [newspaper, St. Anthony, Newfoundland], July 22, 2002.

Fitzgerald, S.M., and L.B. Keith. 1990. Intra– and inter–specific dominance relationships among arctic and snowshoe hares. *Canadian Journal of Zoology* 68: 457–464.

Fortin, C., V. Banci, J. Brazil, M. Crête, J. Huot, M. Huot, R. Lafond, P. Paré, J. Shaefer, and D. Vandal. 2005. National recovery plan for the wolverine (*Gulo gulo*) [Eastern Population]. National Recovery Plan No. 26. Recovery of Nationally Endangered Wildlife (RENEW); Ottawa, Ontario. 33 pp.

Fortin, C., P. Galois, B. Dutil, L. Ponge, and M. Ouellet. 2011. Inventaire de l'herpétofaune dans la région des monts Otish. *Naturaliste Canadienne* 136(1): 122–131.

Foss, K. 2010. Labrador research uncovers oldest black spruce. *Memorial University News*; July 27, 2010.

Foster, J. B. 1961. Life History of the Phenacomys Vole. *Journal of Mammology* 42(2): 181–198.

Garneau, F.X. 1845. Histoire du Canada. Imprimerie de N. Aubin; Québec, QC.

Germain, D., Filion, L., and B. Hétu. 2009. Snow–avalanche regime and related weather scenarios in the Chic–Choc Range, eastern Canada. *Climatic Change* 92(40210): 141–167.

Germain, D., B. Hétu, and L. Filion. 2010. Tree–ring based reconstruction of past snow avalanche events and risk assessment in the Northern Gaspé Peninsula (Québec, QC). In M. Stoffel, M. Bollschweiler, D.R. Butler, and B.H. Luckman.

(eds.), *Tree-ring reconstructions in natural hazards research: a state-of-the-art.* Springer-Verlag; Berlin.

Gervais, C. 1982. La flore vasculaire de la région du Mont Logan, Gaspésie, Québec. *Provancheria* 13. Université Laval; Québec. 63 pp.

Gigon, A. 1987. A Hierarchic Approach in Causal Ecosystem Analysis—The Calcifuge–Calcicole Problem in Alpine Grasslands. *Ecological Studies* 61: 228–244.

Gillett, J.M. 1954. A plant collection from the Mealy Mountains, Labrador, Canada. *Canadian Field-Naturalist* 68: 188–122.

Global Biodiversity Information Facility. 2011. Copenhagen, Denmark. At website: http://www.gbif.org.

Gouvernement du Québec. 2005a. E'weewach Place where the waters come from. Provisory Master Plan. Albanel–Témiscamie–Otish National Park Project. 40 pp. http://www.mddep.gouv.qc.ca/parcs/ato/Plan-en.pdf

Gouvernement du Québec. 2005b. Fact Sheet 5. Zoning for Conservation. E'weewach Place where the waters come from. Provisory Master Plan. Albanel–Témiscamie–Otish National Park Project. http://www.mddep.gouv.qc.ca/parcs/ato/con-ato_en.htm

Gower, C.F., and T. van Nostrand. 1996. Geology of the southeast Mealy Mountains region, Grenville Province, Southeast Labrador. Current research, Newfoundland Department of Natural Resources, Geological Survey, Report Vol. 69(1).

Granberg, H.B. 1994. Mapping heat loss zones for permafrost prediction at the northern/alpine limit of the Boreal forest using high-resolution C–band SAR. *Remote Sensing of Environment* 50(3): 280–286.

Grant, D.R. 1992. *Quaternary geology of the St. Anthony—Blanc–Sablon area.* Newfoundland and Québec. Geological Survey of Canada. 60 pp.

Gray, J.T. 1969. Glacial history of the eastern Mealy Mountains, Southern Labrador. *Arctic* 22(2): 106–111.

Gray, J.T., and R.J.E. Brown. 1979. Permafrost existence and distribution in the Chic-Chocs Mountains, Gaspésie, Québec. *Geographie physique et Quaternaire* 33(3–4): 299–316.

Gray, D.R. 1993. Behavioural adaptations to arctic winter: shelter seeking by arctic hare *(Lepus arcticus). Arctic* 46: 340–353.

Green Mountain Club (GMC). 2011. *Long Trail Guide*, 27th edition. Leahy Press; Montpelier, VT.

Grenfell, W.T. 1911. Labrador. *The Geographic Journal* 37(4): 407–418.

Hanel, C. 2005. Doctor's Brook Rare Plant Survey Final Report. Western Newfoundland Model Forest Report Number 2–221. Corner Brook, NL.

Hanel, C. 2006. Management Plan for Fernald's Milk-vetch (*Astragalus robbinsii* (Oakes) Gray var. *fernaldii* (Rydberg) Barneby) in Newfoundland and Labrador 2006-2011. Corner Brook, NL.

Harrington, F.H., and A.M. Veitch. 1991. Short-term impacts of low-level jet fighter training on caribou in Labrador. *Arctic* 44(4): 318–327.

Harrington, C.R. 2003. Quaternary vertebrates of Québec: A summary. *Géographie physique et Quaternaire* 57(1): 85–94.

Harris, S.K., J.H. Langenheim, F.L. Steele, and M. Underhill. 1964. *AMC Field Guide to Mountain Flowers of New England.* Appalachian Mountain Club Books; Boston, MA.

Harshberger, J.W. 1919. Alpine fell–fields of eastern North America. *Geographical Review* 7(4): 233–255.

Hébert, A. (ed.). 2005. 2005 Status Report: Albanel–Témiscamie–Otish Park Project (E'weeach). Ministere du Développement durable, de l'Environnement et des Parcs (MDDEP); Québec, QC.

Hedderson, T.A., L. Söderström, and G.R. Brassard. 2001. Hepaticae of the Torngat Mountains, northern Labrador, Canada. *Lindbergia* 26: 143–156.

Hétu, B., and J.T. Gray. 2000. Les étapes de la déglaciation dans le nord de la Gaspésie (Québec): Les marges glaciaires des Dryas ancient et récent. *Géographie Physique et Quaternaire,* 54(1): 5–40.

Hilke, C.J. 2007. Visitor impacts, succession, and substrate depth on the granite outcrop "island" communities of Welch Mountain, New Hampshire. Unpublished M.Sc. thesis, Antioch New England Graduate School; Keene, NH.

Hoag, R.W. 1982. The Mark on the Wilderness: Thoreau's Contact with Ktaadn. *Texas Studies in Literature and Language* 24(1).

Hoch, G., and Körner, C. 2003. The carbon charging of pines at the climatic treeline: A global comparison. *Oecologia* 135: 10–21.

Holder, K. R. Montgomerie, and V. L. Friesen. 2004. Genetic diversity and management of Nearctic rock ptarmigan (*Lagopus mutus*). *Canadian Journal of Zoology* 82: 564–575.

Holme, R.F. 1888. A Journey in the Interior of Labrador, July to October, 1887. *Proceedings of the Royal Geographic Society* 10: 189.

Holtmeier, F.K. 2003. *Mountain Timberlines: Ecology, Patchiness, and Dynamics.* Dordrecht; Kluwer, The Netherlands. 437 pp.

Hoving, C.L., R.A. Joseph, and W.B. Krohn. 2003. Recent and historical distributions of Canada lynx in Maine and the Northeast. *Northeastern Naturalist* 10(4): 363–382.

Howard, T. 2009. Vegetation communities of the Adirondack alpine zone. Unpublished manuscript in preparation. New York Natural Heritage Program; Albany, NY.

Howe, N. 2000. *Not Without Peril: 150 Years of Misadventure on the Presidential Range of New Hampshire.* Appalachian Mountain Club Books; Boston, MA.

Hudson, W.D. 1987. The reproductive biology of *Saxifraga Stellaris var. Comosa* on Mt. Katahdin, Maine. Unpublished M.Sc. thesis, University of Vermont; Burlington, VT.

Integrated Taxonomic Information System (ITIS). 2011. At website: http://www.itis.gov

Ives, J.D. 1957. Glaciation of the Torngat Mountains, Northern Labrador. *Arctic* 10(2): 67–87.

Ives, J.D. 1975. Photograph with caption: The Torngat Mountains of Northern Labrador possess a great variety of cirque forms. *Arctic and Alpine Research* 7(4).

(Mealy Mountains and Torngat Mountains, Labrador, Canada). Unpublished M.Sc. thesis, Memorial University; St. John's, Newfoundland. 102 pp.

Nagy, L., and G. Grabherr. 2009. *The Biology of Alpine Habitats.* Oxford University Press; New York, NY. 376pp.

Nagy, L., G. Grabherr, C. Körner, and D.B.A. Thompson. 2003. *Alpine biodiversity in Europe.* Ecological Studies, Vol. 167. Springer; Berlin. 577 pp.

Nalcor Energy. 2009. Labrador–Island Transmission Link. Environmental Assessment Registration Pursuant to the Newfoundland and Labrador Environmental Protection Act. Project Description Pursuant to the Canadian Environmental Assessment Act. January 2009. St. John's, NL.

National Climatic Data Center (NCDC). U.S. Department of Commerce, National Oceanic and Atmospheric Administration. NCDC data archive. Available at: http://www.ncdc.noaa.gov

NatureServe. 2011. NatureServe Explorer: An online encyclopedia of life. Association for Biodiversity Information; Arlington, VA. At website: http://www.natureserve.org/explorer.

Naylor, B. J., J. F. Bendell, and S. Spires. 1985. High density of heather voles, *Phenacomys intermedius* in Jack Pine, *Pinus banksiana,* Forests in Ontario. *Canadian Field–Naturalist* 99: 494–497.

Nelson, F.E.N., and L.E. Jackson. 2003. Cirque forms and alpine glaciation during the Pleistocene, west–central Yukon. *Yukon Exploration and Geology*, 183–198.

Northcott, T. 1975. Long-distance movement of an arctic fox in Newfoundland. *Canadian Field-Naturalist* 89: 464–465.

Odell, N.E. 1933. The mountains of Northern Labrador. *The Geographical Journal* 82(3): 193–210.

Olejczyk, P., and J.T. Gray. 2007. The relative influence of Laurentide and local ice sheets during the last glacial maximum in the eastern Chic-Chocs Range, northern Gaspé Peninsula, Québec. *Canadian Journal of Earth Sciences* 44(11): 1603–1625.

Olofsson, J. 2004. Positive and negative plant–plant interactions in two contrasting arctic–alpine plant communities. *Arctic, Antarctic, and Alpine Research* 36(4): 464–467.

Olofsson, J., J. Moen, and L. Oksanen. 1999. On the balance between positive and negative plant interactions in harsh environments. *Oikos* 86, 539–543.

Ouellet, J.-P., J. Ferron, and L. Sirois. 1996. Space and habitat use by the threatened Gaspé caribou in southeastern Québec. *Canadian Journal of Zoology* 74(10): 1922–1933.

Packer, L., and J.S. Taylor. 2002. Genetic variation within and among populations of an arctic/alpine sweat bee (Hymenoptera: Halictidae). *Canadian Entomologist* 134(5): 619–631.

Palmer, R.S. 1938. Late records of caribou in Maine. *Journal of Mammalogy* 19(1): 37–43.

Palmer, R.S., and W. Taber. 1946. Birds of the Mt. Katahdin Region of Maine. *The Auk* 63(3): 299–314.

Parker, G.R., and S. Luttich. 1986. Characteristics of the wolf (*Canis lupus labradoricus* Goldman) in Northern Québec and Labrador. *Arctic* 39(2): 145–149.

Parks Canada. 1986. Gros Morne National Park. World Heritage Site Nomination No.419. 63 pp. Ottawa, ON.

Parks Canada. 2011. Torngat Mountains National Park of Canada. Parks Canada; Nain, NL. At website: http://www.pc.gc.ca/eng/pn-np/nl/torngats/index.aspx

Parsons, K., and L. Hermanutz. 2006. Conservation of rare, endemic braya species (Brassicaceae): Breeding system variation, potential hybridization. *Biological Conservation* 128: 201–214.

Pauchard, A., C. Kueffer, H. Dietz, C.C. Daehler, J. Alexander, P.J Edwards, J.R. Arévalo, L.A. Cavieres, A. Guisan, S. Haider, G. Jakobs, K. McDougall, C.I. Millar, B.J. Naylor, C.G. Parks, L. J. Rew, and T. Seipel. 2009. Ain't no mountain high enough: Plant invasions reaching new elevations. *Frontiers in Ecology and the Environment* 7: 479–486.

Pauwels, A.M. 2005. Evaluation report: Otish Mountain property. Unpublished report prepared for Xemplar Energy Corp.

Payette, S. 2007. Contrasted dynamics of northern Labrador tree lines caused by climate change and migrational lag. *Ecology* 88(3): 770–780.

Payette, S., and A. Delwaide. 2003. Shift of conifer boreal forest to lichen–heath parkland caused by successive stand disturbances. *Ecosystems* 6: 540–550.

Pease, A.S. 1924. Vascular flora of Coos County, New Hampshire. *Proceedings of the Boston Society of Natural History* 37(3).

Perreault, S. 2007. Dramatic increase in uranium exploration in Québec. *Mining Information Bulletin*; Province of Québec. Québec, QC.

Pickavance, J.R., and C.D. Dondale. 2005. An Annotated Checklist of the Spiders of Newfoundland. *Canadian Field-Naturalist* 119: 254–275.

Pielou, E.C. 1988. *The World of Northern Evergreens.* Comstock Publishing Associates; Ithaca, NY. 174 pp.

Pielou, E.C. 1992. *After Ice Age: The Return of Life to Glaciated north America*. University of Chicago Press; Chicago, IL. 376 pp.

Platnick, N.I. 2011. *World Spider Catalog, Version 12.0.* American Museum of Natural History. http://research.amnh.org/iz/spiders/catalog/.

Pollock, S.L., and S. Payette. 2010. Stability in the patterns of long-term development and growth of the Canadian spruce–moss forest. *Journal of Biogeography* 37: 1684–1697.

Pomerlau, R. 1950. Au sommet de l'Ungava. *Revue de L'Université Laval* 4(5):775-791.

Pope, R. 2005. *Lichens above Treeline: A Hiker's Guide to Alpine Zone Lichens of the Northeastern United States.* University Press of New England; Lebanon, NH.

Preble, E.A. 1899. Description of a new lemming mouse from the White Mountains, New Hampshire. *Proceedings of the Biological Society of Washington* 13: 43–45.

Protected Areas Association. 1993. Towards sustainable development: A protected natural areas strategy for Newfoundland and Labrador part 1 terrestrial component. At website: http://www.nfld.net/paa/towards%20sustainable%20development.pdf

Rajakaruna, N., T.B. Harris, and E.B. Alexander. 2009. Serpentine geoecology of eastern North America: A review. *Rhodora* 111(945): 21–108.

Rankin, D.W., and D.W. Caldwell. 2010. A Guide to the Geology of Baxter State Park and Katahdin. Maine Geological Survey, Bulletin 43, 80 p., 2 maps.

Rayl, N.D. 2012. Black bear movements and caribou calf predation in Newfoundland. Unpublished M.Sc. thesis, University of Massachusetts; Amherst, MA. 243 pp.

Reidy, D.E. 2004. New Hampshire Geology. At website: http://www.nhgeology.org/.

Reisigl, H., and R. Keller. 1987. Alpenpflanzen im Lebensraum: Alpine Rasen, Schutt– und Felsvegetation. Gustav Fischer; Stuttgart, NY.

Rémillard, J. 1978. Lionel Groulx—bibliographie (1964–1979). *Revue d'histoire de l'Amerique française* 32(3): 465–523.

Rhymer, J.M., J.M. Barbay, and H.L. Givens. 2004. Taxonomic relationship between *Sorex dispar* and *S. gaspensis:* inferences from mitochondrial DNA sequences. *Journal of Mammalogy* 85: 331–337.

Riley, G.C. 1962. Stephenville Map–Area, Newfoundland. *Geological Survey of Canada Memoir* 323.

Ring, R.A., and D. Tesar. 1981. Adaptations to cold in Canadian arctic insects. *Cryobiology* 18: 199–211.

Rivers, T. 1997. Lithotectonic elements of the Grenville Province: Review and tectonic implications. *Precambrian Research* 86: 117–154.

Rogerson, R.J. 2012. Labrador Highlands. *Canadian Encyclopedia.* At website: http://thecanadianencyclopedia.com.

Roland, A.E., and M. Zinck. 1998. *Roland's Flora of Nova Scotia.* Nimbus Publishing and Nova Scotia Museum; Halifax, NS.

Rombough, P.J., S.E. Barbour, and J.J. Kerekes. 1978. Life history and taxonomic status of isolated population of Arctic Char, *Salvelinus alpinus,* from Gros Morne National Park, Newfoundland. *Journal of the Fisheries Research Board of Canada* 35(12): 1537–1541.

Rouleau, E., and G. Lamoureux. 1992. *Atlas of the Vascular Plants of the Island of Newfoundland and of the Islands of Saint–Pierre–et–Miquelon.* Fleurbec; Saint Henri–de–Levis, Québec, Canada. 777 pp.

Rousseau, J. 1950. La flore des monts Otish, Québec. *Mémoires et Comptes rendus de la Société royale du Canada* 44(3): Annex F.

Rousseau, J. 1959. Grandeur and décadence des monts Watshish. *Cahiers de géographie du Québec* 3(6): 457–468.

Rune, O. 1954. Notes on the flora of the Gaspe Peninsula. *Svensk Bonisk Tidskrift* 48: 117–138.

Russell, J.E., and R. Tumlison. 1996. Comparison of Microstructure of White Winter Fur and Brown Summer Fur of Some Arctic Mammals. *Arcta Zoologica* 77(4): 279–282.

Sabo, S.R. 1980. Niche and Habitat Relations in Subalpine Bird Communities of the White Mountains of New Hampshire. *Ecological Monographs* 50(2): 241–259.

Samson, I.M., and A.E. Williams-Jones. 1991. C-O-H-N-salt fluids associated with contact metamorphism, McGerrigle Mountains, Québec: A Raman spectroscopic study. *Geochimica et Cosmochimica Acta* 55(1): 169–177.

Savard, J. 2009. *Dynamique de la végétation des sommets alpins de la forêt boréale, Québec (Canada).* Unpublished M.Sc. thesis, Université Laval; Québec.

Schaefer, J.A., and S.N. Luttich. 1998. Movements and activity of caribou, *Rangifer tarandus caribou,* of the Torngat Mountains, Northern Labrador and Québec. *Canadian Field-Naturalist.* 112(3): 486.

Schaefer, J.A., A.M. Veitch, F.H. Harrington, W.K. Brown, J.B. Theberge, and S.N. Luttich. 1999. Demography of decline of the Red Wine Mountains caribou herd. *Journal of Wildlife Management* 63(2): 580–587.

Schaefer, J.A., A.M. Veitch, F.H. Harrington, W.K. Brown, J.B. Theberge, and S.N. Luttich. 2001. Fuzzy structure and spatial dynamics of a declining woodland caribou population. *Oecologia* 126: 507–514.

Schärer, U., T.E. Krogh, and C.F. Gower. 1986. Age and evolution of the Grenville Province in eastern Labrador from U–Pb systematics in accessory minerals. *Contributions to Mineralogy and Petrology* 94(4): 438–451.

Scott, F.W. 1988. Status report on the Gaspé Shrew *Sorex gaspensis* in Canada. Committee on the Status of Endangered Wildlife in Canada; Ottawa, ON.

Scott Wilson Mining (SWM). 2010. Updated technical report on the preliminary assessment of the Renard Project, Québec, QC. NI 43–101 Report.

Scudder, B.E. 1995. *Scudder's White Mountain Viewing Guide.* High Top Press; Bellmore, NY.

Seidel, T.M., D. M. Weihrauch, K. D. Kimball, A. A. P. Pszenny, R. Soboleski, E. Crete, and G. Murray. 2009. Evidence of climate change declines with elevation based on temperature and snow records from 1930s to 2006 on Mount Washington, New Hampshire, U.S.A. *Arctic, Antarctic, and Alpine Research* 41: 362–372.

Selås, V., B.S. Johnsen, and N.E. Eide. 2010. Arctic fox *Vulpes lagopus* den use in relation to altitude and human infrastructure. *Wildlife Biology* 16(1): 107–112.

Shchepanek, M.J. 1973. Botanical investigation of the Otish Mountains, Québec. *Syllogeus* 1(2).

Sibley, D. A. 2001a. *The Sibley Guide to Bird Life & Behavior.* Alfred A . Knopf; New York.

Sibley, D. A. 2001b. *The Sibley Guide to Birds.* Alfred A . Knopf; New York.

Sirois, L., F. Lutzoni, and M.M. Grandtner. 1988. Les lichens sur serpentine et amphibolite du plateau du Mont Albert, Gaspésie, Québec. *Canadian Journal of Botany* 66: 851–862.

Slack, N.G., and A.W. Bell. 1993. *85 Acres: A Field Guide to the Adirondack Alpine Summits.* Adirondack Mountain Club; Lake George, NY. 44 pp.

Slack, N.G., and A.W. Bell. 2006. *AMC Field Guide to New England Alpine Summits,* 2nd edition. Appalachian Mountain Club Books; Boston, MA.

Small, R.J., and L.B. Keith. 1992. An experimental study of red fox predation on arctic and snowshoe hares. *Canadian Journal of Zoology* 70: 1614–1621.

Smith, J.S., T. Bell, and L. Rankin. 2003. Quaternary geology and landscape change, porcupine strand, Labrador. Current Research. Newfoundland Department of Mines and Energy Geological Survey Report 03-1: 293–305.

Smith, N. 2012. Snowy Owl Project web update. Massachusetts Audubon Society; Lincoln, MA. At website: http://massaudubon.org/Birds_and_Birding/snowyowl/intro.php.

Sokoloff, P.C. 2010. Taxonomic status of the narrow endemic *Astragalus robbinsii* var. *fernaldii* (Fernald's Milkvetch—Fabaceae): molecules, morphology, and implications for conservation. Unpublished M.Sc. thesis, University of Ottawa.

Sokoloff, P.C., and L.J. Gillespie. 2012. Taxonomy of *Astragalus robbinsii var. fernaldii* (Fabaceae): molecular and morphological analyses support transfer to *Astragalus eucosmus*. *Botany* 90(1): 11–26.

Spiess, A. 1976. Labrador grizzly (*Ursus arctos* L.): First skeletal evidence. *Journal of Mammalogy* 57(4): 787–790.

Spiess, A., and S. Cox. 1976. Discovery of the skull of a grizzly bear in Labrador. *Arctic* 29(4): 194–200.

Sperduto, D.D., and Cogbill, C.V. 1999. *Alpine and Subalpine Vegetation of the White Mountains, New Hampshire.* New Hampshire Natural Heritage Inventory; Concord, NH.

Sprugel, D.G. 1976. Dynamic structure of wave–regenerated *Abies balsamea* forests in the north-eastern United States. *Journal of Ecology* 64: 889–911.

Squires, S.E., L. Hermanutz, and P.L. Dixon. 2008. Endangered, endemic and almost invisible—Rare Braya on the Limestone Barrens of Newfoundland, Canada. *Endangered Species Update* 25: 41–44.

Squires, S.E., L. Hermanutz, and P.L. Dixon. 2009. Agricultural insect pest compromises survival of two endemic *Braya* (Brassicaceae). *Biological Conservation* 142: 203–211.

Ståhls, G.E. Ribeiro, and I. Hanski. 1989. Fungivorous *Pegomya* flies: spatial and temporal variation in a guild of competitors. *Annales Zoologici Fennici* 26: 103–112.

Steele, F. 1994. *At Timberline: A Nature Guide to the Mountains of the Northeast.* Appalachian Mountain Club Books; Boston, MA.

Steele, F. 1963. A double-flowered form of *Diapensia lapponica*. *Rhodora* 65: 21.

Stevens, K. 1989. Continuation of a metallogenic/exploration study for granite–associate mineralized regions in the Gaspé Peninsula, Québec. Geological Survey of Canada, Open File 2064.

Stone, A., C.W. Sabrosky, W.W. Wirth, R.H. Foote, and J.R. Coulson. 1965. *A catalog of the Diptera of America north of Mexico.* Washington, DC: U.S. Department of Agriculture.

Strateco. 2009. Environmental Impact Statement. Underground Exploration Program. Matoush Property. Strateco Resource, Document 75–752–A–1.

Sutton, J.T., L. Hermanutz, and J.D. Jacobs. 2006. Are frost boils important for the recruitment of arctic–alpine plants? *Arctic, Antarctic, and Alpine Research* 38(2): 273–275.

Thomas, A., V. Jackson, and G. Finn. 1981. Geology of the Red Wine Mountains surrounding area, Central Labrador. Current research, Newfoundland and Labrador Mineral Development Division, Report no. 81–1, 1981, p. 111–120.

Thompson, D.J. 1999. Talus fabric in Tuckerman Ravine, New Hampshire: Evidence for a tongue–shaped rock glacier. *Géographie Physique et Quaternaire* 53(1): 47–57.

Tiffney, W.N.J. 1972. Snow cover and the *Diapensia lapponica* habitat in the White Mountains, New Hampshire. *Rhodora* 74 : 358–377.

Townsend, C.W. 1923. A breeding station of the Horned Lark and Pipit on the Gaspé Peninsula. *The Auk* 40(1): 85–87.

Trant, A.J., R.G. Jameson, and L. Hermanutz. 2011. Persistence at treeline: Old trees as opportunists. *Arctic* 84(3): 367–370.

Trindade, M., T. Bell, and C. Laroque. 2011. Changing climatic sensitivities of two spruce species across a moisture gradient in Northeastern Canada. *Dendrochronologia* 29(2011): 25–30.

Vandal, D. 1984. Ecologie comportementale du caribou du parc des Grand–Jardins. Unpublished M.Sc. thesis, Université Laval; Laval, Québec.

Vellend, M., and M.J. Waterway. 1999. Geographic patterns in the genetic diversity of a northern sedge, *Carex rariflora. Canadian Journal of Botany* 77: 269–278.

Veitch, A.M. 1992. The barren ground bear. *Ursus* 1(4): 5–7, 36.

Veitch, A.M. 1994. Black bear research on the barren–grounds of the Northeastern Labrador Peninsula, 1989–1993. *Osprey* 25: 71–79.

Veitch, A.M., and F.H. Harrington. 1996. Brown bears, black bears, and humans in northern Labrador: An historical perspective and outlook to the future. *Journal of Wildlife Research* 1: 245–250.

Vonlanthen, C. M., A. Bühler, H. Veit, P. M. Kammer, and W. Eugster. 2006. Alpine plant communities: A statistical assessment of their relation to microclimatological, pedological, geomorphological, and other factors. *Physical Geography* 27(2): 137–154.

Wallner, J., and M.J. DiGregorio. 1997. *New England's Mountain Flowers: A High Country Heritage.* Mountain Press Publishing; Missoula, MT.

Wang, L.C.H., D.L. Jones, R.A. MacArthur, and W.A. Fuller. 1973. Adaptation to cold: Energy metabolism in an atypical lagomorph, the arctic hare *(Lepus arcticus). Canadian Journal of Zoology* 51: 841–846.

White Mountain National Forest (WMNF). 2008. *White Mountain National Forest: Monitoring and Evaluation Report.* USDA Forest Service Eastern Region; Laconia, NH.

Whitney, G.G., and R.E. Moeller. 1982. An analysis of the vegetation of Mt. Cardigan, New Hampshire: a rocky, subalpine New England summit. *Bulletin of the Torrey Botanical Club* 109(2): 1977–188.

Wickham, H. F. 1897. A list of Coleoptera from the southern shore of Lake Superior, with remarks on geographical distribution. *Proceedings of the Davenport Academy of Natural Sciences* 6: 125–169.

Wilderness and Ecological Reserves Advisory Council (WERAC) 2007. Annual report 2006–2007. Parks and Natural Areas Division, Department of Environment and Conservation. 6 pp.

Willey, L.L., and M.T. Jones. 2012. Site characteristics of *Sibbaldia procumbens* (Rosaceae) on the Uapishka Plateau, Québec, with notes on the alpine flora. *Rhodora* 114: 21–30.

Williams, E. 1901. A comparison of the floras of Mts. Washington and Katahdin. *Rhodora* 3(30): 160-165.

Yates, D., S. Folsom, and D. Evers. 2010. Maine Appalachian Trail rare mammal inventory from 2006–2008. Natural Resource Report NPS/NETN/NRR— 2010/177. National Park Service; Fort Collins, CO.

Young S.B. 1971. The vascular flora of St. Lawrence Island with special reference to floristic zonation in the Arctic regions. *Contributions from the Gray Herbarium* 201: 11–115.

Young, S.B. 1989. *To the Arctic: An Introduction to the Far Northern World.* Wiley Science; New York, NY.

Zika, P.F., and J.C. Jenkins. 1992. Contributions to the flora of the Adirondacks, NY. *Bulletin of the Torrey Botanical Club* 119(4): 442–445.

Zwinger, A.H., and B.E. Willard. 1996. Land Above the Trees: A Guide to American Alpine Tundra. Johnson Books; Boulder, CO. 448 pp.

Picea mariana	black spruce	**6**, 44, 46, **47**, 54, 59, **60**, 73, 126, 127, 168, **206**, 207, 222, **225**, 226, 228, **245**, 251, 252, 263, 267, 269, **270**, **271**, **277**, 281, 288, 296, 301
Picea rubens	red spruce	59, **60**, 127, 138, 288
Pinguicula vulgaris	common butterwort	73, 163, 168, 177, 193, 194, 240, 268, **294**
Pinus banksiana	jack pine	59, **60**, 269, **289**
Pinus strobus	white pine	73, **126**, 190, **191**
Plantago major	common plantain	206
Platanthera dilatata (*Habenaria dilatata*)	tall white bog orchid (scentbottle)	70, 112
Poa arctica	arctic bluegrass	227
Poa pratensis	Kentucky bluegrass	288
Polygonum viviparum	alpine bistort	**54**, 142, 159, 226
Polystichum scopulinum	mountain holly fern	139
Populus balsamifera	balsam poplar	301
Potentilla spp.	cinquefoil	295
Potentilla nivea	snow cinquefoil	206
Potentilla norvegica	rough cinquefoil	300
Potentilla pulchella (*Potentilla usticapensis*)	pretty cinquefoil (Burnt Cape cinquefoil)	**241**
Potentilla robbinsiana	dwarf mountain cinquefoil (Robbins' cinquefoil)	**102**, 106, 110, **111**, 134, 163
Primula mistassinica	Mistassini primrose	192
Ranunculus acris	tall buttercup (common buttercup)	113
Ranunculus abortivus	kidney-leaved buttercup (little-leaved buttercup)	142
Ranunculus allenii	Allen's buttercup	280, 281
Rhinanthus minor ssp. *groenlandicus*	arctic yellow rattle (Greenland yellow rattle)	113, **115**
Rhodiola rosea (*Sedum rosea*)	roseroot	176, **227**, 240
Rhododendron canadense	rhodora	**126**, 191
Rhododendron groenlandicum (*Ledum groenlandicum*)	Labrador tea	63, 68, 126, 127, 140, 166, 168, 175, 177, 226, 267, 300
Rhododendron lapponicum	Lapland rosebay	**51**, 56, 63, **66**, 73, **103**, 111, 122, **125**, 126, 139, 168, **192**, 193, **237**, 238, 240, 253, **267**
Rhododendron tomentosum	northern Labrador tea	300
Ribes glandulosum	skunk currant	**62**, 253
Rosa nitida	shining rose	**175**, **217**, 218
Rubus arcticus	arctic raspberry	139, 294
Rubus chamaemorus	cloudberry (bakeapple)	59, **67**, 68, 113, **115**, 165, **216**, 217, 228, **252**, 254, 267, **269**
Rumex acetosella	sheep sorrel	113
Salix arctica	arctic willow	142, 192, 193
Salix arctophila	northern willow (arctic-loving willow)	125, 142, 254, **255**, 280
Salix argyrocarpa	Labrador willow (silver willow)	**62**, 112, 140, 217, **253**, 280
Salix brachycarpa	short-capsuled willow (short-fruit willow)	142
Salix chlorolepis	green-scaled willow	139, 140
Salix glauca	gray-leaved willow (grey-leaved willow)	73, 192, 193
Salix glauca var. *cordifolia* (*Salix cordifolia*)	beautiful willow	142
Salix herbacea	snowbed willow (herbaceous	68, **69**, 102, **113**, 142, **145**, **227**,

Triglochin palustris	marsh arrowgrass	193
Trillium undulatum	painted trillium	**126**
Tussilago farfara	coltsfoot	113, **206**
Vaccinium spp.	cranberry	56, 68
Vaccinium angustifolium	early lowbush blueberry (narrow-leaved blueberry)	175, 234
Vaccinium boreale	northern blueberry	234
Vaccinium caespitosum	dwarf bilberry	**69**
Vaccinium myrtilloides (Vaccinum canadense)	velvet-leaved blueberry	139
Vaccinium oxycoccos	small cranberry	**67**
Vaccinium uliginosum	alpine bilberry (bog bilberry, alpine blueberry)	63, **64, 66**, 76, 111, 126, 138, 139, 140, 153, 159, 162, 167, 176, 177, 193, 205, **206**, 207, 217, 226, **253**, 267, 288, 289, 291, 296, **297, 301**
Vaccinium vitis-idaea	mountain cranberry (partridgeberry)	63, **64, 66**, 126, 139, 140, 159, 160, 166, 176, 193, 205, 225, 253, 280, 288, 289
Veratrum viride	green false hellebore	112
Veronica wormskjoldii	Wormskjold's alpine speedwell	70, **72**, 112, 125, 194, **195**, 217, 280
Viburnum edule	squashberry (highbush cranberry, mooseberry)	**62**, 63, **253**
Vicia cracca	bird vetch (cow vetch, tufted vetch)	113, **114**, 294, **295**
Viola labradorica	Labrador violet	193
Woodsia alpina	alpine woodsia	206
Woodsia ilvensis	rusty woodsia	206
Woodsia glabella	smooth woodsia	227, **235**

Invertebrates
Mollusca

Scientific name	Common name	Page reference
Cepaea hortensis	white-lipped snail	**228**, 229
Sphaerium nitidum	arctic fingernail clam	**96**, 178

Arthropods
BUTTERFLIES and MOTHS (Lepidoptera)

Scientific name	Common name	Page reference
Agriades glandon	arctic blue	283
Boloria chariclea	arctic fritillary	**146, 178**, 282
Boloria chariclea montinus	White Mountain fritillary	**94**, 118
Boloria eunomia	bog fritillary	283
Boloria freija	Freija fritillary	146, 259, 283
Boloria polaris	polaris fritillary	283
Celastrina spp.	azures	**259**, 283
Celastrina ladon	spring azure	283
Choristoneura fumiferana	spruce budworm	289
Colias gigantea	giant sulphur	282
Colias palaeno	moorland clouded sulpher	282
Colias pelidne	pelidne sulphur	283
Colias philodice	a sulphur	260
Epelis truncataria	black-banded orange	**178**
Oeneis bore	white-veined arctic	283

Oeneis melissa semidea	White Mountain arctic butterfly	**94**, 118
Oeneis polyxenes	polyxenes arctic	283
Oeneis polyxenes katahdin	Katahdin arctic butterfly	94, 129
Orgyia antiqua	rusty tussock moth	**229**
Papilo canadensis	Canadian tiger swallowtail	283
Papilo zelicaon	anise swallowtail	146
Pieris spp.	a white	260
Pyrgus centaureae	pyrgus skipper	283
Polygonia faunus	green comma	118
Syngrapha spp.	a looper moth	282
Syngrapha u-aureum	golden looper moth	96

CADDISFLIES (Trichoptera)

Scientific name	Common name	Page reference
Limnephilidae	caddisflies	260

GRASSHOPPERS (Orthoptera)

Scientific name	Common name	Page reference
Booneacris glacialis glacialis	wingless White Mountain locust	**94**, 97, **118**, **129**
Neotettix femoratus	short-legged pygmy grasshopper	96
Melanoplus borealis	northern grasshopper	**94**
Melanoplus bruneri	Bruner's grasshopper	**94**
Melanoplus fasciatus	huckleberry grasshopper	94, **146**
Melanoplus gaspesiensis	Gaspé grasshopper (Mont Albert grasshopper)	94, 144

FLIES, MOSQUITOES (Diptera)

Scientific name	Common name	Page reference
Aedes hexodontus	an inland floodwater mosquito	93
Aedes vexans	a woodland floodwater mosquito	93
Culiseta alaskaensis	Alaska snow mosquito	93
Wyeomyia smithii	pitcher plant mosquito	93
Dicranota modesta	a crane fly	94
Nephrotoma penumbra	tiger crane fly	94
Tipula broweri	a crane fly	94
Tipula insignifica	a crane fly	94
Tipula subserta	a crane fly	94
Admontia washingtonae	a tachinid fly	94
Megaselia subobscurata	a scuttle fly	94
Spiniphora slossonae	a scuttle fly	94
Trupheoneura vitrinervis	a scuttle fly	94
Rhamphomyia expulse	a dance fly	94
Rhamphomyia rustica	a dance fly	94
Dolichopus brevicauda	a long-legged fly	94
Neurigona viridis	a long-legged fly	96
Argyra obscura	a gall midge	94
Dirhiza papillata	a gall midge	94
Spilogona argenticeps	an anthomyiid fly	94, 95
Cyrtopogon lyratus	a robber fly	96
Laphria altitudinum	a robber fly	96
Simuliidae	black fly	93, 96, 260, 283
Syrphidae	syrphid fly (hoverfly, flowerfly)	**95**, 117, 118
Diura nanseni	a stonefly (arctic springfly)	96
Aptilotus spatulatus	a lesser dung fly	96

Rudolfina digitata	a lesser dung fly	96
Pegomya ruficeps	anthomyiid fly	98
Pegomya icterica	anthomyiid fly	98
Eristalis tenax	drone fly	**118**

BEES, WASPS, ANTS (Hymenoptera)

Scientific name	Common name	Page reference
Formica sp.	an ant	**95**, 117
Bombus sylvicola	a bumblebee	**95**
Lasioglossum boreale	a sweat bee	96

BEETLES, BUGS (Coleoptera)

Scientific name	Common name	Page reference
Amara hyperborea	a ground beetle	96
Nysius groenlandicus	Greenland seed bug	96
	a ground beetle	**97**
Cicindela spp.	a tiger beetle	**97**
Cicindela longilabris	boreal long-lipped tiger beetle	**97**, 178, 197
Cicindela tranquebarica	oblique-lined tiger beetle	**97**
Cicindela punctulata	sidewalk tiger beetle (punctured tiger beetle)	**97**
Monochamus scutellatus	white-spotted sawyer beetle	**97**, **98**
Elateridae	click beetle	**130**
Cicadidae	cicada	**130**

DRAGONFLIES, DAMSELFLIES (Odonata)

Scientific name	Common name	Page reference
Aeshna spp.	a hawker dragonfly	**96**, 97
Somatochlora spp.	an emerald dragonfly	**96**, 97
Leucorrhinia hudsonica	Hudsonian whiteface dragonfly	**130**

SPIDERS (Araneae)

Scientific name	Common name	Page reference
Aculepeira carbonarioides	a mountain spider	95, 96, 118, 129
Larinioides sp.	a furrow spider	**238**
Lycosidae	a wolf spider	95, **129**, 197
Pardosa sp.	a thin-legged wolf spider	238
Phidippus borealis	boreal jumping spider	**95**
Crustulina sticta	cobweb spider	97

Leeches

Scientific name	Common name	Page reference
Hirudinea	leech	**197**

Vertebrates

Fish

Scientific name	Common name	Page reference
Salmo salar	Atlantic salmon	218

| Salvelinus alpinus | arctic char | 92, **207**, 218, 259 |
| Salvelinus fontinalis | brook trout | **92**, 218, 259 |

Amphibians

Scientific name	Common name	Page reference
Ambystoma laterale	blue-spotted salamander	91, 259
Ambystoma maculatum	spotted salamander	**92, 128**, 129, 258
Anaxyrus americanus (Bufo americanus)	American toad	**91**, **117**, **128**, 156, 178, 196, **197**, **207**, 258, **259**, 271
Anaxyrus americanus copei (Bufo americanus copei)	Hudson Bay toad	**91**, **156**, **259**
Eurycea bislineata	northern two-lined salamander	**129**, 259
Lithobates clamitans	green frog	92, **117**, 163, 178
Lithobates pipiens	northern leopard frog	258, 259
Lithobates septentrionalis (Rana septentrionalis)	mink frog	**91**, **156**, 196, **197**, 207, 258, 259
Lithobates sylvaticus (Rana sylvaticus)	wood frog	**91**, **92**, **117**, **128**, 129, 163, 258, 259, 271
Notophthalmus viridescens	red-spotted newt (eastern newt)	92
Pseudacris crucifer	spring peeper	91, **117**, 163, 258, 259, 271

Reptiles

Scientific name	Common name	Page reference
Thamnophis sirtalis	garter snake	**92**, 258

Birds

Scientific name	Common name	Page reference
Acanthis flammea	common redpoll	89, 271
Anthus rubescens	American pipit	81, **88**, 116, 129, **144**, 155, 258, 271, 282
Aquila chrysaetos	golden eagle	87, 271
Asio flammeus	short-eared owl	241
Bubo scandiacus	snowy owl	89, 144, 155, 248
Bucephala islandica	Barrow's goldeneye	89, 257
Buteo jamaicensis	red-tailed hawk	87
Buteo lagopus	rough-legged hawk	89, 258, 282
Calcarius lapponicus	Lapland longspur	89, 258
Calidris minutilla	least sandpiper	282
Catharus bicknelli	Bicknell's thrush	**89**, 116
Catharus minimus	gray-cheeked thrush	**89**, 258, 271
Contopus cooperi	olive-sided flycatcher	282
Corvus corax	raven	**87**, 116, 294
Eremophila alpestris	horned lark	**88**, 144, 196, 241, 258
Eremophila alpestris alpestris	northern horned lark	88
Euphagus carolinus	rusty blackbird	282
Falcipennis canadensis	spruce grouse	88, **90**, 282
Falco peregrinus	peregrine falcon	87, 282
Falco rusticolus	gyrfalcon	89
Gavia immer	common loon	257
Haliaeetus leucocephalus	bald eagle	87, 196
Histrionicus histrionicus	harlequin duck	282
Junco hyemalis	dark-eyed junco (slate-colored junco)	89, **90**, 116, **155**, 258
Lagopus lagopus	willow ptarmigan	**87**, 88, 155, 195, **196**, 218, **258**, 282, 297

Lagopus mutus (*Lagopus muta*)	rock ptarmigan	81, **87**, 88, 155, 171, 178, 195, **196**, **202**, **207**, 218, 229, 282
Lagopus mutus welchi	Newfoundland rock ptarmigan	88
Larus spp.	gull	258
Loxia curvirostra	red crossbill	90
Loxia leucoptera	white-winged crossbill	**90**
Melospiza georgiana	swamp sparrow	90
Oenanthe oenanthe	northern wheatear	89
Passerculus sandwichensis	savannah sparrow	**90**, 196
Perisoreus canadensis	gray jay (Canada jay)	**90**, **116**, 258
Picoides arcticus	black-backed woodpeckers	90
Plectrophenax nivalis	snow bunting	89, 258
Poecile hudsonicus	boreal chickadee	90
Regulus calendula	golden-crowned kinglet	90
Regulus satrapa	ruby-crowned kinglet	90, 258
Setophaga coronata (*Dendroica coronata*)	yellow-rumped warbler	89, 116, 258
Setophaga magnolia (*Dendroica magnolia*)	magnolia warbler	89
Setophaga striata (*Dendroica striata*)	blackpoll warbler	89, 282
Somateria mollissima	eider	282
Spinus pinus (*Carduelis pinus*)	pine siskin	90
Sterna hirundo	common tern	258, 282
Tringa melanoleuca	greater yellowlegs	90, 196
Troglodytes hiemalis	winter wren	90
Turdus migratorius	American robin	87, 258
Zonotrichia albicollis	white-throated sparrow	**89**, 116, **155**, 196, 258
Zonotrichia leucophrys	white-crowned sparrow	**89**, 258, 271, 297

Mammals

Scientific name	Common name	Page reference
Alces americanus (*Alces alces*)	moose	85, 86, **116**, 144, **195**, **207**, 218, **228**, 241, 255, 257, 270, 281
Canis latrans	coyote	86, 144, 154, 178, 196, 218
Canis lupus	gray wolf (timber wolf)	83, 115, 154, 255, 257, 270, 281
Canis lupus lycaon	eastern wolf	83
Canis lupus beothucus	Newfoundland wolf (Beothuk wolf)	83
Castor canadensis	beaver	**86**, 116, 257, 271
Condylura cristata	star-nosed mole	87
Erethizon dorsatum	porcupine	**86**, 116, 128, 136, 257
Gulo gulo	wolverine	86, 115, 134, 154, 155, 255, 256, 257, 271, 272, 308
Lepus americanus	snowshoe hare	**83**, 84, 116, 257, 281
Lepus arcticus	arctic hare	81, 82, **83**, 134, 171, **177**, 178, 195, **196**, **202**, 207, 218, 228
Lontra canadensis	river otter	257, 271
Lynx canadensis	Canada lynx	83, 154, **218**, 255, 257, 270, 281
Marmota monax	groundhog (woodchuck)	86, 116, 257, **295**
Martes americana	American marten	**79**, **82**, 116, 129, 154, **164**, 165, 255, 257, 270, 271
Martes americana atrata	Newfoundland marten	82, 178
Microtus chrotorrhinus	rock vole	84, 129, 257
Microtus pennsylvanicus	meadow vole	87, 177, **178**, 195, **196**
Microtus pennsylvanicus terraenovae	Newfoundland meadow vole	207
Mustela erminea	ermine (short-tailed weasel)	86, 282

Photographers

The following photographers graciously contributed their photographs to this book:

(MA)	Marilyn Anions
(TB)	Trevor Bell
(NC)	Noah Charney
(LSC)	Laura Siegwart Collier
(GD)	Guillaume de Lafontaine
(WD)	Bill DeLuca
(WLD)	Lawson Dickson
(DD)	Daniel Dubie
(MD)	Martin Dubois
(CE)	Charley Eiseman
(JF)	Jon Feldgajer
(MF)	Mollie Freilicher
(DG)	Daniel Germain
(JG)	Julia Goren
(BG)	Ben Griffith
(AH)	Arthur Haines
(JH)	Joe Hardenbrook
(LH)	Luise Hermanutz
(IATNL)	International Appalachian Trail – Newfoundland & Labrador
(PJ)	Patrick Johnson
(LJ)	Ludovic Jolicoeur
(MJ)	Mike Jones
(WK)	Will Kemeza
(KK)	Ken Kimball
(JK)	Jerry Kobalenko
(MK)	Monica Kopp
(MLL)	Michael Lederer
(ML)	Mathieu Lemieux
(JPM)	Jean-Philippe Martin
(JM)	Jonathan Mays
(SM)	Scott Melvin
(MR)	Mark Rainey
(NR)	Nathaniel Rayl
(JS)	Jim Salge
(BT)	Brian Tang
(JW)	James Wall
(AW)	Alison Whitlock
(LW)	Liz Willey
(DW)	Daren Worcester
(PW)	Paul Wylezol
(DY)	Derek Yorks

Photographs are labeled with photographer's initials throughout the book.
Unless otherwise noted, plate photographs were taken by Mike Jones and Liz Willey

Contributors

Mike Jones is currently the Massachusetts State Herpetologist. His years with the Appalachian Mountain Club included two as Hutmaster at Lakes the Clouds on Mount Washington and four as winter caretaker at Carter Notch. Mike holds a Ph.D. in biology from the University of Massachusetts, serves on the board of Beyond Ktaadn, and is a 2001 graduate of the Center for Northern Studies.

Liz Willey worked throughout the mountains of New England for the Appalachian Mountain Club, U.S. Geological Survey, and Mount Washington Observatory before co-founding Beyond Ktaadn with Mike Jones and Will Kemeza. Liz holds an engineering degree from MIT and a Ph.D. in biology & wildlife from UMass Amherst. She serves on the faculty of Antioch University New England.

Jean-Philippe Martin spends most of his free time climbing, hiking or skiing. He completed a Ph.D. at the Université du Québec à Montréal, where his research focused on mass wasting processes in eastern mountains, the evolution of the alpine environment, and the relationships between geomorphology and ecology. He is completing a post-doc at Brock University.

Daniel Germain is a professor of Geography at the Université du Québec à Montréal, where his major research interests include alpine geomorphology (such as coarse debris landforms, talus slopes, snow avalanches, rock glaciers, etc.) Over the last few years he has worked in the Appalachians, northern Québec, Carpathian Mountains (Romania) and the French Alps.

Kenneth Kimball is the research director for the Appalachian Mountain Club. Ken has been a lead researcher on alpine plant endangered species restoration, northeastern alpine ecosystems, and the impacts of climate change, air pollution and wind power development on eastern U.S. mountains.

Neil Lareau earned his Ph.D. in Atmospheric Science at the University of Utah. His research focuses on topics of mountain meteorology including the formation and break-up of "cold-air pools" in mountain valleys and the structure of storm tracks across mountainous regions. Neil previously worked as a weather observer at the Mount Washington Observatory.

Noah Charney is an impassioned naturalist and conservationist. His publications include scientific articles on amphibians, birds, mammals, and mangroves, as well as an award-winning field guide to invertebrate tracking that he co-authored with Charley Eiseman. Noah is currently a Bullard Fellow at Harvard University.

Paul Wylezol is the Chairperson of the International Appalachian Trail (IAT) and International Appalachian Trail Newfoundland & Labrador (IATNL). He owns and operates a beachfront inn and guiding business just north of Gros Morne National Park, Newfoundland.

Claudia Hanel is a botanist with the Newfoundland and Labrador Department of Environment and Conservation. Her searches for rare flora have allowed her to explore many of the mountains and limestone barrens of Newfoundland and Labrador.

Luise Hermanutz is a conservation biologist whose research includes impacts of climate change on tundra and treeline in Labrador, and preservation of endangered plants on the Limestone Barrens of the Northern Peninsula of Newfoundland, Canada. She is a Professor in the Department of Biology at Memorial University and serves on several federal and provincial governmental committees that focus on protected areas establishment. She has just co-authored "Our Plants, Our Land," a book documenting the use of plants by Nunatsiavut (Labrador) Inuit. Luise lives in Portugal Cove, Newfoundland.

Guillaume de Lafontaine is a plant ecologist integrating paleoecology, ecology and genetics to study the origin and dynamics of forest tree species (such as white spruce) and plant communities. Guillaume's research has examined the population genetics of Canadian and European forest tree species, and has incorporated the use of macrofossil and soil charcoal analysis to reconstruct past vegetation and fire histories.

Marilyn Anions has lived, trekked, and canoed all over Canada. She has especially enjoyed working and living in many national parks (including St. Lawrence Islands, Wood Buffalo, Prince Albert, Gros Morne, and Mealy Mountains). Marilyn is currently the Director of Science at NatureServe Canada.

Michael Lederer has led and organized numerous wilderness mountaineering expeditions since the late 1970s. He is focused on exploring seldom-visited ranges or peaks in all seasons. Michael is a member of the American Alpine Club, and was awarded the Expedition Flag (NY Section) for his exploration of Labrador's English Mountains. Michael is an attorney in New York City.

Andrew Trant is an Assistant Professor of Ecology at the University of Waterloo. Andrew's Ph.D. work focused on how vegetation in the Mealy Mountains, Labrador, is responding to climate change by looking at shifts in disturbance regimes and life history strategies.

Laura Siegwart Collier earned her Ph.D. in the Department of Biology at Memorial University, St. John's, Newfoundland, Canada. Her research focused on the impacts of climate change on berry plant growth and fruit production in treeline and tundra plant communities of northern Labrador.

Julia Goren is an avid hiker, alpine enthusiast, and the coordinator of the Adirondack High Peaks Summit Steward Program in the Adirondack Mountains of New York. She earned a M.Sc. in Environmental Studies from Antioch University New England in Keene, NH and has served on the board of the Waterman Alpine Stewardship Fund since 2008.

Matt Burne is the Vice President and co-founder of the Vernal Pool Association, which promotes the study and protection of vernal pools in New England. He is also the Conservation Director for the Walden Woods Project in Lincoln, Massachusetts. Matt was the Vernal Pool Ecologist for the Massachusetts Endangered Species Program where he co-authored the *Field Guide to the Animals of Vernal Pools*. He holds a M.Sc. from the University of MA.

Iona Woolmington holds a degree in Earth Sciences from Dartmouth College in New Hampshire, and has spent several seasons working for the Appalachian Mountain Club as a naturalist in the White Mountain huts. She enjoys drawing cartoon stories about the natural world in her free time and searching for wolverine in the wilds of eastern Canada.

Will Kemeza resides in Harvard, Massachusetts with his family. Will teaches high school English at Concord-Carlisle High School and worked seasonally with the Appalachian Mountain Club, Green Mountain Club, and Randolph Mountain Club. He travels to the mountains whenever possible, now, increasingly, with children (students and his own sons).

Charley Eiseman is a freelance naturalist, conducting plant and wildlife surveys throughout New England. He is the lead author of the award-winning *Tracks & Sign of Insects and Other Invertebrates* and writes an insect-themed blog called "BugTracks." He holds a M.Sc. from the University of Vermont's Field Naturalist Program.